AF389522

Catalysts and Processes for H$_2$S Conversion to Sulfur

Catalysts and Processes for H$_2$S Conversion to Sulfur

Editor

Daniela Barba

MDPI • Basel • Beijing • Wuhan • Barcelona • Belgrade • Manchester • Tokyo • Cluj • Tianjin

Editor
Daniela Barba
Industrial Engineering
University of Salerno
Fisciano (SA)
Italy

Editorial Office
MDPI
St. Alban-Anlage 66
4052 Basel, Switzerland

This is a reprint of articles from the Special Issue published online in the open access journal *Catalysts* (ISSN 2073-4344) (available at: www.mdpi.com/journal/catalysts/special_issues/H2S_Conversion).

For citation purposes, cite each article independently as indicated on the article page online and as indicated below:

LastName, A.A.; LastName, B.B.; LastName, C.C. Article Title. *Journal Name* **Year**, *Volume Number*, Page Range.

ISBN 978-3-0365-3137-3 (Hbk)
ISBN 978-3-0365-3136-6 (PDF)

Contents

About the Editor

Daniela Barba

Daniela Barba was born in 1984 and graduated with Laude in Chemical Engineering (2010) at the University of Salerno, discussing a thesis entitled "Screening of Catalysts for H_2S abatement from Biogas to feed Molten Carbonate Fuel Cells". From 2010 to 2014, she collaborated on a research project, agreed between the Proceed Laboratory and ENEA (Casaccia, Rome), focusing on the study and preparation of catalysts for the removal of H_2S from Biogas by selective oxidation of sulfur at low temperatures. Subsequently, she started a PhD Course in Industrial Engineering, by studying an innovative process for the abatement of H_2S from refinery streams by catalytic oxidation of sulfur and hydrogen. In March 2018, she received the title of Doctor of Philosophy in Industrial Engineering. She is also a Research Fellow at the University of Salerno, and she is working on the study of a process involving zero emissions of SO_2 concerning the catalytic removal of sulfur acid compounds (H_2S, COS, CS2) at low temperatures. She is the author of several book chapters and 33 publications in international journals and the proceedings of international conferences.

Preface to "Catalysts and Processes for H$_2$S Conversion to Sulfur"

Today, more stringent regulations on SOx emissions and growing environmental concerns have led to considerable attention on sulfur recovery from hydrogen sulfide (H$_2$S). Hydrogen sulfide is commonly found in raw natural gas and biogas, even if a great amount is obtained through sweetening of sour natural gas and hydrodesulphurization of light hydrocarbons. It is highly toxic, extremely corrosive and flammable, and for these reasons, its elimination is necessary prior to emission in atmosphere. There are different technologies for the removal of H$_2$S, the drawbacks of which are the high costs and limited H$_2$S conversion efficiency. The main focus of this Special Issue will be on catalytic oxidation processes, but the issue is devoted to the development of catalysts able to maximize H$_2$S conversion to sulfur minimizing SO$_2$ formation, pursuing the goal of "zero SO$_2$ emission". The Special issue welcomes short communications, original research papers, and review papers concerning the formulation of novel catalysts for the optimization of the traditional processes or for new technologies. Submissions are welcome in the following areas:

Preparation, physical–chemical characterization of novel catalysts;

Optimization of the catalyst formulation and operating conditions; and

Study of the reaction mechanism and deactivation phenomena.

Daniela Barba
Editor

Editorial

Catalysts and Processes for H$_2$S Conversion to Sulfur

Daniela Barba 

Department of Industrial Engineering, University of Salerno, Via Giovanni Paolo II 132, 84084 Fisciano, Italy; dbarba@unisa.it

1. Definitions, State of the Art, Challenges

Hydrogen sulfide is one of the main waste products of the petrochemical industry; it is produced by the catalytic hydrodesulfurization processes (HDS) of the hydrocarbon feedstocks, and it is a byproduct from the sweetening of sour natural gas and from the upgrading of heavy oils, bitumen, and coals. It is a toxic gas and is classified as hazardous industrial waste. The exploitation of hydrogen sulfide as fuel using conventional combustion technologies is forbidden and criminalized by the more stringent environmental policies due to its deleterious effect like SO$_2$ formation, which is the main factor responsible for acidic precipitation.

There are different technologies for the removal of hydrogen sulfide, but they are characterized by high costs and limited H$_2$S conversion efficiency. The main purification treatments for H$_2$S removal comprise absorption, adsorption, membrane separation, and catalytic processes. Among these, adsorption and catalytic oxidation could be considered interesting methods to carry out the desulfurization thanks to their simplicity, efficiency, and low cost. Materials with high surface area and large pore volume, such as activated carbons, zeolites, mesoporous silica, and metal-organic frameworks, are generally used for the adsorption process [1].

The catalytic methods are among the most attractive as they allow the conversion of highly hazardous hydrogen sulfide into a nontoxic, marketable product, elementary sulfur. Basic processes for hydrogen sulfide-to-sulfur conversion are the direct oxidation of H$_2$S into elementary sulfur and the low-temperature reduction of sulfur dioxide [2].

Today, hydrogen sulfide is usually removed by the well-known Claus process, which is mainly used in refineries and natural gas processing plants for the treatment of rich-H$_2$S gas streams. The Claus process is the dominant technology to produce sulfur but it is not economically profitable because the hydrogen is lost as water. It is worth noting that the vast majority (about 94%) of the 8.1 million metric tons of sulfur produced in the United States in 2020 was synthesized using the Claus process [2,3].

Recently, there has been a growing interest in the utilization of H$_2$S as feedstock for hydrogen generation. Thermochemical cycles have been proposed by many researchers in order to obtain hydrogen and sulfur from hydrogen sulfide such as electrolysis, photolysis, plasmolysis, and their many variants [4]. An interesting alternative could be to produce sulfur and hydrogen simultaneously by the thermal catalytic decomposition of H$_2$S, even if the amount of energy requested to achieve extremely high temperatures, a low hydrogen yield, and the need for subsequent separation stages represent the main drawbacks to an industrial application.

Therefore, the challenge is to realize the H$_2$S removal in a one-reaction step in the presence of an active catalyst very selective to sulfur. The choice of the catalyst plays a fundamental role in assuring a high grade of H$_2$S removal with a lower selectivity to SO$_2$. In this regard, many efforts are addressed to the identification of active materials at the lower admissible temperature, in order to improve the H$_2$S abatement by reducing operational costs and so optimizing the desulfurization technology.

Citation: Barba, D. Catalysts and Processes for H$_2$S Conversion to Sulfur. *Catalysts* **2021**, *11*, 1242. https://doi.org/10.3390/catal11101242

Received: 5 October 2021
Accepted: 13 October 2021
Published: 15 October 2021

Publisher's Note: MDPI stays neutral with regard to jurisdictional claims in published maps and institutional affiliations.

2. Special Issue

I would particularly like to thank all the authors who contributed their excellent papers to this Special Issue covering significant aspects of this topic accompanied by a variety of novel approaches. The contributions represent interesting and innovative examples of the current research trends in the field of H_2S removal from liquid and gas streams.

I also wish to thank the Editorial Staff of *Catalysts* for their help to organize this issue and in particular Vivian Niu for the support, assistance, and encouragement.

This Special Issue is particularly devoted to the preparation of novel powdered/ structured supported catalysts and their physical–chemical characterization, the study of the aspects concerning stability and reusability, as well as the phenomena that could underlie the deactivation of the catalyst.

This Special Issue comprises 7 articles, 1 communication, and 1 review regarding the desulfurization of sour gases and fuel oil, as well as the synthesis of novel adsorbents and catalysts for H_2S abatement. In the following, a brief description of the papers included in this issue is provided to serve as an outline to encourage further reading.

Chen et al. investigated porous carbonaceous materials for the reduction of H_2S emission during swine manure agitation. Two biochars, highly alkaline and porous, made from corn stover and red oak were tested. The authors verified the possibility of using surficial biochar treatment for short-term mitigation of H_2S emissions during and shortly after manure agitation [5].

Bao et al. used the waste solid as a wet absorbent to purify the H_2S and phosphine from industrial tail gas. The reaction mechanism of the simultaneous removal of H_2S and phosphine by manganese slag slurry was investigated. The best efficiency removal of both H_2S and phosphine was obtained by the modified manganese slag slurry [6].

The desulfurization of sour gases was studied by Duong-Viet et al. in carbon-based nanomaterials in the form of N-doped networks by the coating of a ceramic SiC. The chemical and morphological properties of the nano-doped carbon phase/SiC-based composite were controlled to get more effective and robust catalysts able to remove H_2S from sour gases under severe desulfurization conditions such as high GHSVs and concentrations of aromatics as sour gas stream contaminants [7].

Li et al. carried out the oxidative desulfurization of fuel oil for the removal of dibenzothiophene by using imidazole-based polyoxometalate dicationic ionic liquids. Three kinds of catalyst were synthesized and tested under different conditions [8]. The catalytic performance of the catalysts was studied under different conditions by removing the dibenzothiophene from model oil. The authors identified a catalyst with an excellent DBT removal efficiency under optimal operating conditions.

The removal of H_2S and SO_2 at low temperatures was investigated by Ahmad et al. in eco-friendly sorbents from raw and calcined eggshells. They studied the effect of relative humidity and reaction temperatures. The best adsorption capacities for H_2S and SO_2 were obtained at a high calcination temperature of eggshell [9].

Zulkefli et al. prepared a zinc acetate supported with commercial activated carbon for the capture of H_2S by adsorption. The optimization conditions for the adsorbent synthesis were carried out using RSM and the Box–Behnken experimental design. Several factors and levels were evaluated, including the zinc acetate molarity, soaking period, and soaking temperature, along with the response of the H_2S adsorption capacity and the surface area [10].

Vanadium-sulfide-based catalysts supported on ceria were used for the direct and selective oxidation of H_2S to sulfur and water at a low temperature. Barba et al. performed a screening of catalysts with different vanadium loadings in order to study the catalytic performance in terms of H_2S conversion and SO_2 selectivity. The effect of the temperature, contact time, and H_2S inlet concentration was studied in relation to the catalyst that has exhibited the highest H_2S removal efficiency and the lowest SO_2 selectivity [11].

H_2S adsorption was studied in relation to a novel kind of hydrochar adsorbent derived from chitosan or starch and modified by CuO-ZnO. Zang et al. investigated the formation of CuO-ZnO on hydrochar, the effect of the hydrochar species, the adsorption temperature, and the adsorption mechanism [12]. A review concerning the different technologies of the gas-based phase for the direct catalytic oxidation of H_2S to sulfur was the object of study by Khairulin et al. [2]. The development of catalysts for the direct oxidation reaction (e.g., metal oxides, nanocarbon materials) and the discussion of the data concerning the Claus process and its recent adaptations were widely analyzed. Furthermore, the authors presented the results of basilar investigations obtained at the Institute of Catalysis where an industrial installation for H_2S removal from gas streams was located.

I hope that the topics presented in this issue will inspire readers to further investigate new materials and solutions to significantly reduce the presence of pollutants such as H_2S, SO_2 and other sulfur-based compounds, thereby pursuing the objective of "zero emissions" in the atmosphere.

Funding: This research received no external funding.

Conflicts of Interest: The author declares no conflict of interest.

References

1. Yang, J.H. Hydrogen sulfide removal technology: A focused review on adsorption and catalytic oxidation. *Korean J. Chem. Eng.* **2021**, *38*, 674–691. [CrossRef]
2. Khairulin, S.; Kerzhentsev, M.; Salnikov, A.; Ismagilov, Z. Direct Selective Oxidation of Hydrogen Sulfide: Laboratory, Pilot and Industrial Tests. *Catalysts* **2021**, *11*, 1109. [CrossRef]
3. Sulfur. Available online: https://pubs.usgs.gov/periodicals/mcs2021/mcs2021-sulfur.pdf (accessed on 9 September 2021).
4. Zaman, J.; Chakma, A. Production of hydrogen and sulfur from H_2S. *Fuel Process. Technol.* **1995**, 159–198. [CrossRef]
5. Chen, B.; Koziel, J.A.; Białowiec, A.; Lee, M.; Ma, H.; Li, P.; Meiirkhanuly, Z.; Brown, R.C. The Impact of Surficial Biochar Treatment on Acute H_2S Emissions during Swine Manure Agitation before Pump-Out: Proof-of-the-Concept. *Catalysts* **2020**, *10*, 940. [CrossRef]
6. Bao, J.; Wang, X.; Li, K.; Wang, F.; Wang, C.; Song, X.; Sun, X.; Ning, P. Reaction Mechanism of Simultaneous Removal of H_2S and PH_3 Using Modified Manganese Slag Slurry. *Catalysts* **2020**, *10*, 1384. [CrossRef]
7. Duong-Viet, C.; Nhut, J.-M.; Truong-Huu, T.; Tuci, G.; Nguyen-Dinh, L.; Pham, C.; Giambastiani, G.; Pham-Huu, C. Tailoring Properties of Metal-Free Catalysts for the Highly Efficient Desulfurization of Sour Gases under Harsh Conditions. *Catalysts* **2021**, *11*, 226. [CrossRef]
8. Li, J.; Guo, Y.; Tan, J.; Hu, B. Polyoxometalate Dicationic Ionic Liquids as Catalyst forExtractive Coupled Catalytic Oxidative Desulfurization. *Catalysts* **2021**, *11*, 356. [CrossRef]
9. Ahmad, W.; Sethupathi, S.; Munusamy, Y.; Kanthasamy, R. Valorization of Raw and Calcined Chicken Eggshell for Sulfur Dioxide and Hydrogen Sulfide Removal at Low Temperature. *Catalysts* **2021**, *11*, 295. [CrossRef]
10. Zulkefli, N.N.; Masdar, M.S.; Wan Isahak, W.N.R.; Abu Bakar, S.N.H.; Abu Hasan, H.; Sofian, N.M. Application of Response Surface Methodology for Preparationof $ZnAC_2$/CAC Adsorbents for Hydrogen Sulfide (H_2S) Capture. *Catalysts* **2021**, *11*, 545. [CrossRef]
11. Barba, D.; Vaiano, V.; Palma, V. Selective Catalytic Oxidation of Lean-H_2S Gas Stream to Elemental Sulfur at Lower Temperature. *Catalysts* **2021**, *11*, 746. [CrossRef]
12. Zang, L.; Zhou, C.; Dong, L.; Wang, L.; Mao, J.; Lu, X.; Xue, R.; Ma, Y. One-Pot Synthesis of Nano CuO-ZnO Modified Hydrochar Derived from Chitosan and Starch for the H_2S Conversion. *Catalysts* **2021**, *11*, 767. [CrossRef]

Review

Direct Selective Oxidation of Hydrogen Sulfide: Laboratory, Pilot and Industrial Tests

Sergei Khairulin [1,*], Mikhail Kerzhentsev [1], Anton Salnikov [1] and Zinfer R. Ismagilov [1,2]

[1] FRC, Boreskov Institute of Catalysis SB RAS, 630090 Novosibirsk, Russia; ma_k@catalysis.ru (M.K.); salnikov@catalysis.ru (A.S.); zri@catalysis.ru (Z.R.I.)
[2] FRC of Coal and Coal Chemistry, Institute of Coal Chemistry and Chemical Materials Science SB RAS, 630090 Novosibirsk, Russia
* Correspondence: sergk@catalysis.ru; Tel.: +73-833-306219

Abstract: This article is devoted to scientific and technical aspects of the direct catalytic oxidation of hydrogen sulfide for the production of elemental sulfur. It includes a detailed description of the Claus process as the main reference technology for hydrogen sulfide processing methods. An overview of modern catalytic systems for direct catalytic oxidation technology and known processes is presented. Descriptions of the scientific results of the Institute of Catalysis of the SB RAS in a study of the physical and chemical foundations of the process and the creation of a catalyst for it are included. The Boreskov Institute of Catalysis SB RAS technologies based on fundamental studies and their pilot and industrial testing results are described.

Keywords: gas purification; hydrogen sulfide; direct catalytic oxidation; fluidized catalyst bed; hydrogen sulfide removal facilities

check for **updates**

Citation: Khairulin, S.; Kerzhentsev, M.; Salnikov, A.; Ismagilov, Z.R. Direct Selective Oxidation of Hydrogen Sulfide: Laboratory, Pilot and Industrial Tests. *Catalysts* **2021**, *11*, 1109. https://doi.org/10.3390/catal11091109

Academic Editor: Daniela Barba

Received: 21 July 2021
Accepted: 31 August 2021
Published: 15 September 2021

Publisher's Note: MDPI stays neutral with regard to jurisdictional claims in published maps and institutional affiliations.

1. Introduction

According to modern classification, hydrogen sulfide is a highly hazardous substance which contributes significantly to the pollution of the atmosphere. The destruction of vegetation, the death of aqueous flora and fauna, an increase in the incidences of cancer and diseases of the respiratory tract, and "acid" rain are typical direct consequences of the release of hydrogen sulfide and sulfur dioxide into the atmosphere. The main sources of hydrogen sulfide emissions into the atmosphere and water include mining and the processing of sulfurous natural gas and oil, coal gasification, and biomass processing [1–4].

In fact, 40% of global gas reserves currently identified as viable, i.e., more than 70 trillion nm^3, are "acidic", and more than 10 trillion Nm^3 contain more than 10% H_2S [5].

To date, the estimated overall flow rate of the produced and processed sulfuric gas is about 100 billion m^3/year, and its contribution to the global mining of natural gases is 10–15%. At the same time, up to 60% of global sulfur production depends on the H_2S in these sulfuric gases, and there is a steady increasing trend in the share of sulfur obtained in this manner in the global balance of elementary sulfur production [6].

Another typical example characterizing the overall situation is the disposal of sulfurous oil-associated gases formed during the extraction of sulfur oil. The total flow rate of deposits located in the densely populated areas of the Volga-Ural oil and gas province is up to 1 billion m^3/year. The involvement of such gases in the fuel and energy balance will save up to 1 million tons/year of fuel. However, the high hydrogen sulfide content (1–6%) precludes their use as hydrocarbon fuel supplied to the population, industrial enterprises, and as raw materials for the synthesis of chemical products.

At present, the torching of such gases leads to the contamination of the atmosphere with toxic sulfur di- and tri- oxide, sulfuric acid, products of incomplete burning of hydrocarbons, and carcinogenic soot in amounts of up to one million tons per year. The average fraction of the incinerated associated oil gas in Russia was 24.4% in 2013 [7].

The ecological effects of burning are significantly worsened due to the flare disposal of hydrogen sulfide-containing oil-associated gases (OAG). The burning of one billion nm^3 of OAG results in atmospheric emissions of up to 60,000 tons of highly toxic H_2S, SO_2 and SO_3, soot, carbon monoxide, and up to 3 million tons of carbon dioxide, as well as, which is no less important, the loss of hundreds of millions of cubic meters of hydrocarbons, raw materials for oil and gas chemistry. For example, the qualified processing of 1000 m^3 of associated gas produces 820 m^3 of dry gas, 200 kg of a wide fraction of light hydrocarbons, and up to 61 kg of stable gasoline [8].

Given the global relevance of these problems, a wide range technologies which make use of sulfurous compounds have been implemented; however, the strengthening of environmental protection requirements dictates the need to create new technologies. These technologies must be highly efficient with a wide range of purified gases, and must minimize environmental damage while maximizing the yield valuable products. Such technologies should also meet the requirements of compactness and ease of process control.

To this end, catalytic methods are the most attractive, as they allow the conversion of highly hazardous hydrogen sulfide into a nontoxic, marketable product, i.e., elementary sulfur. Basic processes for hydrogen sulfide-to-sulfur conversion are the direct oxidation of H_2S into elementary sulfur and the low-temperature reduction of sulfur dioxide.

Due to the relevance of the aforementioned problems, this paper describes attempts to develop and improve the processes of purification and processing of hydrogen sulfide-containing gases. At present, three main categories of methods for cleaning gases from hydrogen sulfide can be distinguished:

- adsorption methods
- absorption methods
- catalytic methods

The general feature of the first two methods is that they are essentially ways to concentrate hydrogen sulfide from a purified gas, and must operate jointly with sulfur production plants using the Claus method. This process is currently the only large-tonnage method which is able to obtain sulfur from highly concentrated hydrogen sulfide-containing gas streams. It is characterized by:

- multistage operation
- insufficient environmental safety, due to the presence of a high-temperature furnace in the technological chain, which is a source of toxic byproducts
- a limited range of applications (thus, it is impossible to treat gases with hydrogen sulfide contents below 20 vol.% or gas streams with flow rates below 1000 Nm^3/h).

Therefore, as a supplement or alternative to the Claus process, direct selective catalytic oxidation of hydrogen sulfide to elemental sulfur is currently being explored.

2. Direct Selective Oxidation in the Liquid Phase. RedOx Processes

One means by which to purify gases from hydrogen sulfide is oxidation to elemental sulfur using oxygen in solutions of complex compounds of metals with wide variation of the pH of the medium.

The process proceeds at a rapid rate in a wide range of temperatures at pressures of 5–50 atm and provides a high degree of gas purification from hydrogen sulfide. Especially noteworthy are the processes developed by Wheelabrator Clean Air Systems, Inc (Pittsburgh, PA, USA). (ARI–Lo-Cat I®, ARI–Lo-Cat-II®), Shell Oil Company (Houston, TX, USA), and Dow Chemical (SulFerox®), as well as those based on the process of the direct oxidation of hydrogen sulfide in a solution of iron (3+) chelate complexes [9,10].

In SulFerox®, the reagent used was characterized by increased stability and low capital and operating costs. Reagent costs are 80–100$ per 1 t of hydrogen sulfide. Available data show that in the process of gas purification, up to 50–80% of methyl mercaptan and 30–60% of carbonyl sulfide can be removed from the initial content.

The SulFerox® process uses a new composition of a complexone, which is similar to EDTA (ethylenediamine-tetraacetate). However, the concentration of iron in the absorbent used is significantly higher (up to 3 wt.%) than in Lo-Cat processes (up to 0.5 wt.%, Figure 1) [11]. The first installation put into operation had a capacity of 120,000 m^3/day of the gas containing 4.5% hydrogen sulfide and 57% carbon dioxide at a pressure of 20 atm. The largest installation was launched in 1992 in Denver City, Texas. At this unit, 1500 ppmv of hydrogen sulfide in carbon dioxide gas at a pressure of 20 atm was reduced to 20 ppmv. The Sulferox process is currently the object of the greatest amount of research. From 1990 to 1995, Shell designed, built, and constructed more than 20 installations for the cleaning of various technological gases.

Direct Treatment Process for H₂S Removal from Sour Gas Streams

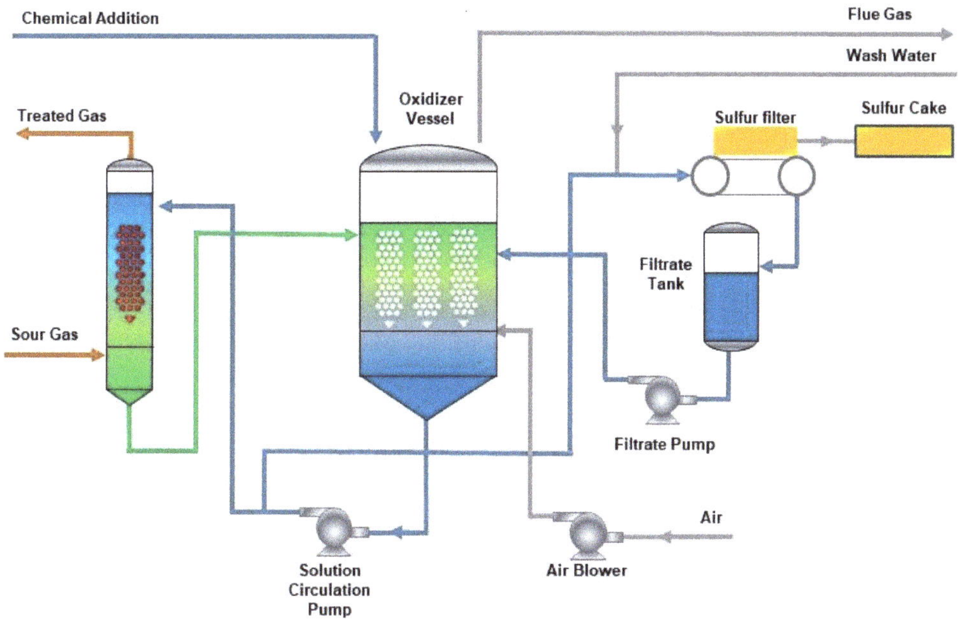

Figure 1. Schematic of the Lo-Cat process (adapted from [11]).

In the SulFerox® process, the concentration of iron compounds is significantly higher than in the ARI–Lo-Cat I®, ARI–Lo-Cat-II® processes. This fact explains the broader introduction of the ARI–Lo-Cat processes in gas cleaning operations. In the literature, information was found on the creation of only a few technological complexes for the purification of gases which simultaneously yield elementary sulfur, as opposed to the conventional procedure of amine absorption coupled with the Claus process. The Volga Research Institute of Hydrocarbon Fuels (JSC VNIIUS, Kazan, Russia) developed the Serox-2 process for cleaning gas flows from hydrogen sulfide with solutions of iron complexes to obtain elemental sulfur. The process is an analog of the "LO-CAT" process; its main distinctive feature is the composition of the absorbent with low corrosion activity with respect to carbon steel and high stability under the conditions required for the purification of gases. The process is implemented according to a standard two-step procedure, i.e., sulfur foam filtration and purification of hydrocarbon gas with a residual H$_2$S content of no more than 20 mg/m^3 (National Standard 5542-87) [12].

As more than 40 years of field testing experience shows, the process of cleaning gases from hydrogen sulfide with chelate complexes (iron salts of EDTA) has some disadvantages that limit its application for gases with a hydrogen sulfide content of more than 1–2 g/m^3.

Due to the need to use a working solution with a pH no higher than 8.5–9.0, the applicability of this solution with respect to hydrogen sulfide is limited, leading to the need to increase the rate of its circulation through the absorber (i.e., energy consumption for pumping increases).

The formation of a side product in the oxidation of hydrogen sulfide–thiosulfate is inevitable, which necessitates the use of a reagent such as EDTA.

A substantial technological problem is the separation of sulfur from the resulting pulp. Although to date, automatic filter machines (automatic filter presses and drum vacuum filters) have been developed, the complexity of their operation and their high costs make the cleaning process economically costly. Additionally, the low quality of the resulting elemental sulfur makes its commercialization difficult.

3. Claus Process

The most widely-used procedure for the large-scale reprocessing of highly-concentrated gases is the Claus method [13], which consists of several steps (Figure 2). The feed for the Claus process is acid gases.

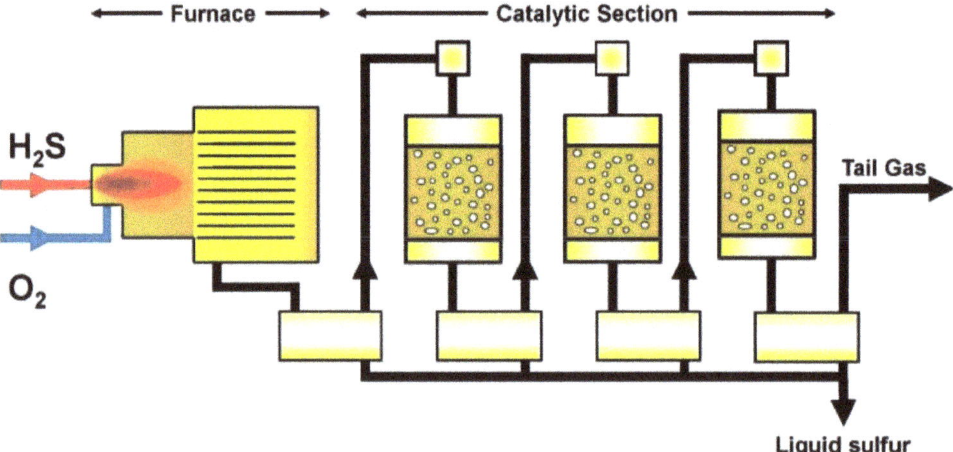

Figure 2. Schematic of Claus installation (adapted from [13]).

The Claus process is the dominant technology to produce gas (regenerated) sulfur. It is worth noting that the vast majority (about 94%) of the 8.1 million metric tons of sulfur produced in the United States in 2020 was synthesized using the Claus process [14].

The term "acid gases" is used to designate gases obtained after the absorptive treatment of hydrocarbon raw materials. Some typical characteristics of acid gases of various origins are shown in Table 1.

Table 1. Typical characteristics of acid gases of various origins.

Process	H_2S Content, Vol.%	Other Gas Components
Purification of gases from oil processing (MEA treatment)	90–98	Carbon dioxide, hydrogen, methane
Purification of natural and oil-associated gas (MEA or DEA treatment)	10–70	Carbon dioxide, water vapor, hydrocarbons C_1–C_6

As a rule, acid gases formed in the process of hydrotreating oil fractions are characterized by rather low flow rates ($\leq$1000 nm^3/hour) and high contents of hydrogen sulfide, the concentration of which, depending on the efficiency of the primary cleaning unit, usually exceeds 90%, compared with acid gases produced by a gas processing plant (for example, the power of only one technological line at the Astrakhan GPP is over 15,000 nm^3/hour).

4. The Thermal Stage of the Claus Process. Process Conditions. Chemical Reactions Proceeding in the System

The thermal stage of the Claus process largely determines the efficiency of the process as a whole, because at this stage, the main part (up to 70%) of the target product, i.e., elemental sulfur, is produced.

Upon mixing the acid gas with air (at the same stoichiometric hydrogen sulfide:oxygen ratio used in Reaction 1), a gas stream containing H_2S, O_2, N_2, CO_2, H_2O, sometimes hydrocarbons, and in some cases NH_3, HCN, etc. is formed, which is fed to the Claus furnace. Accordingly, during H_2S oxidation in the furnace, in addition to the main reactions [15–17]:

$$H_2S + 0.5O_2 \rightarrow 0.5S_2 + H_2O \tag{1}$$

$$H_2S + O_2 \rightarrow H_2 + SO_2 \tag{2}$$

$$H_2 + 0.5SO_2 \rightarrow H_2O + 0.25S_2 \tag{3}$$

$$H_2S \leftrightarrow H_2 + 0.5S_2 \tag{4}$$

$$CH_4 + 2S_2 \leftrightarrow CS_2 + 2H_2S \tag{5}$$

$$CO + H_2O \leftrightarrow CO_2 + H_2 \tag{6}$$

$$CO + 0.5S_2 \leftrightarrow COS \tag{7}$$

$$CH_4 + 0.5O_2 \rightarrow CO + 2H_2 \tag{8}$$

$$CO + H_2S \leftrightarrow COS + H_2 \tag{9}$$

Since the oxidation of hydrogen sulfide is an exothermic reaction, the temperature in the Claus furnace can reach 1200–1500 °C, while the minimum critical value of the temperature for sustaining a steady flame in the furnace is 1050 °C. The factor determining the temperature of the flame in the standard implementation of the process is the concentration of hydrogen sulfide in the acid gas. Despite significant progress in developing burner devices for the combustion of hydrogen sulfide-containing mixtures, the optimal conditions for the stable operation of the flame furnace are those with a hydrogen sulfide content in the feed gas of $\geq$60 vol.%. Technical approaches for maintaining sustainable operation in Claus furnaces are examined in [18] where, along with the results of calculations carried out using the Gibbs energy minimization method, the experimental data are in good compliance with the results of the calculations.

The following methods are considered:

- The reheating of the initial gas streams, acid gas and air: Even at a concentration of hydrogen sulfide in the acid gas of 40 vol.%, it is necessary to heat initial gas streams to 300 °C to reach the lower threshold of the stable operation of the Claus furnace, i.e., 1050 °C. In practice, as the experience of operating Claus installations at the Orenburg GPP shows, considering the essential heat loss, the preheating temperature can be as high as 600 °C.

- The use of oxygen-enriched air as an oxidant: Even at hydrogen sulfide concentrations in the acid gas of 50%, the required oxygen concentration in the supplied air should be at least 50 vol.% in order to reach the lower threshold of the stable operation of the Claus furnace.

- Supply of hydrocarbon fuel gas to the flame furnace: A supply of fuel gas at 25–30% of the acid gas flow rate with a high H_2S concentration will not provide the necessary temperature in the furnace to maintain stable operation. The heat of the combustion of hydrogen sulfide is utilized by heating chemically purified water, with water vapor production in a waste heat boiler. The hot gas passes through the boiler tubes and heats the water therein to boiling point. The gas cooled in the boiler is sent to the condenser, where it is cooled further to about 150 °C.

5. Catalytic Stage of the Claus Process. Catalysts Used

Gases from the Claus furnace condenser located after the waste heat boiler containing mainly H_2S, SO_2, N_2, CO_2, H_2O, COS, CS_2, CO, H_2, and traces of sulfur are further passed to the main catalytic stage. The process is usually carried out in the adiabatic fixed beds of the granular catalyst, in which, in addition to the Claus reaction, hydrolysis reactions of sulfur-organic compounds also proceed [17,19–22]. Catalytic convertors:

$$2H_2S + SO_2 \leftrightarrow 2H_2O + 0.5S_6 \tag{10}$$

$$COS + H_2O \rightarrow CO_2 + H_2S \tag{11}$$

$$CS_2 + 2H_2O \rightarrow CO_2 + 2H_2S \tag{12}$$

$$2H_2S + SO_2 \leftrightarrow 2H_2O + 0.5S_6 \tag{13}$$

The most commonly used catalyst for the Claus process is aluminum oxide with various modifications. The production of catalysts for the Claus process reaches hundreds of thousands of tons; the big players in this market are BASF [23], producing Claus catalysts Activated Alumina–DD-431, Promoted Alumina DD-831, EURO SUPPORT (previously Kaiser Alumina), and their successors LaRoche and UOP [24], producing catalysts S-2001/ESM-221, S-501/ESM-251, Axens [25] alumina catalysts CR 3-7, CR-400, CR-3S.

A new generation of Claus catalysts based on titanium dioxide [26] is now being actively implemented. The CRS-31 catalyst of French companies Rhone Poulenc and Elf Aquitaine (the current name is Axens Procatalyse, Paris, France) has gained broad recognition. Experience with its industrial use has revealed high stability for a long time in the presence of oxygen, high activity in the Claus reaction, and COS and CS_2; see also [25,27].

At the Boreskov Institute of Catalysis SB RAS, the activities of the oxides of various metals in the Claus reaction were comparatively studied [27,28]. Twenty-one oxides were investigated, of which nine were stable. These metal oxides can be arranged according to activity per surface area as follows:

$$V_2O_5 \gg TiO_2 > Mn_2O_3 > La_2O_3 > CaO > MgO > Al_2O_3 > ZrO_2 \gg Cr_2O_3$$

The surface activity of vanadium pentoxide is 16 times higher than that of titanium dioxide and 73 times higher than that of γ-alumina. However, pure V_2O_5 has a low value of specific surface area, and, in connection with the activity per unit of mass, it is inferior to TiO_2 and approximately equivalent to Al_2O_3. Furthermore, vanadium pentoxide is not very effective in the hydrolysis reaction of sulfur-organic compounds. However, its use in mixed catalytic systems is promising. Based on V_2O_5 at the Boreskov Institute of Catalysis SB RAS, the ICT-27-36 catalyst was developed. This catalyst is characterized by high activity in Claus and hydrolysis reactions, high stability at operation in oxygen-containing mixtures, and high mechanical strength [29].

6. Claus Process. Enhancement. Oxygen Enrichment

It should be noted that the enriched oxygen in the air supplied in the thermal stage of the Claus process is obtained using COPE® Technology (Kingswinford, UK, The Claus Oxygen Based Enhancement, Figure 3), developed by GOAR, Allison & Associates, LLC [30,31]. This technology is used in installations for sulfur production; its main advantage is the possibility of increasing in the power of the Claus process without incurring significant additional expenses.

Based on experiments and calculations, it was shown that an increase in the concentration of O_2 in the air, i.e., to 30 vol.% (a low degree of enrichment), could increase the

Claus installation capacity by 25%. It is proposed to transport liquid oxygen in cryogenic tanks without on-site cryogenic or membrane separation installations. This configuration is optimal, giving rise to sulfur production of up to 50 tons/day. The average degree of air enrichment with oxygen (C_{O2} = 40–45 vol.%) will increase the installation capacity by 75%. In this case, the furnace must be equipped with additional nozzles to supply oxygen. The anticipated production of such a plant is 100 tons sulfur/day. With an increase in oxygen concentration up to 100%, the daily production of sulfur can be increased by 150%. However, this will require significant changes in the structure of the flame furnace or the combustion of hydrogen sulfide using the "Sure" Double Combustion Process technology developed by Lurgi [32]. The process is conducted as follows: the burning of hydrogen sulfide with pure oxygen is carried out in a two-section furnace; the reaction products subsequently enter the sulfur condenser and the water condenser, and in recycling mode are then fed to the inlet of the torch.

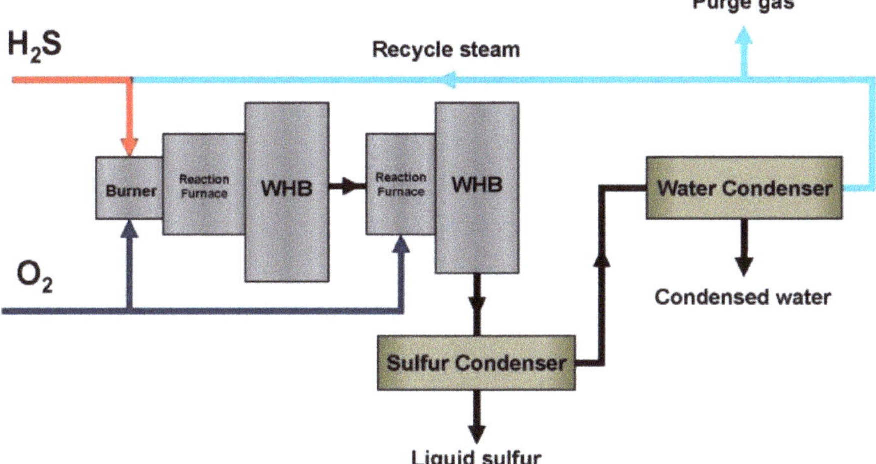

Figure 3. COPE® Technology (Kingswinford, UK) flow sheet diagram (adapted from [28]).

It should be emphasized that the cost of reconstruction (capital investment) of existing installations for the transition to the Cope® technology is only 5–25% of the construction cost of a new installation with increased capacity.

7. PROClaus Process

In the proposed concept of the modification of the Claus process, the first and second stages are standard: the combustion of hydrogen sulfide of acid gas in a flare furnace with further catalytic conversion of a mixture of hydrogen sulfide and sulfur dioxide in the first catalytic reactor. The main distinguishing feature of the PROClaus process is the use of a specially developed catalyst for sulfur dioxide reduction comprising the oxides of Fe, Co, Ni, Cr, Mo, Mn, Se, Cu, and Zn, in the presence of which, in a temperature range of 200–380 °C, there is almost complete sulfur dioxide conversion into elemental sulfur. Additional reducing agents are the products of side reactions proceeding in the high-temperature Claus furnace, i.e., CO and H_2 [33–35].

Furthermore, sulfur is separated from the gas stream containing hydrogen sulfide as the primary reagent, and the gas flows to the direct catalytic oxidation reactor which is filled with a Hi-Activity catalyst [36]. The Hi-Activity catalyst is a modified form of the KS-1 catalyst previously developed in the Azerbaijan Institute of Oil and Chemistry containing iron, zinc, and chromium oxides as the main components [37].

The calculated value of the total extracted sulfur from the gas is 99.2% using the three-reactor scheme and 99.5% for the four-reactor one. There characteristics were confirmed in laboratory studies of the concept of the process. However, the tests at the industrial level ended unsuccessfully: the hypothetical level of sulfur extraction was not observed, as the sulfur dioxide reduction catalyst did not achieve the proposed rate of SO_2 conversion into elementary sulfur. According to Alkhazov [38], in the process of laboratory studies, a factor of inhibition of the catalyst activity by sulfur vapor coming with the gas flow after the condenser of the first catalytic stage was not taken into account.

At the same time, according to the company JACOBS [39], the EUROCLAUS process was implemented on an industrial scale using the concept of catalytic reduction of SO_2 with the subsequent oxidation of hydrogen sulfide to elementary sulfur. In the EUROCLAUS process, an additional bed of the reduction catalyst is loaded into the Claus catalytic converter.

8. SuperClaus Process

The main distinguishing feature of the modification of Claus technology known as the SuperClaus process is the supply of substoichiometric air in the thermal stage (Figure 4). Such a method results in the reaction mixture composition after the second catalytic converter containing predominantly hydrogen sulfide at a concentration of 0.8–3.0 vol.%, with trace amounts of sulfur dioxide. Such a mixture passes to the third sequential reactor filled with a catalyst for the direct catalytic oxidation of hydrogen sulfide [40,41].

According to data provided by Jacobs Comprimo, the technology licensee from the beginning of the first industrial demonstration of the process in 1988, over 190 installations using the SuperClaus® process are currently operating or are under construction, with a total capacity of up to 1200 tons of sulfur per day.

According to the authors, the catalyst provides the sulfur yield in the third converter at a level of 85%, and the total sulfur yield is 99–99.5% [42]. It should be noted that industrial experience shows the inconsistency of the real and expected results. Thus, in the SuperClaus® process, at the stage of the selective oxidation of hydrogen sulfide under industrial conditions, the sulfur yield does not exceed 80–83%, and the achieved total degree of sulfur extraction in SuperClaus® industrial installations is 98–98.6% [43], instead of the declared 99–99.5%. However, a sulfur yield of 85% at the stage of selective oxidation of hydrogen sulfide in treatment of tail gases of the Claus process is currently recognized as the best modern level for technologies using the direct heterogeneous catalytic oxidation of hydrogen sulfide.

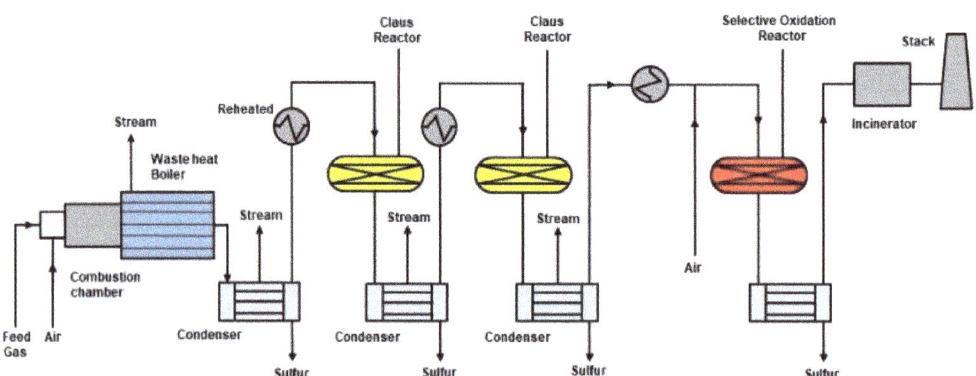

Figure 4. SuperClaus® process flow sheet diagram (adapted from [28]).

9. Modifications of the Claus Process

Research attempts have been made to optimize the Claus process (notably, the catalytic part). It has been proposed that the interaction of sulfur dioxide and hydrogen sulfide be carried out in the fluidized catalyst bed. Studies were carried out in a cylindrical reactor with an internal diameter of 0.1 m and a height of 0.86 m. Spherical active alumina of the Kaiser S-501 brand with an effective diameter of 195 μm was used as a catalyst. The maximum concentration of the reagents were H_2S-1300 ppmv and SO_2-650 ppmv, and the test temperatures were 100–150 °C, that is, below the dew point of the sulfur. The fluidization number was varied in the range of 2.2–8.8. It was shown that the observed conversion of sulfur compounds was ~96% in the initial period, although this decreased with an increase in sulfur sediments to 60% in 16 days of continuous operation. As the main advantages of the method, the authors note catalyst loading was reduced by up to 50% compared to the three-reactor scheme of the Claus process. The developed method could be considered as an alternative to the known processes of purification of tail gases based on sulfur condensation (CBA, Sulfreen) followed by the regeneration of catalytic material, and not a fundamentally new process for replacing the catalytic stages of the Claus process [44].

A process for the purification of hydrocarbon gas with hydrogen sulfide contents of 2.3–5 vol.% and carbon dioxide of 3–5 vol.% is proposed. The initial gas also contains from 40 ppmv to 90 ppmv benzene, toluene from 45 ppmv to 220 ppmv, xylene from 20 ppmv to 150 ppmv, carbon sulfoxide (COS) from 25 ppmv to 70 ppmv, heavy hydrocarbons (to C_{50}), mercaptans from 15 ppmv up to 50 ppmv. The overall gas processing complex includes the amine treatment installation and the "classic" three-reactor scheme of the catalytic conversion of sulfur dioxide and hydrogen sulfide into elementary sulfur. In the proposed procedure, the gas stream coming from the amine treatment unit is split in a ratio of 75%/25%, and the larger flow enters the first zone, a specially developed furnace, while the smaller stream enters a flushing column after the catalytic converters. In the flushing column, at the interaction of the gas flow components with a caustic soda solution, selective absorption of carbon dioxide occurs, and the stream enriched in hydrogen sulfide flows into the second zone of the thermal stage of the total process line. Such technological approaches are specifically used to extend the lower limit of the range of hydrogen sulfide concentrations in the initial gas stream to 30%. Furthermore, according to the authors' statements, this method is an alternative to the COPE process, while the complex as a whole will ensure the following characteristics of the purified gas: the content of hydrogen sulfide in the purified gas is 4 ppmv, the carbon dioxide content is not higher than 1.7 vol.%; and the content of organic sulfur compounds is not higher than 60 ppmv. It should be noted that the proposed procedure was conceived via computer simulations, and did not undergo any testing on the pilot or experimental levels [45].

There are also proposals to increase the degree of hydrogen sulfide conversion by its removal from the tail gas using reagents based on triazines in order to neutralize residual H_2S [46].

Researchers from Politecnico Di Milano developed a rather interesting concept, i.e., the simultaneous disposal of H_2S and CO_2 [47].

With regard to the gasification process of coal, they proposed the joint utilization of acid gas components by reacting carbon dioxide and hydrogen sulfide according to the following Equation:

$$2H_2S + CO_2 \rightarrow H_2 + CO + S_2 + H_2O \tag{14}$$

In this case, carbon dioxide is used as a "soft" oxidizer.

At the same time, Pirola and co-authors [47] demonstrated the results of a comparative analysis, where the superiority of the AG2STM process (Acid gas to SynGas) is shown in comparison to the traditional Claus process.

Thus, several essential problems were solved:

- The generation of additional synthesis gas

- Complete recovery of hydrogen sulfide in the form of elementary sulfur
- The utilization of carbon dioxide.

It should be noted that this work was performed on a computer simulation level, and and that the concept has not undergone laboratory and pilot testing.

10. Modern Trends in the Field of Hydrogen Sulfide Treatment with the Formation of Elemental Sulfur. Direct Heterogeneous Catalytic Oxidation of Hydrogen Sulfide to Elemental Sulfur

The process has some significant advantages, the main of which are:

- the single-step characteristic and continuity;
- "soft" conditions (T = 220–280 °C) due to the use of highly active catalysts, which allow for the oxidation of hydrogen sulfide directly in the composition of hydrocarbon.

It should be noted that the apparent advantages of the direct oxidation process are the main reason to consider the technologies using Reaction (15) as an alternative to Claus technology [48,49].

11. Chemism of the Process of Direct Catalytic Oxidation of Hydrogen Sulfide

In the process of direct H_2S oxidation, the following reactions can proceed [50]:

$$H_2S + 1/2O_2 \rightarrow H_2O + 1/2S_2 \tag{15}$$

$$H_2S + 1/2O_2 \rightarrow H_2O + 1/6S_6 \tag{16}$$

$$H_2S + 1/2O_2 \rightarrow H_2O + 1/8S_8 \tag{17}$$

$$H_2S + 3/2O_2 \rightarrow H_2O + SO_2 \tag{18}$$

$$H_2S + 1/2SO_2 \rightarrow H_2O + 3/4S_2 \tag{19}$$

$$H_2S + 1/2SO_2 \rightarrow H_2O + 1/4S_6 \tag{20}$$

$$H_2S + 1/2SO_2 \rightarrow H_2O + 3/16S_8 \tag{21}$$

The allotropic form of sulfur S_2 is stable in the temperature range of 100–900 °C. The characteristic temperature range for the reaction of direct oxidation of hydrogen sulfide is usually 100–300 °C; in this range, sulfur is present as S_6 and S_8. The so-called reverse Claus reaction accompanies the oxidation of hydrogen sulfide:

$$3S + 2H_2O \rightarrow 2H_2S + SO_2 \tag{22}$$

At reduced temperatures which are typical for direct oxidation, sulfur chains predominantly consisting of six or eight atoms are formed. With an increase in temperature to 800 °C, hydrogen sulfide oxidation proceeds mainly with the formation of sulfur in the form of S_2.

In the temperature range of 25–727 °C, the equilibrium constant of the hydrogen sulfide oxidation reaction with oxygen to elemental sulfur is, on average, 10 orders of magnitude higher than that of the reaction of oxidation with sulfur dioxide. Consequently, the probability of the formation of elemental sulfur in Equations (15)–(18) is higher than by Equations (19)–(21) [50].

The thermodynamic features of the reaction of the direct oxidation of hydrogen sulfide are presented in the form of temperature dependence (Figure 5) [51].

The reaction of direct H_2S oxidation can proceed with selectivity achieving 100% target product at low temperatures; with an increase in temperature above 200 °C, the selectivity significantly decreases. The chemical equilibrium is determined by the Claus reaction, i.e., the only reversible reaction of the system. If a catalyst with high activity for the reaction (15) is selected that is practically unaffecting the Claus reaction rate (21), then a super-equilibrium sulfur yield (100%) can be attained [52]. Therefore, the use of TiO_2- and Al_2O_3-based catalysts in this process is ineffective. An increase in pressure in the system favorably affects the yield of elemental sulfur and increases the selectivity, even at elevated temperatures.

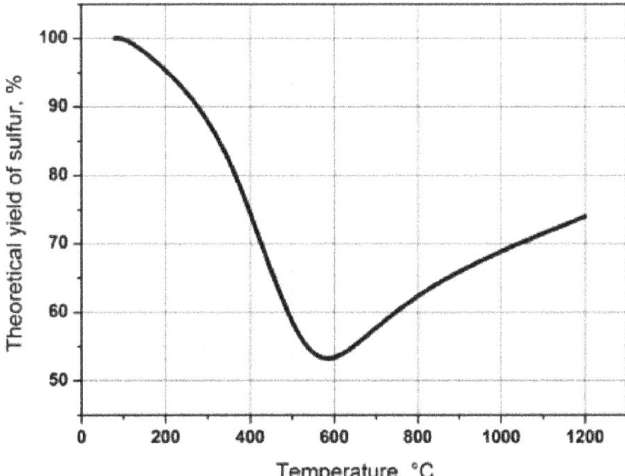

Figure 5. The dependence of the equilibrium sulfur yield on the temperature in the reaction of direct oxidation of hydrogen sulfide at atmospheric pressure (adapted from [51]).

12. Main Types of Catalysts Used in the Process of Direct Heterogeneous Oxidation of Hydrogen Sulfide. Industrial Processes. Brief Description of the Most Common Catalysts for the Hydrogen Sulfide Oxidation Reaction with Oxygen to Elementary Sulfur

Specific requirements associated with the particular features of the reaction are imposed on catalysts for the process of the direct oxidation of hydrogen sulfide into elemental sulfur. In terms of selecting a catalyst, the thermodynamics of the process should be taken into account, as well as the possibility of the homogeneous evolution of the process through the radical-chain mechanism at elevated temperatures and the condensation of sulfur in catalyst pores at low temperatures.

Catalysts for the direct oxidation of hydrogen sulfide to sulfur are used in the temperature range of 200–350 °C. The sulfur dew point determines the lower limit of the temperatures. The upper limit is due to the possibility of the reactions of sulfur and hydrogen sulfide oxidation to sulfur dioxide, which leads to a significant drop in the reaction selectivity.

Despite considerable efforts devoted to the direct catalytic oxidation of hydrogen sulfide described in the literature, the scope of catalytic systems for this process is somewhat limited. These, first of all, are activated carbons and artificial zeolites [53,54], as well as natural bauxites, traditionally used as catalysts for this process [55].

However, the most promising systems are individual metal oxides or mixtures of transition metal oxides due to their apparent advantages, i.e., high mechanical strength, thermal stability, and relative cheapness. It should be noted that oxides are used both in a bulk state and in the supported form. This is confirmed by the fact that all commercial processes for sulfur production from H_2S through its direct catalytic oxidation, such as Catasulf® of BASF (Ludwigshafen, Germany), BSR/SELECTOX® of Unocal Company (California, CA, USA), Modop® of Mobil Oil (Panama City, Panama), etc., are based on the application of heterogeneous multicomponent oxide catalysts.

13. Activated Carbon. Catalysts Based on Activated Carbon

As demonstrated above, active carbon (AC) simultaneously acts as the adsorbent of hydrogen sulfide and the catalyst for the oxidation of the latter to sulfur, as H_2S is transformed into sulfur which accumulates in AC pores upon purification.

Microporous ACs have been well investigated as adsorbents/catalysts for the periodical partial oxidation of hydrogen sulfide at temperatures below 150 °C [56–66]. They demonstrate high activity and selectivity under these conditions. As shown, a relatively large volume of large pores is required for the oxidation process to occur, whereas smaller pores serve for adsorptive desulfurization processes. Elemental sulfur is mainly accumulated in pores <12 Å, and initially in small and later in larger ones. It has been shown that there is maximum adsorption in cases where the pore size is maximally close to the size of the adsorbent molecules [67,68]. Hence, efficient carbon materials should have the optimal pore structure with a good volume of both micro- and meso- pores, and a relatively narrow pore size distribution to ensure high selectivity for sulfur. Nevertheless, the complete picture of the effect of AC pore structure on H_2S selective oxidation is not quite clear yet.

Primavera and co-authors [69] investigated the effect of water vapor on adsorbent efficiency. It was found that a relative moisture content of about 20% facilitates an enhanced reaction rate. The reaction rate drops dramatically when the moisture content is decreased, and less significantly when increased. It is assumed that HS^- ions are generated in water, being readily oxidized to sulfur with oxygen.

Surface chemistry has a significant effect on catalyst efficiency; therefore, AC-based sorbents/catalysts undergo modification with various reagents, such as metal salts [70] and alkaline [63,64,71] or oxidative (permanganate) additives [62], by the introduction of heteroatoms, such as nitrogen, oxygen, and phosphorus [72], and also by thermal treatment and controlled surface oxidation [73–75].

When AC is treated with nitric acid, oxygenated groups (C=O, C–O, and C–O–) are generated. Modified ACs contain charged oxygen particles, have higher catalytic activities, and may oxidize to 1.9 g of H_2S per g of catalyst, which is much higher than literature data for carbon catalysts [58,62,76,77].

The dynamic adsorption capacity of AC is reduced, as high temperatures decrease adsorption efficiency and selectivity for sulfur because COS and SO_2 are formed. In order to improve the capacity for sulfur and catalytic activity at high temperatures, AC modified with metal oxides is used [62]. At 180 °C and in the absence of water vapor, catalytic activity is varied in series, i.e., Mn/AC > Cu/Ac > Fe/AC > Ce/AC > Co/AC, being reduced to between 142 mg and 6 mg of H_2S/g for Mn/AC and V/AC, respectively. The major reaction product is elemental sulfur, which forms on active sites (carrier and coal micropores). When these pores are blocked with sulfur, the catalyst is deactivated.

When CO and CO_2 are present in the gas, a side product, COS, appears [78,79]. The impregnation of AC with sodium hydroxide facilitates hydrogen sulfide conversion [71,80–82], as NaOH improves H_2S dissociation to form hydrosulfide ion (HA^-) followed by its oxidation to S, SO_2, and H_2SO_4. Hydroxyl groups (OH^-) on the carbon surface enable binding SO_2 with COS due to an ion-dipole interaction between OH^- and COS.

Reaction conditions for the selective catalytic oxidation of hydrogen sulfide (temperatures, ratios of O_2/H_2S, and volumetric flow rate) have a significant effect on the activity and selectivity of AC-based catalysts [61]. Herewith, the microporosity and relatively small pore volume limit adsorptive capacity, with values of 0.2–0.6 g of H_2S/g for AC treated with alkalis and 1.7–1.9 g of H_2S/g for AC with oxygenated groups on the surface; sulfur saturation of the catalyst requires its frequent replacement. Other drawbacks of such adsorbents/catalysts preventing their wide applications are connected with the trend to spontaneously ignite upon hydrogen sulfide adsorption on alkaline AC and limited regeneration possibilities.

Sun and co-authors [83] describe the synthesis and properties of nitrogen-doped mesoporous carbon. This material shows a high concentration of catalytically active sites and a large pore volume. When nitrogen content is 8%, adsorptive capacity values of 2.77 g of H_2S/g at 30 °C and relative moisture content of 80% were achieved. The presence of pyridine nitrogen explains the elevated capacity. Nitrogen atoms located at the facets of graphite cavities have a high electron acceptor capacity, which facilitates the adsorption of oxygen atoms and therefore facilitates the oxidative reaction. Furthermore, the presence of pyridine active sites on the surface increases the basicity of the aqueous layer therein and simplifies H_2S dissociation to form HS^- ions. The nitrogen content plays a key role, affecting the basicity and thus the concentration of HS^-.

14. Catalysts Based on Carbon Nanotubes

Nanocarbon materials, i.e., carbon nanotubes (CNT) and carbon nanofibers (CNF), have recently attracted considerable research attention [84–86]. In particular, due to their lack of microporosity, diverse structures (the outer or inner diameter and the number of graphene layers), and rich surface chemistry (heteroatoms and structural defects), they are more promising than microporous AC, in which quite a few micropores substantially increase the role of diffusion. In particular, the tubular morphology of CNT ensures a special reactivity with liquid and gas reagents when passing through small tubes. For example, the so-called confinement effect [84] should be mentioned. Moreover, the chemical inertness of CNT species avoids problems of sulfation.

Metal oxides, alkaline agents, and heteroatoms are often used to modify CNT. According to the data of Nhut and co-authors [87], Ni_2S-modified CNT has a high capacity for sulfur (1.8 g of H_2S per catalyst) in a trickle-bed reactor. Active sites of Ni_2S are located inside the tube due to the confinement effect, and condensed water acts as the conveyor track, transferring elemental sulfur from the inner graphene layers to the outer ones in multilayer CNT, from where it is desorbed from the active phase. This mechanism ensures a high rate for hydrogen sulfide removal without any deactivation for 70 h. A substantial free catalyst volume makes it possible to save the resulting sulfur. However, because of the hydrophobic properties of Na_2S/CNT, condensed water is required to maintain high activity, which complicates reactor design and production.

Multiwalled CNT modified with Na_2CO_3 also make it possible to achieve capacity values of 1.86 g of H_2S/g catalyst at a low temperature (30 °C), which is approximately four times higher than commercial AC (0.48 g of H_2S/g catalyst) [88]. As in the case of NiS_2/CNT, a high capacity for sulfur is ensured by the presence of a large free volume formed by voids between CNT aggregates. In addition, introducing Na_2CO_3 increases the hydrophilicity and alkaline properties of CNT as an adsorbent. Alkaline properties promote the sorption and dissociation of H_2S into HS^- ions in the aqueous layer. The gradual deactivation of the catalyst is linked to a decrease in pH upon sodium sulfate formation and the blocking of catalyst pores with sulfur.

Hydrogen sulfide oxidation over multiwalled CNT decorated with tungsten sulfide was investigated in [89,90]. The metal content in the catalysts was 4.7–4.9%. The catalyst activity was examined compared to WS_2/AC and WS_2 catalysts using single-walled CNT under the following conditions: 5000 ppmv of H_2S, 20% of water vapor, a volumetric flow rate of 5000 h^{-1}, O_2/H_2S = 2, and a temperature of 60 °C. As shown, the catalyst over multiwalled tubes displayed the highest activity. The catalyst activity has been shown to increase with increased metal content but to cease when the latter is over 15%. When the volumetric flow rate is increased, the conversion degree naturally decreases. Upon an increase in temperature to between 70 °C and 180 °C, there is a high degree of conversion of hydrogen sulfide (at 180 °C), i.e., close to 10%, which is stable for 10 h, in contrast to the process performed at lower temperatures. This is related to the fact that sulfur is removed from catalyst pores more quickly at a high temperature, i.e., close to the melting point.

Macroscopical nitrogen-doped CNT (N-CNT) were developed by Ba al [91] for hydrogen sulfide oxidation at high temperatures (>180 °C) with heavy mass flow rates, WHSV = 0.2 − 1.2 h^{-1}. As demonstrated, H_2S conversion increases with nitrogen content, which is associated with a simultaneous increase in the concentration of active oxygen sites. Correspondingly, when the temperature was 250 °C, the degree of H_2S conversion and selectivity were 91% and 75%, respectively. When the catalyst is deposited onto a spongy carrier, SC, process indicators are substantially improved: conversion degree and selectivity reach 90% after 120 h of operation at 190 °C and high WHSV values.

A recent paper by Chizari and co-authors [92] investigates the activity of N-CNT catalysts formed as spherical granules with a diameter of around 5 nm. The test conditions were as follows: temperature of 210–230 °C, H_2S concentration of 1%, O_2 content of 2.5%, water level of 30%, and gas hourly space velocity (GHSV) of 2400 h^{-1}. As shown, the N-CNT catalyst was more efficient in terms of hydrogen sulfide removal compared to the

Fe$_2$O$_3$/SiC catalytic agent: the conversion degree reached 100% and selectivity was around 80% at 210 °C compared to the deposited oxide catalyst, for which H$_2$S conversion degree under these conditions was only 30%.

15. Carbon Nanofiber-Based Catalysts

As in the case of nanotubes, the main advantages of carbon nanofiber-based catalysts are related to the high thermal conductivity of the latter, chemical inertness, and the lack of ink-bottle pores where elemental sulfur may settle [93]. Furthermore, the presence of pores as microcavities between nanofibers increases the sulfur capacity of the material.

Using CNT for H$_2$S selective oxidation at high temperatures (>180 °C) has been investigated more widely than CNT-based catalysts. The latter are more promising from the standpoint of using a high excesses of oxygen to stoichiometry [94]. Herewith, the catalytic characteristics might be quite different depending on the nature of the initial catalyst over which the synthesis of nanofibers was carried out.

When water is absent, nanofibers produced over a Fe-Ni catalyst [95] with a structure of multilayered CNT have the highest selectivity for sulfur. The selectivity for sulfur is maintained at a level of 90%, whereas H$_2$S conversion degree decreases to 65% after 25 h of the reaction. The most highly active CNT samples were obtained using a Ni-Cu catalyst. After 25 h of reaction, hydrogen sulfide conversion degree and selectivity for sulfur were 95% and 70%, respectively. Compared to those species, nanofibers grown on Ni-catalyst displayed low activity because of sulfur deposits. In order to improve catalytic characteristics, these fibers were modified by treatment with HNO$_3$ or NH$_3$ [91]. As determined, acid treatment improved catalyst stability and selectivity for sulfur due to the the partial removal of nickel from CNF. In contrast, ammonia treatment reduced selectivity. As noted, the presence of 40% of water vapor improved the characteristics of the procedure, achieving a conversion degree of 70% and a selectivity of 89%.

Shinkarev and co-authors [96] investigated the process kinetics of selective hydrogen sulfide oxidation over CNT. The proposed kinetic model matched well with experimental results across a broad temperature range (155–250 °C) with hydrogen sulfide, oxygen, and water vapor concentrations of 0.5–2 vol.% and 0.25–10 vol.%, and 0–35 vol.%, respectively. The findings may be used when modeling processes and reactor designs for H$_2$S selective oxidation using nanofiber-based catalysts.

Chen and co-authors [97] systematically investigated H$_2$S selective oxidation over acrylonitrile-derived CNF impregnated with Na$_2$CO$_3$. The capacity for sulfur over these catalysts was shown to be 0.10–0.81 g of H$_2$S/g. First of all, the pore structure affected the sulfur capacity. Additionally, unlike other nitrogen-doped carbon materials, the concentration of nitrogenated functional groups almost did not affect the characteristics of the H$_2$S oxidation process. As demonstrated by analysis data, the prevalent product, i.e., elemental sulfur, was deposited in larger pores, whereas H$_2$SO$_4$ was generated in smaller ones.

The effect of temperature and water on H$_2$S selective oxidation over CNF-based macroscopic catalysts was analyzed by Coelho and co-authors [98]. Carbon nanofibers were grown over a graphite fiber substrate. The active phase was NiS$_2$. The catalyst demonstrated very high selectivity and stability at 60 °C owing to its stability to sulfur deposits removed therefrom through the presence of water and the hydrophobic properties of the catalyst. The efficiency of H$_2$S removal using catalysts based on new nanocarbon materials, i.e., CNT and CNF, was shown to be much higher, and material doping with nitrogen improved the purification process characteristics to a greater extent.

This research demonstrates that carbon materials are highly efficient during direct H$_2$S oxidation and the sorption of sulfur compounds.

Liu and co-authors [99] described the synthesis and study of a catalyst for the catalytic oxidation of H$_2$S to S at room temperature. The catalyst was activated carbon with supported iron and cerium oxides. The introduction of ceria was a positive factor, increasing catalytic activity due to the improved oxidation of Fe^{2+} to Fe^{3+} by redox-pair Ce^{4+}/Ce^{3+}. Also, the sorption capacity increased significantly. The adsorption-catalytic parameters of the system were investigated at a relative humidity of 80%, an oxygen content of 10 vol.%, a temperature of 30 °C and a space velocity of 7440 h^{-1}. The time of continuous stable operation with the sulfide conversion close to 100% was 71 h, and the value of the adsorption capacity was 820 mg S/g catalyst, which significantly exceeded this indicator for KNa/AC systems. It was found that the obtained sulfur is mainly precipitated inside the pore volume of the AC, but that some also formed on the AC surface.

Note that, depending on the nature of the process occurring on activated carbon during gas purification, the requirements for its porous structure may be different. For an adsorption process, carbons with narrow pores are required, the surface of which should have minimal catalytic activity. For catalytic oxidation of hydrogen sulfide, a wide-pore carbon is needed with a large total pore volume and, naturally, high catalytic activity. Large pores are needed to accumulate the resulting sulfur, in which up to 120% sulfur relative to the mass of the carbon can be adsorbed [100].

In connection with the development of technologies for the production of nanoscale carbon fibers, recently, the use of these materials and catalysts based on them in the reaction of partial oxidation of hydrogen sulfide to sulfur has attracted much interest [87,94,101]. It has been shown that carbon nanofibers make it possible to increase both the catalyst activity and sulfur resistance to deposition on the catalyst surface at low temperatures compared to conventional catalysts.

The issue of the use or regeneration of spent AC sorbents/catalysts deserves special consideration. Standard (industrial) processes for solving this problem are:

1. burning out sulfur at elevated temperatures
2. treatment of catalyst/sorbent with steam, with resulting hydrogen sulfide formation
3. washing catalyst with an organic solvent, effectively dissolving sulfur.

Obviously, the second and third options are the most acceptable for carbon materials, because, when exposed to oxygen at high temperatures, destruction (combustion) of the carbon matrix will inevitably occur.

16. Zeolite Catalysts for Direct Oxidation of Hydrogen Sulfide

Along with activated carbon, zeolites also can be used as adsorbents with catalytic properties for the oxidation reaction of hydrogen sulfide with molecular oxygen [102,103]. For low concentrations of hydrogen sulfide, the activity of zeolites NaX, NaY, NaA decreases with time; however, after a certain time (stabilization time), the fall in activity is terminated. The stabilization time decreases with a decrease in the hydrogen sulfide concentration in the gas and does not depend on the temperature of the process. At temperatures below 300 °C, the degree of transformation of H_2S does not depend on its initial concentration.

The oxidation of hydrogen sulfide on various zeolites was detailed studied in detail by Z. Dudzik and co-authors. In [103], the reaction of the direct oxidation of hydrogen sulfide on the sodium form of faujasite was examined. As shown, when the oxygen pulse is supplied to the activated zeolite NaX, on which hydrogen sulfide was preadsorbed, the sample became intensely paramagnetic, and the electronic paramagnetic resonance method allowed the registration of a sulfur biradical -*S-(S)-S*. The measurement of catalytic activity showed that sodium faujasite is an effective catalytic system for direct catalytic oxidation of hydrogen sulfide to elementary sulfur, partially paramagnetic sulfur. The degree of hydrogen sulfide removal from the initial gas flow gradually drops and reaches a constant value during the reaction. In this case, the level of stationary activity is directly proportional to the temperature of the process to temperatures of about 150 °C; the further temperature rise leads to the intensive formation of an unwanted by-product-sulfur dioxide.

Lee and co-authors [104], studied oxides of transition metals supported on NaX zeolite as catalysts for the oxidation of hydrogen sulfide in gaseous products of coal gasification. Coal gasification gases can contain H_2, CO, H_2S, CO_2, O_2, and H_2O. The authors identified the influence of the nature of transition metal oxides on the catalyst activity and selectivity to sulfur formation.

It was shown that the catalyst based on vanadium oxide showed the maximum activity (X_{H2S} = 70%) and selectivity (S = 80%). To characterize the composition of coal gasification gases [104–106] introduced the term "Reducing Power" [Equation (I)]:

$$Reducing\ Power = \frac{[H_2S] + [CO] + [H_2]}{[O_2] + [H_2O] + [CO_2]} \tag{23}$$

The strong dependence of the activity and selectivity of catalysts on the "Reducing Power" of coal gasification gases was established. In their work, the authors concluded that the vanadium oxide catalyst could be effectively used to remove H_2S from gases of coal gasification.

17. Catalysts Based on Sic

Recently, catalytic systems based on new materials are being intensively developed. One of these materials is SiC-silicon carbide. Catalysts supported on silicon carbide are proposed to be used in highly exothermal reactions such as partial oxidation of hydrogen sulfide to sulfur. Recently, several works were published offering catalysts-metal oxides on silicon carbide [105–107].

The use of SIC as support of hydrogen sulfide oxidation catalysts has several advantages:

1. The chemical inertness of the material allows the use of catalysts in aggressive media, providing high stability of catalysts;
2. High SiC thermal conductivity (150 W/m·K) compared to alumina (15 W/m·K) ensures a uniform temperature distribution in the catalyst bed and prevents local overheating of the catalyst;
3. SiC-based catalysts can be used to remove H_2S from highly concentrated gases (>2 vol.%);
4. The meso- and macroporous SiC structure allows the use of catalysts for the oxidation of hydrogen sulfide at temperatures below the dew point or in the presence of excess water.

Nguyen and co-workers [105] investigated Fe_2O_3-based catalysts supported on γ-Al_2O_3 and SiC in the oxidation reaction of 1 vol.% H_2S in the presence of 30 vol.% H_2O. It was shown that the SiC catalyst had much higher activity in the hydrogen sulfide oxidation reaction compared with the alumina-based catalyst.

The Fe_2O_3/SiC catalyst showed high activity in the hydrogen sulfide oxidation reaction and selectivity to the formation of elemental sulfur in excess of oxygen and in the presence of water vapor. To determine the nature of the active component, Fe_2O_3/SiC, FeS_2/SiC, and $FeSO_4$/SiC catalysts were synthesized [107]. It was shown that at H_2S conversion close to 100%, the selectivity to sulfur on these catalysts decreases in the following sequence:

$$FeSO_4/SiC > Fe_2O_3/SiC > FeS_2/SiC$$

The catalyst containing sulfate groups on the surface showed selectivity to sulfur of about 100% at 240 °C, whereas the formation of SO_2 was observed on the other catalysts in noticeable quantities. For example, on the FeS_2/SiC catalyst, the selectivity to sulfur was about 60%.

Keller and co-authors [106] proposed to use for the oxidation of hydrogen sulfide in the presence of water vapor NiS_2/SiC catalyst supported on mesoporous SiC. To avoid the catalyst deactivation in the reaction conditions, it was proposed to use a binary catalyst containing the hydrophobic SiC support and the hydrophilic layer of SiO_2 located in the support pores of the carrier. The transformation of the initial NiS_2 to nickel oxysulfide which has high activity in the hydrogen sulfide oxidation reaction explains the high activity of the proposed catalyst.

The mechanism of the catalyst deactivation in the absence of water vapor was proposed and an explanation of the high stability of the catalyst in the presence of water was found (Figure 6). According to the authors, the catalyst has a hydrophilic layer in SiC pores. Under the reaction conditions, in the presence of water vapor, the water film is formed on the hydrophilic layer, which delivers/transfers the resulting elemental sulfur to hydrophobic parts of the SiC support, where its deposition and the subsequent transition to the gas phase occurs. Thus, the active component remains available for reagents, and the catalyst is not deactivated (Figure 6a).

In the reaction medium without vapor water, such a film is not formed. Therefore, sulfur deposition occurs mainly on the active component of the catalyst, which leads to the capsulation of the active component and deactivation of the catalyst (Figure 6b).

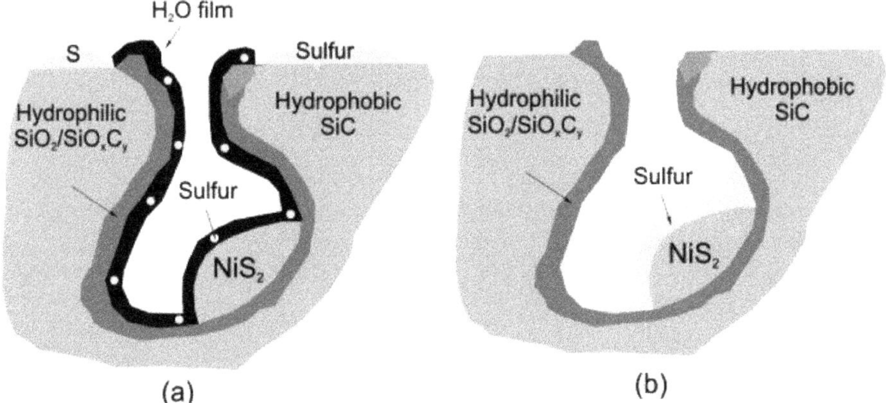

Figure 6. The tentative mechanism of sulfur deposition on the surface of a catalyst based on SiC: (**a**) in the presence of water vapor; (**b**) without water vapor (adapted from [106]).

Thus, SiC catalysts are promising for use in the reaction of partial oxidation of H$_2$S to sulfur. Several SiC-based catalysts were proposed. These are mainly iron oxide systems and a nickel sulfide-based catalyst. However, these catalytic systems are not optimal for the hydrogen sulfide oxidation reaction, and the development of the composition of an active component simultaneously active and selective in the reaction required additional research. The most optimal catalytic systems for the oxidation reaction of hydrogen sulfide to sulfur are transition metal oxides or a combination of oxides.

18. Transition Metal Oxides

Catalysts based on metal oxides are most widely used and studied in the continuous process of H$_2$S selective oxidation. Their main feature is that they provide a stable operation with different H$_2$S/O$_2$/H$_2$O ratios. Also, undoubted advantages of oxide catalysts are high mechanical and thermal stability, availability, wider ranges of hydrogen sulfide concentrations, and space velocities; therefore, their productivity is much higher than that, for example, of carbon-based catalysts [108].

This group of catalysts is promising for the direct oxidation of hydrogen sulfide due to the previously indicated reasons: high mechanical and thermal stability and availability.

Catalysts for gas-phase oxidation of hydrogen sulfide to elementary sulfur are known on bulk alumina or alumina with additives of titania (5.0–15.0 wt.%). The catalysts have high activity, and selectivity in a temperature range of 160–230 °C: the total conversion of hydrogen sulfide into sulfur and sulfur dioxide is 80–100% depending on the temperature range studied [109].

In the practice of gas-phase oxidation of hydrogen sulfide with air oxygen to elementary sulfur, the use of titania in the form of a mixture of its rutile (5–50 wt.) and anatase (50–95 wt.%) modifications as a catalyst is described. In the presence of the catalyst of the specified composition, it is possible at a space velocity of 3000 h^{-1}, a temperature of

230–280 °C, an initial H_2S concentration of 3 vol.% with the stoichiometric H_2S/O_2 ratio to provide ~98–100% conversion of hydrogen sulfide [110]. However, the catalyst of the specified composition has extremely low mechanical strength. The introduction into the catalyst of strengthening additives of magnesium oxide in an amount 0.3–1.0 wt.% and alumina slightly increases the mechanical strength (the catalyst attrition rate decreases by two times). More significant strengthening (4 to 6 times) is achieved by deposition of the active component on the faience aluminosilicate support.

The complete kinetic data relating to the oxidation of hydrogen sulfide to elementary sulfur on metal oxides can be found in the works of V.I. Marshnyova and Davydov A. A. [111], who studied more than twenty individual metal oxides under the following standard conditions for all samples:

$$T = 250\,°C$$

The concentration of reagents:

$$C_{H2S} = 0.5\ \text{vol.\%}$$

$$C_{O2} = 0.25\ \text{vol.\%}$$

It was shown that for the kinetic region, an activity row of individual oxides is as follows:

$$V_2O_3 > Mn_2O_3 > CoO > TiO_2 > Fe_2O_3 > Bi_2O_3 > Sb_6O_{13} > CuO > Al_2O_3 = MgO = Cr_2O_3$$

Representing the conversion of hydrogen sulfide in the form of three reactions: the Claus reaction (I), the total oxidation reaction (II), and the reaction of partial oxidation (III) [Equations (22)–(24)], Alkhazov and coauthors [112] found the following patterns for individual oxides.

$$2H_2S + SO_2 \leftrightarrow 2H_2O + 3/nS_n \tag{24}$$

$$2H_2S + 3O_2 \rightarrow 2H_2O + 2SO_2 \tag{25}$$

$$2H_2S + 3O_2 \rightarrow 2H_2O + 2/nS_n \tag{26}$$

It was shown that for the kinetic region, an activity row of individual oxides is as follows:

$$V_2O_5 > Mn_2O_3 > CoO > TiO_2 > Fe_2O_3 > Bi_2O_3 > Sb_6O_{13} > CuO > Al_2O_3 = MgO = Cr_2O_3$$

Whereas for the total conversion of hydrogen sulfide to sulfur and sulfur dioxide (II + III) these oxides can be arranged in the following row [111]:

$$V_2O_5 > Bi_2O_3 > Fe_2O_3 = CoO > Mn_2O_3 > Sb_6O_{13} = TiO_2 > CuO > Cr_2O_3 > Al_2O_3 > MgO$$

The maximum stationary activity in all reactions (I-III) is observed for vanadia V_2O_5. Analyzing the data on selectivity, Davydov and co-authors [111] concluded that the most selective catalysts for the process (III) are V_2O_5, MgO, and Mn_2O_3, while the oxides of Bi, Fe, and Cu are catalysts for deep oxidation of hydrogen sulfide to SO_2 (II).

The above series of activities are significantly different from similar ones given by T.G. Alkhazov and N.S. Amirgulyan [37] who studied the catalytic properties of metal oxides of the IV period to select the optimal catalyst for the partial oxidation of hydrogen sulfide. According to their data, the catalytic activity of individual oxides in the reaction of direct selective oxidation of hydrogen sulfide to elementary sulfur at temperatures of 280–300 °C decreases in the following sequence:

$$Co_3O_4 > V_2O_5 > Fe_2O_3 > Mn_2O_3 > CuO > TiO_2 > ZnO > NiO > Cr_2O_3$$

They also give the activity row of these oxides in the reaction of deep H_2S oxidation to sulfur dioxide:

$$Co_3O_4 > V_2O_5 > NiO > ZnO > CuO > Fe_2O_3 > TiO_2 > Cr_2O_3 > Mn_2O_3$$

Unfortunately, it is impossible to determine the causes of discrepancies, since in [39], absolute values of the specific rates and the conditions for conducting experiments are not given: the ratio of reagents, the size of the catalyst pellet, etc.

Batygina and co-authors [113], studied the catalytic activity of transition metal oxides deposited on γ-Al_2O_3 under the conditions listed below:

- The content of hydrogen sulfide in the feed, vol.% 20;
- Gas hourly space velocity, h^{-1} 7200;
- Hydrogen sulfide/oxygen polar ratio 2/1;
- Temperature range of testing, °C 200–300;
- The geometric shape of catalysts spherical granules;
- Active component individual oxides of cobalt;
- manganese, chromium;
- Vanadium;
- The active component content, wt.% 0.1–0.6.

It was shown that with the stoichiometric ratio of reagents, the activity of metal oxides decreased in the following sequence:

- Co > V > Fe = Cr > Mn > γ-Al_2O_3 (at T > 250 °C);
- V > Fe = Cr > Co > Mn > γ-Al_2O_3 (at T < 250 °C).

CeO_2-based catalysts are potentially suitable for H_2S-selective oxidation, but their practical application is limited due to the problem of sulfate formation. Shape-specific CeO_2 nanocrystals (rods, cubes, spheres and nanoparticles) with well-defined crystal facets and hierarchically porous structure were successfully synthesized and used as model catalysts to study the structure-dependent behavior and reaction mechanism for H_2S selective oxidation over ceria-based catalysts. It is deduced that the defect sites and base properties of CeO_2 are intrinsically determined by the surface crystal facets. Among the nanocrystals, CeO_2 nanorods with well-defined [110] and [100] crystal facets exhibits superb catalytic activity and sulfur selectivity. The high reactivity for H_2S selective oxidation is attributed to the high concentration of surface oxygen vacancies which are beneficial for the conversion of lattice oxygen to active oxygen species. Besides, the presence of hierarchically porous structure of CeO_2 nanorods hinders the formation of SO_2 and sulfate, ensuring good sulfur selectivity and catalyst stability. Through a combined approach of density-functional theory (DFT) calculations and in situ DRIFTS investigation, the plausible reaction mechanism and nature of active sites for H_2S selective oxidation over CeO_2 catalysts have been revealed. Thus, morphology engineering can be one of the effective methods in boosting the H_2S conversion [114].

The comparison of catalytic activity and selectivity, taking into account the stability of oxides, allowed Ismagilov and co-authors [115] to find that iron oxide is the most effective catalyst for the reaction of oxidation of hydrogen sulfide to elementary sulfur. Bulk iron oxide catalysts demonstrate high activity at 250 °C, providing almost 100% conversion of hydrogen sulfide at sufficiently high selectivity [112]. It was also shown that the method of iron oxide preparation does not significantly affect the conversion of hydrogen sulfide and the selectivity of the process. The effects of additives of K, Cr, Ag, Ti, V, Mn, and anions: Cl^-, SO_4^{2-}, PO_4^{3-} in the amount of 1–5% to the initial catalyst on catalytic properties were investigated. In particular, it was demonstrated that the introduction of chromium ions leads to a decrease in the activity, and vanadium ions and SO_4^{2-} to a decrease in the selectivity. A particular feature of the studied catalysts is their ability to oxidize hydrogen sulfide in the presence of hydrocarbons of natural gas without their involvement in the reaction. A significant advantage of these catalysts is their ability to selectively oxidize hydrogen sulfide under an over stoichiometric O_2/H_2S ratio.

An increase in the selectivity of the iron oxide catalyst at elevated temperatures can be achieved by deposition of the active component on an alumina support. The most active catalyst at a temperature of 300 °C is the catalyst of the following composition: 0.5 wt.% Fe_2O_3/Al_2O_3 [116]. 34A mixed catalytic system was investigated as a catalyst for partial oxidation of hydrogen sulfide, which consists of iron and titanium oxides (anatase modification). It is proposed to use this system in the process of two-stage oxidation of hydrogen sulfide (at the concentrations of hydrogen sulfide over 5 vol.%). When conducting the process in a temperature range of 200–300 °C, it is possible to remove hydrogen sulfide in the form of elementary sulfur with an efficiency close to 98 99% [117].

It is necessary to emphasize the work creating a catalyst for the direct oxidation of hydrogen sulfide made in VEG-Gasinstitut and the University of Utrecht (The Netherlands) under the general scientific leadership of Professor J. Geus [118–120]. The overall goal of these works is to create an effective and highly selective catalyst for the direct oxidation of hydrogen sulfide by using alumina with a low specific surface area (α-modification). It was experimentally shown (Figure 7) that the selectivity of the iron oxide supported on to α-alumina had higher selectivity than the catalyst on γ-alumina even in conditions of a significant excess of oxygen in the reaction mixture. The authors' explanation of the observed results is as follows. By the use of α-alumina, it is possible to synthesize the catalyst in which the active component evenly covers the support surface and prevents the diffusion of reagents to the surface of alumina, which has been shown to actively catalyze the reverse Claus reaction-the interaction of sulfur and water vapor to form hydrogen sulfide and sulfur dioxide. This iron catalyst was specially designed for the created SUPERCLAUS®process.

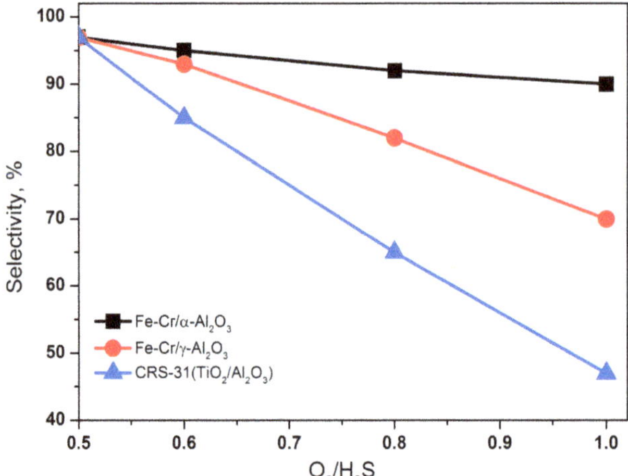

Figure 7. The dependence of the selectivity of the process of direct H_2S oxidation on the ratio O_2/H_2S in the initial mixture (adapted from [118]).

The same authors undertook interesting attempts [121] to create effective catalytic systems based on various composite systems and, in particular, alloys of type Hastelloy-X and Inconel with fillers-SUPERCLAUS®commercial catalysts (the amount of additive did not exceed 2 wt.%). The simultaneous use of such materials as catalysts and reactor construction materials was proposed, that is, the effective combination of construction and catalytic properties. It was shown that such systems in the future could make a serious competition to classic reactors with bulk catalysts. Particular emphasis was bestowed on the possibility of using such structures (combined reactor-catalyst) to carry out the hydrogen sulfide oxidation process at a high H_2S content, given the high heat engineering characteristics of the developed materials. However, it was indicated that the upper limit of the content of hydrogen sulfide for the effective operation of these systems should not exceed 10 vol.%.

New possibilities extending the range of the application of the technology of direct H_2S oxidation are provided by monolithic honeycomb catalysts, which possess several technological advantages over granulated catalysts (most important of them low pressure drop), especially for the gases with low excessive pressure and for purposes when the pressure loss is unacceptable.

For the first time, such studies were undertaken by the research team under the guidance of Professor Z.R. Ismagilov. Successful pilot and experimental-industrial tests of the direct oxidation process in reactors with monolithic catalysts of the honeycomb structure to purify the tail gases of the Claus process and geothermal steam are reported [122–124].

Laboratory studies of the catalysts for direct oxidation of hydrogen sulfide in the form of monolithic catalysts of the honeycomb structure were also reported by Italian scientists [125]. Monoliths from cordierite (9 channels) from 10 to 50 mm long, 6 mm wide, and 6 mm high with 226 channels per square inch (CPSI) were used as a substrate for coating. The commercial ceria-zirconia composition (EcoCat) having the initial solids content of 40 wt.% was deposited on cordierite. The active phase (V_2O_5) deposition was carried out from an aqueous solution of ammonium metavanadate (NH_4VO_3). The authors reported that at a temperature of 200 °C, contact time >200 ms and initial H_2S content 500 ppmv, high conversion of H_2S (90%) and a very low selectivity toward SO_2 (3%) were obtained.

Similar data are given by Eom and co-authors [126], where the results of studies on the use of selective catalytic oxidation to remove hydrogen sulfide from landfill gas using monolithic catalysts of the honeycomb structure are described. The efficiency of removing H_2S at a temperature of 200 °C was the highest for the V/TiO_2 catalyst obtained by incipient wetness impregnation. The optimal content of vanadium is 10% by weight. In addition, it was shown that the selectivity to sulfur and minimization of the formation of SO_2 substantially depends on the O_2/H_2S ratio. It is shown that with increasing the number of CPSI, the honeycomb catalyst productivity can be significantly increased. The efficiency of H_2S removal also increases with an increase of the specific surface (m^2/m^3). The analysis of the long-term operation of a honeycomb catalyst at the cleaning of landfill gas with the composition including CH_4 and CO_2 (typical components) showed that the purification degree is more than 90%. In addition, the catalyst's performance can be restored by thermal regeneration at sufficiently "soft" conditions (400 °C, 3 h in airflow).

19. Description of Modern Industrial Methods Based on the Process of Direct H_2S Oxidation

The Catasulf® process of the German company BASF (Ludwigshafen, Germany) [127] is based on the reaction of the oxidation of the acid gas containing 5–15% H_2S (I) in the tubular reactor 1 (Figure 8). The tube space of reactor 1 is filled with a special highly selective catalyst, which is a mixture of aluminum, nickel, and vanadium oxides, and the inter-tube space is cooled with a high-boiling liquid silicon coolant (II), which, circulating, transfers the removed heat to the refrigerator 2. Gases emerging from reactor 1, (III) are cooled in the sulfur condenser 3 and fed into the adiabatic reactor 4, where there is a further interaction of hydrogen sulfide with sulfur dioxide. The resulting sulfur is separated in the second consecutive condenser 5. Removing the sulfur in the first stage is 94%, after the adiabatic reactor up to 97.5%. It is supposed that by increasing the number of stages, it is possible to attain the degree of sulfur extraction of 99.99%. At the oil refining plant, Ludwigshafen (Germany), the only Catasulf® industrial installation is operated.

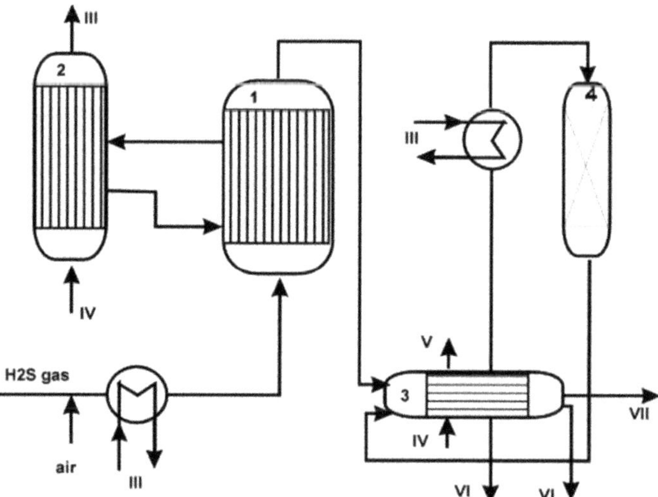

Figure 8. Schematic of Catasulf® process Catasulf®, perhaps, the only "active" large-capacity technology to obtain sulfur based on the reaction of the direct oxidation of hydrogen sulfide (adapted from [127]).

The Sulfatreat® DO process of the company M-I Swaco [128,129] is a technology for the purification of associated oil gases from hydrogen sulfide by direct oxidation. The process is carried out at a temperature of 175 °C, a pressure of up to 65 bar and allows the treatment of gases containing up to 3 vol.% hydrogen sulfide. The catalyst is a mixture of metal oxides of transitional valence, promoted with alkali metals oxides. The results of tests of the experimental installation showed that the H_2S concentration in the purified gas does not exceed 950 ppmv. The degree of sulfur removal, in this case, was more than 88%, with only a slight conversion of the hydrocarbon part.

Other industrial processes based on this reaction are designed exclusively for cleaning tail gases of existing sulfur-producing installations. One of the most common methods is BSR/Selectox® of Unocal and Ralph M. Parsons companies. In this process, the exhaust gases of the Claus installation are reduced in a catalytic reactor by synthesis-gas formed in a special generator by a steam reforming of natural gas. Then the resulting gas is cooled, mixed with air, and hydrogen sulfide is subjected to selective oxidation to sulfur at a temperature of 200–230 °C. The use of a special vanadium oxide catalyst Selectox-67 allows attaining the selectivity of oxidation to sulfur of ca. 100%. The Beavon-Selectox process ensures the degree of sulfur removal of 98.5–99.5% at a relatively low installation cost (about 50–60% of the cost of a Claus installation [130]. The first such installation was launched in 1978 in Germany.

Somewhat later, other technologies similar to the Beavon-Selectox process were developed. Among them are the most famous MODOP® of Mobil Oil Corp (Dallas, TX, USA) [131]. and SUPERCLAUS® of Comprimo BV Company (Dallas, TX, USA) [42]. They differ from the Beavon-Selectox process by using other catalysts (CRS-31 for MODOP and a special highly selective iron oxide catalyst for the SUPERCLAUS® process), and by the fact that the oxidation of hydrogen sulfide is carried out in two stages. In addition, in the SUPERCLAUS® process, it is possible to use the H_2S direct oxidation stage without the hydrogenation stage due since in the Claus installation, the process is carried out with an excess of hydrogen sulfide (this allows, among other things, protect the catalyst in the Claus reactors from sulfation). These processes provide the total degree of sulfur removal of 99.3–99.5%. The first two MODOP® (Dallas, TX, USA) installations were put into operation in Germany in 1983 and 1987 and two SUPERCLAUS® installations in 1988 in Germany and 1989 in Holland [132–134].

The advantage of the processes described above is the possibility to supply air for the oxidation of hydrogen sulfide in a small excess compared to stoichiometry, which simplifies the control of the process in the conditions of variation of the composition and flow rate of the reaction mixture.

However, the use of direct heterogeneous catalytic oxidation of hydrogen sulfide is significantly limited because of intense heating of a fixed catalyst bed due to high heat generation. At the Boreskov Institute of Catalysis SB RAS, the technology of direct oxidation of hydrogen sulfide in a reactor with a fluidized catalyst bed was developed, which is largely free of these shortcomings.

A research program was implemented under which the effects of temperature and concentration of components on the kinetic parameters of the direct hydrogen sulfide oxidation process were studied. The oxidation of hydrogen sulfide in the composition of hydrocarbon-containing mixtures, the kinetic parameters of the hydrogen sulfide process for various catalytic systems, and the elementary stages of the process were investigated, and the activities of a wide range of supported oxide catalysts in the target reaction were measured.

The main results of the research are the following:

A wide range of supported catalysts meeting the requirements for catalytic systems operating in the reactor with a fluidized bed by their structural and mechanical characteristics (i.e., high mechanical strength and thermal stability) were synthesized and characterized [113,135,136]. Honeycomb catalysts have also been developed for the H_2S oxidation process [137–139].

When studying the regularities of the reaction on the magnesium-chromium oxide catalyst, it was found that the following equation could describe by reaction kinetics [Equation (24)] [140]:

$$W = \kappa \cdot \left(C_{H_2S}\right)^m \cdot \left(C_{O_2}\right)^n \tag{24}$$

The orders of m and n have similar values close to 0.5.

This value of the observed order of the hydrogen sulfide oxidation reaction indirectly indicates that the first elementary stage of the process is the dissociative adsorption of hydrogen sulfide on the catalyst's surface. The activation energy is significantly lower than the value found for alumina, and it is about 8.1 kcal/mol [140].

The effect of hydrocarbons in the composition of the gas mixture on the parameters of the reaction of the direct oxidation of hydrogen sulfide was studied, which is a fundamental issue in developing the scientific foundations for the purification of hydrogen sulfide containing fossil fuels.

As can be seen from the results presented in Figure 9, the temperature areas of the effective action of catalysts for the selected reactions are sufficiently separated, that is, in the temperature range of 220–260 °C, where the sulfur yield achieved is close to 100% propane oxidation reaction proceeds at a low rate [141].

Since kinetic data is quite formal and does not give unequivocal information about the mechanisms involved in the process, attempts have been made to study the elementary reaction stages using spectral methods [142–144]. To this end, three systems were selected:

- Baseline magnesium-chromium oxide catalyst $MgCr_2O_4/\gamma\text{-}Al_2O_3$
- Iron oxide catalyst $Fe_2O_3/\gamma\text{-}Al_2O_3$
- γ-alumina $\gamma\text{-}Al_2O_3$.

FTIR spectroscopy of the adsorbed CO revealed that all the catalysts had both Lewis and Broensted acid sites on the surface (Figure 10). However, the nature, strength, and number of sites varied according to the type of catalyst. H_2S adsorption and the formation of intermediates occurred on Lewis acid sites, as confirmed by the disappearance of LAS bands after H_2S adsorption.

The adsorption of H_2S on the surface of catalysts (Figure 11) led to the formation of two types of surface species: sulfates (I) at 1100 cm^{-1} and (II) registered at higher frequencies 1264 and 1342 cm^{-1}, corresponding to organic sulfates. The sulfates of type (I) formed on γ-alumina at 100 °C, while type (II) formed at 250 °C. The formation of these two species was detected at much lower temperatures on $Fe_2O_3/\gamma\text{-}Al_2O_3$ and $MgCr_2O_4/\gamma\text{-}Al_2O_3$ due to their higher oxidative activity.

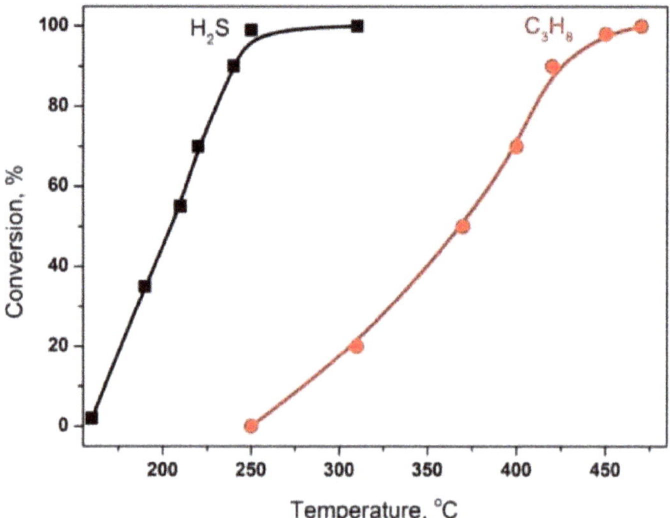

Figure 9. Results of laboratory experiments on the separate oxidation of propane and hydrogen sulfide over a $MgCr_2O_4/\gamma\text{-}Al_2O_3$ catalyst (Catalyst: $MgCr_2O_4/\gamma\text{-}Al_2O_3$; residence time: 0.8 s; C_{H2S}: 30 vol.%; C_{C3H8}: 15 vol.%).

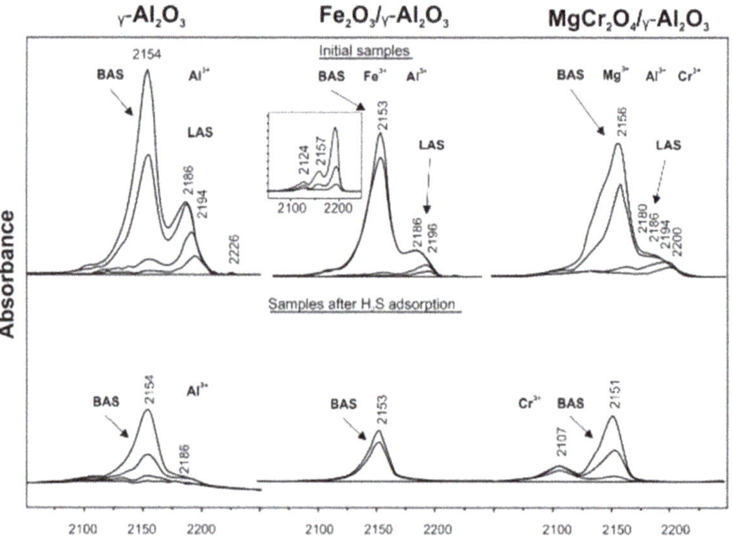

Figure 10. IR spectra of adsorbed CO: initially and after H_2S adsorption.

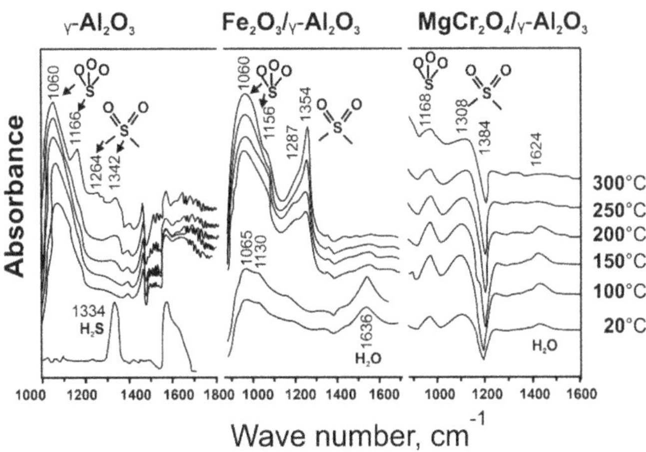

Figure 11. IR spectra of samples of different catalysts after adsorption of 20 torr of H_2S at various temperatures.

The DRS study (Figure 12) revealed that various types of elemental sulfur, i.e., S_4–S_8, formed on the catalyst surface during the reaction depending on the nature of the catalyst.

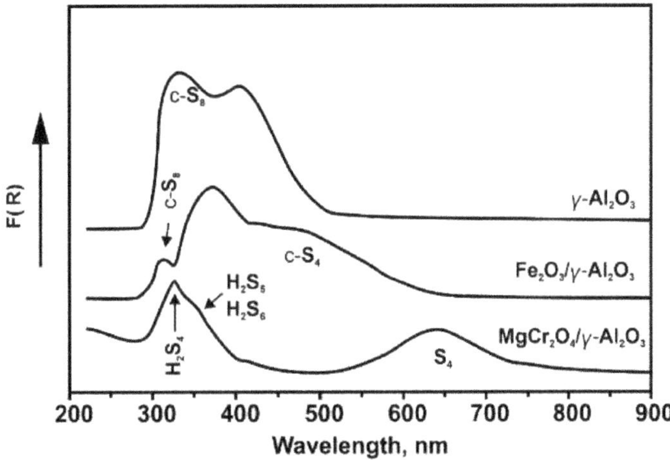

Figure 12. Different sulfur species formed on the surface of catalysts after H_2S oxidation, as detected by UV DRS (30 torr H_2S, 1 h, T = 250 °C).

Based on the data obtained, the reaction mechanism for the direct oxidation of hydrogen sulfide can be represented by the schematic depicted in Figure 13.

In the first stage, the hydrogen sulfide is adsorbed on the surface of the catalyst. The adsorption can occur (i) through the participation of the LACs and the sulfur atom of the hydrogen sulfide molecule, (ii) through the participation of the BACs and the sulfur atom with the formation of hydrogen bonds, or (iii) through the participation of a particular catalyst center, e.g., surface oxygen and the proton of the H_2S molecule. The adsorption on Lewis acid centers leads to the greatest activation of the hydrogen sulfide molecule.

Next, the hydrogen sulfide molecule adsorbed on the Lewis acid center can interact with a neighboring oxygen atom of the catalyst or a hydroxyl group. This process can lead to the dissociation of hydrogen sulfide molecules to form a hydroxyl group or water.

The oxygen of the catalyst surface oxidizes the formed surface particles to form surface SO_2 groups, which, upon interacting with the hydrogen sulfide molecule from the gas phase or with an adsorbed hydrogen sulfide molecule, will yield the final reaction products, i.e., elementary sulfur and water, via the surface Claus reaction.

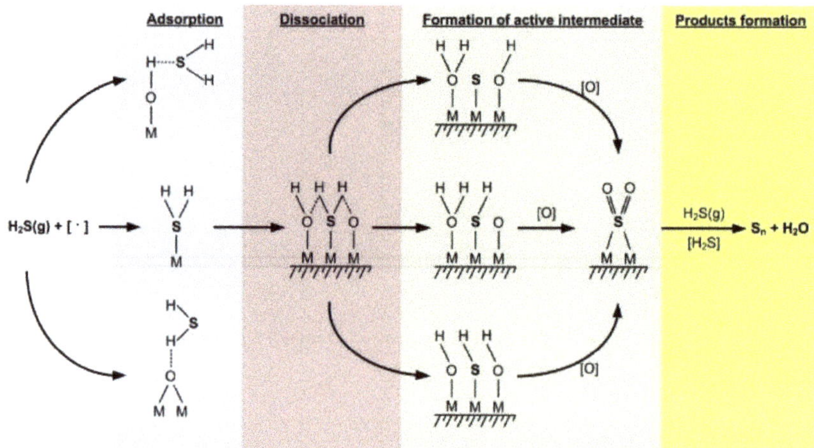

Figure 13. Proposed mechanism of H_2S oxidation on oxide catalysts.

A raw hydrogen sulfide-containing gas is supplied to the reactor with a fluidized catalyst. Simultaneously, oxygen (or air) is fed into the catalyst bed via a separate flow. Before the gas stream supply, the catalyst bed is heated to initiate the catalytic reaction. The excessive heat of the exothermic reaction of H_2S oxidation is efficiently removed by a heat-exchanger in the fluidized bed. The bed temperature is maintained within the preset range (280–320 °C) with high uniformity by regulating the amount of heat removed from the bed with a heat-exchange agent.

The technology was successfully tested on a pilot and industrial scale in Russia's largest sour gas fields, refineries, and gas processing plants.

20. Developments of the Boreskov Institute of Catalysis SB RAS Regarding the Creation of Processes of Heterogeneous Catalytic Oxidation of Hydrogen Sulfide for the Treatment of Various Gases

At the Boreskov Institute of Catalysis SB RAS, various technologies for direct catalytic oxidation of hydrogen sulfide have been developed.

Using data from the development of highly exothermic catalytic processes, in particular, processes of combustion of organic fuels in fluidized catalyst bed reactors [145–148], a new technology using highly concentrated hydrogen sulfide-containing gas was proposed, the essence of which consists of a reaction conducted in a fluidized bed catalytic reactor-Modification 1 (Figure 14).

Due to moderate temperatures (250–320 °C) applied in this technology, no hydrocarbon cracking reactions were observed. Thus, hydrogen formation seems unlikely. This assumption was confirmed by the results of the pilot and industrial tests, that showed:

1. The preservation of qualitative and quantitative composition of hydrocarbons, and
2. The absence of hydrogen in the reaction products after the reactor.

The primary source, which is potentially dangerous from the viewpoint of explosion safety, is the catalytic reactor in which the formation of hydrogen sulfide mixtures at explosive concentrations (i.e., 4.3–45.5 vol.% in the air) is possible. However, this problem is minimized by the following factors:

1. Hydrogen sulfide is almost completely removed on the first three dm of the catalyst bed; that is, its concentration drops significantly, i.e., to below the explosion limit.
2. The reaction proceeds solely on the surface of the catalyst, so the transition of the process into the reactor volume proceeding according to the homogeneous chain (explosive) mechanism is excluded. Thus, the catalyst bed acts essentially as an effective flame arrester.

A feed of hydrogen sulfide-containing gas is supplied to the reactor via a fluidized catalyst bed. Simultaneously, oxygen (or air) is fed into the catalyst bed via a separate flow. Before the gas stream supply, the catalyst bed is heated to initiate the catalytic reaction. The excessive heat of the exothermic reaction of H_2S oxidation is efficiently removed by a heat-exchanger in the fluidized bed. The bed temperature is maintained within the preset range (280–320 °C) with high uniformity by regulating the amount of heat removed from the bed with a heat-exchange agent [49,149–153].

At the same time, there is a treatment problem, i.e., a low pressure drop is required in the reactor, for gases with low concentrations of hydrogen sulfide, such as the tail and ventilation gases of various chemical industries, as well as the purification of energy carriers, such as oil-associated gases and geothermal steam where the pressure loss is extremely undesirable.

To solve these problems, the reaction was conducted in the reactor with a monolithic catalyst with a honeycomb structure (process modification 2, Figure 15). Such catalysts have some advantages, in particular a low pressure drop and a high ratio of the outer surface area to volume [154]. Technologies have been tested on a pilot and experimental industrial scale, and their abilities to clean various gases containing hydrogen sulfide have been demonstrated (Tables 2 and 3, Figures 16–20).

Table 2. Pilot and experimental industrial tests of the technology of direct catalytic oxidation of hydrogen sulfide (Modification 1-fluidized bed) [140].

#	Location Object H_3S Content	Operation Conditions		Year	H_2S, %
		Scale	Gas Supply		
1	Astrakhan sour gas field Natural gas $C_{(H2S)}$ = 27 vol.%	Pilot	up to 50 nm^3/h	1987	98
2	Astrakhan sour gas field Natural gas $C_{(H2S)}$ = 27 vol.%	Pilot	up to 50 nm^3/h	1988	98
A3	Astrakhan sour gas field Natural gas $C_{(H2S)}$ = 27 vol.%	Pilot	up to 20 nm^3/h	1991	98
4	Ufa Refinery Hydrodesulfurization gas $C_{(H2S)}$ = 70 vol.%	Pilot	up to 50 nm^3/h	1990	98
5	Shkapovo GPP Acid gas from amine unit $C_{(H2S)}$ = 65 vol.%	Semi-industrial	up to 350 nm^3/h	1995	98
6	Bavly oil field Acid gas from amine unit $C_{(H2S)}$ = 65 vol.%	Semi-industrial	up to 70 nm^3/h of acid gas	2004–2009	99.5

Table 3. Pilot and experimental industrial tests of the technology of direct catalytic oxidation of hydrogen sulfide (Modification 2-Honeycomb Catalyst) [140].

#	Location Object H$_3$S Content	Operation Conditions		Year	H$_2$S, %
		Scale	Gas Supply		
1	Novo-Ufimsky Refinery Tail gas of Claus process $C_{(H2S)}$ = 2 vol.%	Pilot	up to 20 nm^3/h	1989-1990	98
2	Astrakhan GPP Tail gas of Claus process $C_{(H2S)}$ = 2 vol.%	Pilot	up to 20 nm^3/h	1991	98
3	Orenburg GPP Gases of zeolites regeneration $C_{(H2S)}$ = 2 vol.% $C_{(RSH)}$ = 5 vol.%	Pilot	up to 20 nm^3/h P up to 0.5 MPa	1990	98
4	Kamchatka peninsula Geothermal steam $C_{(H2S)}$ < 1 vol.% $C_{(H2O)}$ > 99 vol.%	Fixed bed Pilot	up to 0.5 tn. steam/h P up to 1.0 MPa	1989-1990	99.9 2500 h of continuous operation
5	Novo-Ufimsky Refinery Tail gas of Claus process $C_{(H2S)}$ = 2 vol. %	Semi-industrial	up to 7000 nm^3/h	1994	98

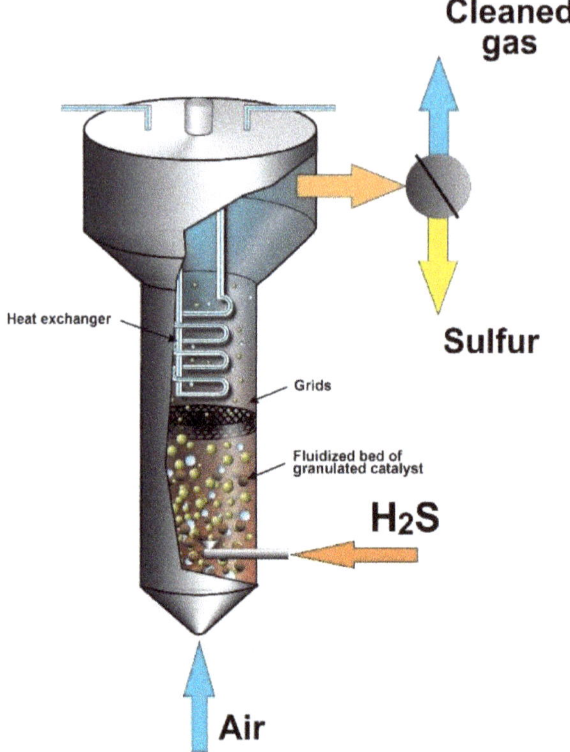

Figure 14. Direct catalytic oxidation in a reactor with a fluidized catalyst bed. Basic engineering concept.

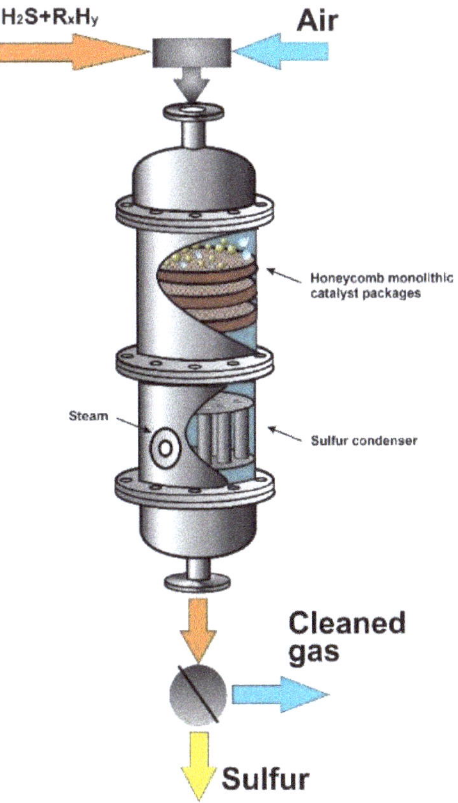

Figure 15. Direct catalytic H_2S oxidation in a reactor via a monolithic catalyst with a honeycomb structure.

Figure 16. Pilot Plant at the Ufa Refinery.

Figure 17. Mutnovskoe deposit of geothermal steam.

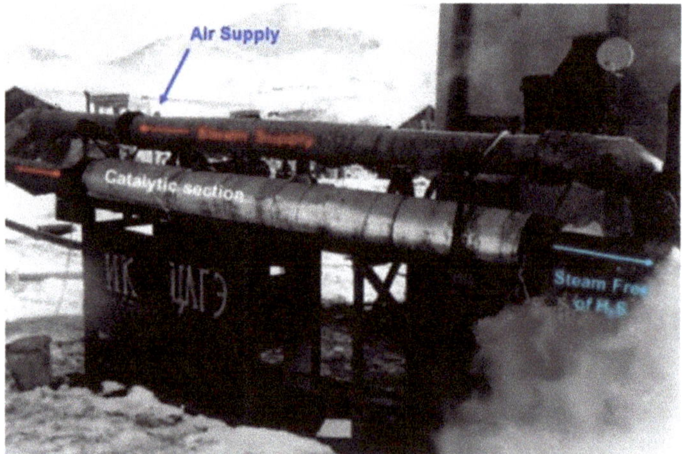

Figure 18. Pilot plant for H$_2$S removal from geothermal steam.

Figure 19. Catalytic segment after 2500 h of continuous operation.

Figure 20. Semi-industrial installation for the direct oxidation of hydrogen sulfide via monolithic catalyst with a honeycomb structure. Tail-gas of the Claus process.

21. Installation for H_2S Recovery from Acid Gas after the Amine Treatment of Oil-Associated Gases at Bavly Gas Shop of PJSC Tatneft

The use of associated petroleum gas (APG) is strictly regulated according to the legislation implemented by the Russian government on 8 November, 2012 (#1143, edited on 17 December 2016), which states that "Regarding peculiarities of the cost calculation for the negative environmental impact during emission of pollutants generated upon combustion using flare facilities and/or associated petroleum gas scattering". The legislation also includes a statement on the peculiarities of cost calculation for a negative environmental impact due to the emission of pollutants generated by facilities using a combustion flare and/or associated petroleum gas scattering.

Additionally, APG is a source of the propane-butane fraction for petroleum chemistry companies in Russia. This fraction is often in short supply. In order to address the issue of the primary removal, sorption facilities for amine treatment have been developed to remove hydrogen sulfide and transport hydrocarbon components to the appropriate sites of further treatment. However, the problem is addressed only partially, as the hydrogen sulfide released is burnt with flares.

Typical examples of the implementation of such an approach are the Bavlinsky gas workshop, PJSC Tatneft, Shkapovskiy, and Tuymazinskiy gas processing plants of PJSC ANK Bashneft.

In 2011, an industrial installation with a fluidized catalyst bed for the removal of hydrogen sulfide from acid gases from the amine treatment of oil-associated gases was created and put into operation by the Boreskov Institute of Catalysis SB RAS [155–158] at the PJSC Tatneft Bavly gas shop (Figures 21–23).

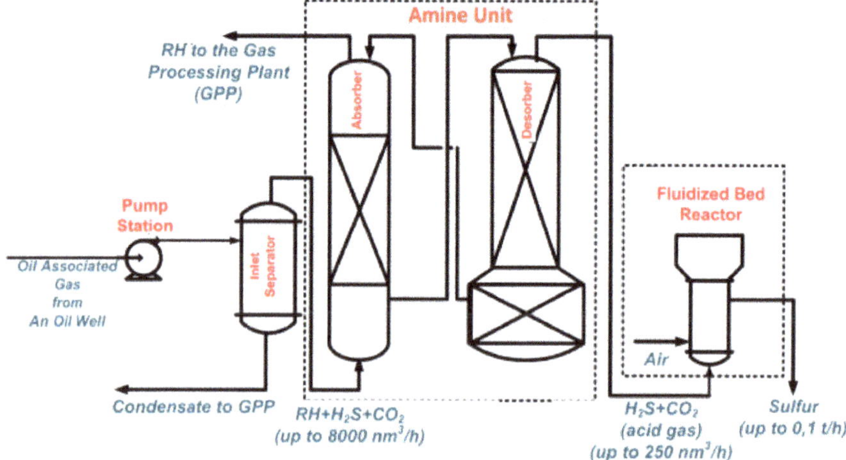

Figure 21. Bavly gas shop of PJSC Tatneft. Purification of associated oil gas. Amine treatment and direct oxidation.

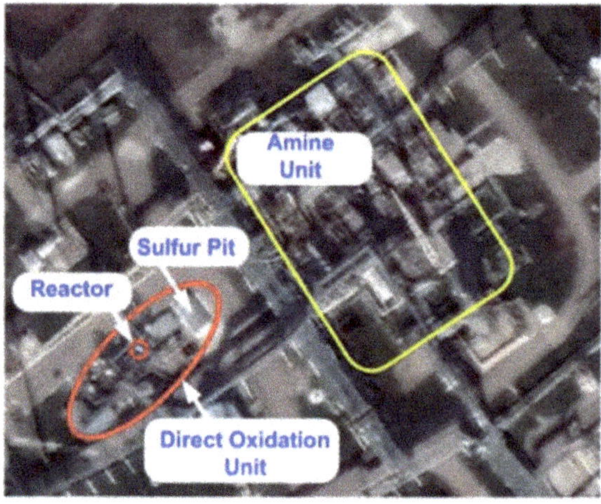

Figure 22. Bavly gas shop of PSC Tatneft. Purification of oil associated gas. Amine treatment and direct oxidation. Source: Satellite photo.

Figure 23. Industrial unit with a capacity of acid gas of up to 250 nm^3/h, which has been in continuous operation since 2011. The hydrogen sulfide content is 30–65 vol.%.

The main feature of the initial feed is the extreme instability of the input gas parameters; see Figure 24.

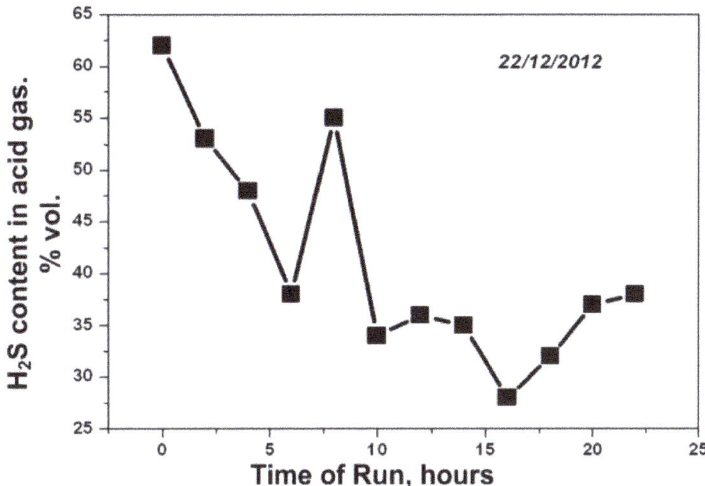

Figure 24. Fluctuations of H$_2$S content in the feed gas subjected to purification.

However, the developed computer control system made it possible to rapidly adjust the air and coolant flows to maintain the preset temperature in the catalytic reactor.

The quality of the resulting sulfur (Figure 25) surpassed the Russian National Standard #127.1-93 (commercial grade sulfur 9990).

The main results of the operation of the installation in the Bavly gas shop are given below:

- Over 1 billion m^3 of purified gas produced
- 6000 tons of hydrogen sulfide converted to elementary sulfur
- Emission of 12,000 tons of sulfur dioxide and sulfuric acid (340 railway tanks) into the atmosphere prevented;
- Environmental damage amounting to about 2.9 billion rubles avoided
- One-stage technology with computer control providing stable operation with variable parameters in terms of the acid gas (for example, hydrogen sulfide content).

Figure 25. Sulfur produced at the Bavly gas shop of PJSC Tatneft.

22. Facility for the Purification of Gases Caused by Blowing-off Sour Crude Oil

Strict limitations for the hydrogen sulfide content in oil for pipeline transport have been in place in Russia since 2002. Herewith, the mass fraction of hydrogen sulfide is limited to within 20–100 ppmv (GOST P 51858-2002. Oil. General technical conditions). When purifying 200 g of oil an hour, about 0.1 t of H_2S, or, on a yearly basis, 800 t of H_2S, are generated. This is particularly relevant because of the short period of transition (2019–2020) regarding technical regulations put in place by the Eurasian Economic Union "Regarding the safety of oil prepared for transportation or use" (TR EAES 045/2017), limiting hydrogen sulfide levels to 20 ppmv. The blowing-off process of H_2S with purified gas (mainline natural gas) is used for oils from the fields in the Volga Ural oil and gas province (Nurlatskoye, Aznakaevskoye, and Aznakaevskoye) with hydrogen sulfide levels up to 600 ppmv.

In this case, a H_2S-enriched hydrocarbon flow is formed which then undergoes amine treatment. Meanwhile, the concentrated hydrogen sulfide should be disposed of using the most reasonable method. To this end, the Boreskov Institute of Catalysis SB RAS, in collaboration with specialists from JSC SHESHMAOIL and JSC VNIIUS, constructed an industrial unit (Table 4).

The unit is scheduled to be used on an industrial basis in 2021 [159].

Table 4. Main characteristics of the installation for the removal of H_2S from gases originating from the blowing off of sour crude oil, designed for JSC SHESHMAOIL.

#	Parameters	Value
1	Acid gas flow rate after amine unite to the direct oxidation unit, nm^3/hour	to 110
2	H_2S concentration in acid gas, vol. %	75–90
3	Diameter of the fluidized bed reactor, m	0.52
4	Catalyst loading, kg	185
5	Sulfur yield, tons/hour	0.13

Minirefineries and GPPs or plants with a capacity of recycled hydrocarbon sulfurous raw materials of up to 3 million tons of oil per year for refineries and up to 80 million m^3 gas per year for GPPs are worthy of special consideration. Such enterprises are becoming rather numerous in the CIS countries. Their main goal is to localize the production of high-quality motor fuels in regions which are distant from large oil and gas processing centers. The low capacity of such production does not allow the creation of full-size hydrogen sulfide utilization units based on the Claus process, and the hydrogen sulfide formed as a result of the primary processing processes is usually burned off.

To solve this issue, at the JSC Condensate (Republic of Kazakhstan), an installation for hydrogen sulfide removal with sulfur production was built. Investors recognized the compact direct oxidation plant as the most rational way to solve the problem from a technical and economical viewpoint.

The installation has successfully passed commissioning and is ready to begin permanent operations.

At present, a plant which will use the hydrogen sulfide formed in the hydrocracking process is being created at the Ust-Luga Complex of PJSC NOVATEK. The technology was selected as a result of a vote, as it proved to be superior to those proposed by other licensers. The acid gas flow rate to the direct oxidation unit after the amine unit is about 170 nm^3/h.

The present status of the technology is as follows:

- The basic design of the technology has been finalized.
- The design and working documentation have been presented.
- The various apparatus units have been fabricated (Figure 26);
- The block of the plant has been delivered to the customer (Figure 27);
- The technology achieves the direct catalytic oxidation of hydrogen sulfide via the use of acid gases. It is an alternative to the Claus process (MTU-0.5 Mini Plant, Republic of Kazakhstan).

Figure 26. Reactor Block.

Figure 27. Cooling Block.

23. Unit for the Direct Oxidation of Hydrogen Sulfide as a Component of the Associated Petroleum Gas

Another application for sour associated gas is as low-debit flows with a capacity of 1000 nm³/h. On the one hand, these flows are environmental pollution sources and may be used with a compact method of purification for the autonomic regeneration of heat energy and electric power for travel heaters, the power supply for gas turbine units, etc.

SMP specialists Neftegaz JSC, BIC, in collaboration with TatNIINentefemash JSC and VNIIUS JSC, developed a production unit to selectively remove hydrogen sulfide directly from APG (Figures 28 and 29) [160–162]. The unit has undergone a complete cycle of industrial tests and is ready for industrial application.

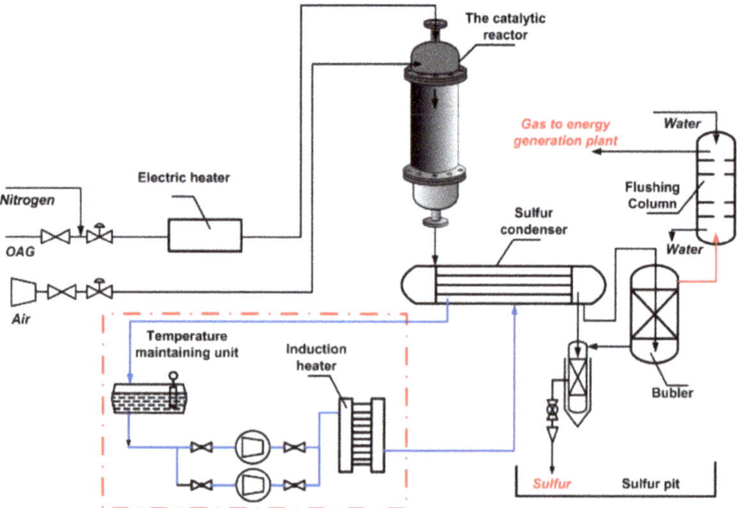

Figure 28. Flow-sheet diagram of the purification plant. Direct catalytic oxidation of hydrogen sulfide.

Figure 29. Industrial APG purification plant by direct catalytic oxidation.

The most important indicator of the process is its selectivity with respect to the hydrocarbon part of the purified gas. In this regard, a technique was developed to study the composition of the hydrocarbon part of the gas which is able to precisely identify individual components based on GC analysis.

The results are shown in Table 5.

As shown in the given data, hydrocarbon components are preserved during gas purification, and the purified gas can be used to generate thermal and electrical energy with minimal damage to the environment.

Table 5. GC analysis of initial and purified gas.

#	Compound	Initial Feedstock Gas, %Vol.	Purified Gas, %Vol.
1	H_2S	1.50	<50 ppmv
2	Water	0.69	2.030
3	He	0.05	0.04
4	Hydrogen	0.006	0.004
5	Oxygen	0.04	0.92
6	CO_2	4.70	4.56
7	Nitrogen	39.82	41.00
8	Ethane	9.60	9.60
9	Methane	25.60	24.22
10	Propane	9.96	9.80
11	iso-Butane	2.02	1.96
12	n-Butane	3.45	3.34
13	neo-Pentane	0.003	0.003
14	iso-Pentane	1.23	1.19
15	n-Pentane	0.85	0.81
16	Hexanes	0.32	0.31
17	Heptanes	0.07	0.07
18	Octanes	0.10	0.09

The preliminary results of techno economic analysis are given below (Table 6). For the sake of comparison, an existing Claus plant now in operation at the Minibay GPP was selected (See Figures 30 and 31).

Table 6. Comparison of the characteristics of the Claus Pant and a direct oxidation plant with the same capacities [163,164].

#	Parameters	Direct Oxidation Unit	Three Stage Claus Unit Minibay Gas Processing Plant
1	Acid gas (H_2S+CO_2) supply, nm^3/h	1050	1050
2	H_2S content, %vol.	80	80
3	Air supply, nm^3/h	2000	2000
4	Sulfur production Annually, ton	10.000	10.000
5	Dimensions of the main units	Calculation Fluidized bed reactor: Diameter = 1.5 m Height = 6 m Fixed bed reactor Diameter = 2.5 m Height = 6 m	Direct data Thermal stage furnace Diameter = 2.5 m Length = 7 m Catalytic converters (3 pieces) Diameter = 2.5 m Length = 4 m
6	Catalyst load, ton	Calculation Fluidized bed reactor-2 Fixed bed reactor-5	Direct data Total: 18
7	Sulfur cost Arbitrary units, estimation	1	2.5

As shown in the data in Table 6, the installation using the direct oxidation process is significantly more compact, primarily due to the use of a reactor with a fluidized catalyst bed, where the actual target process is effectively combined with the simultaneous removal of excess heat. The required temperature of the direct oxidation process is adjusted by changing the flow rate of the coolant through a heat exchanger placed in the catalyst bed according to fluctuations in the H_2S content in the feed gas. With such technology, capital costs are significantly reduced thanks to the need of fewer parts in the process chain and the significantly lower metal weight. The operational costs are also reduced due to lower energy consumption and the reduced number of required service personnel, which ultimately leads to a decrease in the cost of the final product, i.e., elementary sulfur. The absence of a flame furnace increases the environmental friendliness of the process due to the absence of the formation of toxic side products which occur due to high-temperature interactions of H_2S with CO_2-carbonyl sulfide and carbon disulfide.

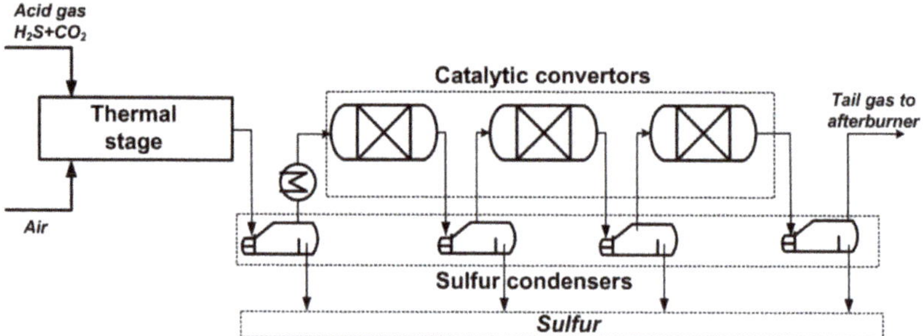

Figure 30. Schematic diagram of the Claus plant operating at the Minibay GPP.

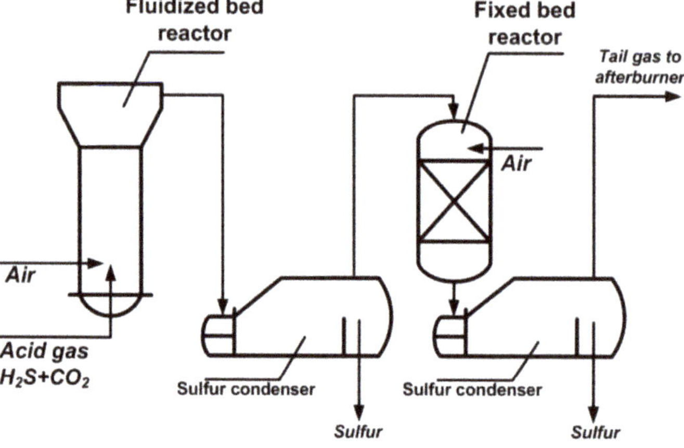

Figure 31. Schematic diagram of the alternative direct oxidation plant.

24. Conclusions

An overview of various technologies based on the direct catalytic oxidation of hydrogen sulfide to obtain elementary sulfur is given. Such technologies, primarily their gas-phase version, have obvious advantages, including:

- continuity of the process that allows simultaneous gas purification and the production of a commodity, i.e., elemental sulfur;
- "soft" conditions for implementing the process (T = 220–280 °C) due to the use of a highly active catalyst.

Data on the Claus process and its modern modifications, as the dominant technology for the conversion of hydrogen sulfide into elemental sulfur, are given. The results of research on the development of various catalysts for a direct oxidation process are described. It is shown that catalysts based on transition metal oxides are the most promising.

Oxide catalysts have indisputable advantages over other potential systems, including high thermal stability, low cost of raw materials, and potential for large-scale production, making them optimal in terms of the quality/price ratio, which is a significant indicator for the technical and economic efficiency of commercial processes. This observation is confirmed by the widespread use of Jacobs iron catalysts in SuperClaus installations.

This review also described the results of fundamental studies of the direct catalytic oxidation of hydrogen sulfide, carried out at the Institute of Catalysis SB RAS, on the basis of which industrial installations for hydrogen sulfide removal from gas streams were created.

The industrial facility in the Bavlinskiy gas shop of the PJSC Tatneftegazpererabotka is now in continuous operation.

Several other facilities have been developed and constructed and are now beginning operations:

- An installation for the purification of blow-off gases of high-sulfur crude oil
- An installation for the direct oxidation of hydrogen sulfide as an alternative to the conventional Claus Process
- An installation for the direct oxidation of hydrogen sulfide in the composition of oil-associated gases

The developed technology, in combination with amine treatment, provides:

- The production of commercial products, i.e., fuel gas and sulfur that correspond to technical standards (GOST 5542-87 and GOST 127.1-93, respectively)
- Extended operational range by H_2S content in comparison with Claus units
- Substantial improvement of the environmental situation by avoiding hazardous emissions and the production of waste materials.

Author Contributions: Conceptualization, methodology and formal analysis, S.K.; writing, original draft preparation, S.K., M.K., and A.S.; writing, review and editing, S.K., M.K., and A.S.; supervision, Z.R.I. All authors have read and agreed to the published version of the manuscript.

Funding: The work was carried out with financial support from the Ministry of Science and Higher Education of RF within the State Assignment for the Boreskov Institute of Catalysis SB RAS (Project No. AAAA-A21-121011390010-7).

Institutional Review Board Statement: Not applicable.

Informed Consent Statement: Not applicable.

Conflicts of Interest: The authors declare no conflict of interest.

Abbreviations

AC	Activated carbon
APG	Associated petroleum gas
BAS	Broensted acid sites
CNF	Carbon nanofibers
CNT	Carbon nanotubes
DEA	Diethanolamine
EDTA	Ethylenediaminetetraacetic acid
DRS	Diffuse reflectance spectra
FRC	Federal Research Center
FTIR	Furier transform infrared
GHSV	Gas hourly space velocity
GPP	Gas processing plant
JSC	Joint-stock company
k	Rate constant
LAS	Lewis acid sites
LLC	Limited liability company
MEA	Monoethanolamine

Nm^3	Normal cubic meters
OAG	Oil-associated gases
ppmv	Part per million by volume
PJSC	Public joint-stock company
SB RAS	Siberian Branch of the Russian Academy of Sciences
W	Reaction rate
WHSV	Weight hourly space velocity
WHB	Waste heat boiler

References

1. Hendrickson, R.G.; Chang, A.; Hamilton, R.J. Fatalities from Hydrogen Sulfide. *Am. J. Ind. Med.* **2004**, *45*, 346–350. [CrossRef] [PubMed]
2. Chou, C.H.S.J. *Hydrogen Sulfide: Human Health Aspects*; World Health Organization: Geneva, Switzerland, 2003.
3. Ministry of Justice of the Russian Federation. *Hygienic Standard 2.2.5.3532-18*; Ministry of Justice of the Russian Federatio: Moscow, Russia, 2018.
4. Chief State Sanitary Doctor of the Russian Federation. *Hygienic Standard 2.1.6.3492-17*; Chief State Sanitary Doctor of the Russian Federation: Moscow, Russia, 2017.
5. Lallemand, F.; Lecomte, F.; Streicher, C. Highly sour gas processing: h$_2$s bulk removal with the sprex process. In Proceedings of the International Petroleum Technology Conference, Doha, Qatar, 21 November 2005.
6. Faramawy, S.; Zaki, T.; Sakr, A.A.-E. Natural Gas Origin, Composition, and Processing: A Review. *J. Nat. Gas Sci. Eng.* **2016**, *34*, 34–54. [CrossRef]
7. Ponkratov, V.V.; Pozdnyaev, A.S. Tax and Regulatory Incentives to Improve Utilization of Associated Petroleum Gas (APG). *Russ. Econ. Taxes Law* **2014**, *5*, 88–94.
8. World Energy Outlook. Available online: http://www.worldenergyoutlook.org/ (accessed on 9 September 2021).
9. Mokhatab, S.; Poe, W.A. *Handbook of Natural Gas Transmission and Processing*, 2nd ed.; Gulf Professional Publishing: Houston, TX, USA, 2012.
10. Merichem Company. *LO-CAT®Process for Cost Effective Desulfurization of All Types of Gas Streams*; Merichem Company: Houston, TX, USA, 2013.
11. Hydrocarbon Processing. Available online: http://hydrocarbon.processengineer.info/sulferox-process-by-shell.html (accessed on 9 September 2021).
12. Mazgarov, A.M. Liquid Phase Oxidations of Mercaptans and Hydrogen Sulfide with Metal Phthalocyanine Catalysts and Development of Pro-Cesses for Desulfurization of Hydrocarbon Feed. Ph.D. Thesis, Kazan University of Chemical Technology, Kazan, Russia, 2004.
13. Mokhatab, S.; Mak, J.Y. Gas Processing Plant Automation. In *Handbook of Natural Gas Transmission and Processing*, 2nd ed.; Elsevier Inc.: Amsterdam, The Netherlands, 2019.
14. Sulfur. Available online: https://pubs.usgs.gov/periodicals/mcs2021/mcs2021-sulfur.pdf (accessed on 9 September 2021).
15. Westbrook, C.K.; Dryer, F.L. Simplified Reaction Mechanisms for the Oxidation of Hydrocarbon Fuels in Flames. *Combust. Sci. Technol.* **1981**, *27*, 31–43. [CrossRef]
16. Zarei, S. Life Cycle Assessment and Optimization of Claus Reaction Furnace through Kinetic Modeling. *Chem. Eng. Res. Des.* **2019**, *148*, 75–85. [CrossRef]
17. Zarei, S. Exergetic, Energetic and Life Cycle Assessments of the Modified Claus Process. *Energy* **2020**, *191*, 116584. [CrossRef]
18. Zare Nezhad, A.B.; Hosseinpour, N. Evaluation of Different Alternatives for Increasing the Reaction Furnace Temperature of Claus SRU by Chemical Equilibrium Calculations. *Appl. Therm. Eng.* **2008**, *28*, 738–744. [CrossRef]
19. Tong, S.; Dalla Lana, I.G.; Chuang, K.T. Kinetic Modelling of the Hydrolysis of Carbonyl Sulfide Catalyzed by Either Titania or Alumina. *Can. J. Chem. Eng.* **1993**, *71*, 392–400. [CrossRef]
20. Tong, S.; Dalla Lana, I.G.; Chuang, K.T. EFfect of Catalyst Shape on the Hydrolysis of COS and CS$_2$ in a Simulated Claus Converter. *Ind. Eng. Chem. Res.* **1997**, *36*, 4087–4093. [CrossRef]
21. Tong, S.; Dalla Lana, I.G.; Chuang, K.T. Kinetic Modeling of the Hydrolysis of Carbon Disulfide Catalyzed by Either Titania or Alumina. *Can. J. Chem. Eng.* **1995**, *73*, 220–227. [CrossRef]
22. Mendioroz, S.; Munoz, V.; Alvarez, E.; Palacios, J.M. Kinetic Study of the Claus Reaction at Low Temperature Using γ-Alumina as Catalyst. *Appl. Catal. Gen.* **1995**, *132*, 111–126. [CrossRef]
23. Claus Catalysts and Tail Gas Treatment Solutions for Sulfur Recovery. Available online: https://catalysts.basf.com/files/literature-library/BASF_CAT-001731_SRU_Bruschuere_AS-viewing.pdf (accessed on 9 September 2021).
24. Scirè, S.; Fiorenza, R.; Bellardita, M.; Palmisano, L. Titanium Dioxide (TiO$_2$) and Its Applications Catalytic Applications of TiO$_2$. *Met. Oxides* **2021**, *21*, 637–679.
25. Claus Catalysts. Available online: https://www.eurosupport.com/media/cache/20170925050616BrochureClaus.pdf (accessed on 9 September 2021).
26. Available online: https://www.axens.net/solutions/catalysts-adsorbents-grading-supply/claus-catalysts (accessed on 9 September 2021).

27. Ismagilov, Z.R.; Kuznetsov, V.V.; Okhlopova, L.B.; Tsickza, L.T.; Yashnik, S.A. *Oxides of Titanium, Cerium, Zirconium, Yttrium and Aluminum. Properties, Applications and Methods of Production*; Publishing House of SB RAS: Novosibirsk, Russia, 2010.
28. Marshnyova, V.I.; Mokrinsky, V.V. Catalytic Activity of Metal Oxides in Hydrogen Sulfide Reactions with Oxygen and Sulfur Dioxide. *Kinet. Catal.* **1988**, *29*, 989–993.
29. Bukhtiyarova, G.A. Development of a Polyfunctional V-Mg-Ti-Ca Catalyst for the Claus Process. Ph.D. Thesis, Boreskov Institute of Catalysis, Novosibirsk, Russia, 1999.
30. Khudenko, B.; Gitman, G.M.; Wechsler, T. Oxygen Based Claus Process for Recovery of Sulfur from H_2S Gases. *J. Environ. Eng.* **1993**, *119*, 1233–1251. [CrossRef]
31. HeGarly, W.P.; Davis, R.; Kammiller, R. Claus Plant Capacity Boosted by Oxygen Enrichment Process. *Technol. Oil Gas J.* **1985**, *30*, 39–41.
32. El-Bishtawi, R.; Haimour, N. Claus Recycle with Double Combustion Process. *Fuel Process. Technol.* **2004**, *86*, 245–260. [CrossRef]
33. Kwong, K.V. Process for Direct Reduction of Sulfur Compounds to Elemental Sulfur in Combination with the Claus Process. U.S. Patent US6214311B1, 4 October 2001.
34. Rameshni, M.; Street, R. PROClaus: The New Standard for Claus Performance. In *Sulfur Recovery Symposium*; Brimstone Engineering Services: Calgary, AB, Canada, 2001.
35. Jin, Y.; Yu, Q.; Chang, S. Catalyst for the Reduction of Sulfur Dioxide to Elemental Sulfur. U.S. Patent 5494879, 27 February 1996.
36. Alkhazov, T.; Meissner, R.E., III. Catalysts and Process for Selective Oxidation of Hydrogen Sulfide to Elemental Sulfur. U.S. Patent 5603913, 18 February 1997. [CrossRef]
37. Alkhazov, T.G.; Amirgulyan, N.S. Catalytic Oxidation of Hydrogen Sulfide on Iron Oxides. *Kinet. Catal.* **1982**, *23*, 1130–1134.
38. Borsboom, J.; Van, N.P.F. Process for the Removal of Sulphur Compounds from Gases. U.S. Patent 6800261, 5 October 2004.
39. Berben, P.H.; Borsboom, J.; Geus, J.W.; Lagas, J.A. Process for Recovering Sulfur from Sulfur-Containing Gases. U.S. Patent 4988494, 29 January 1991.
40. Geus, J.W.; Teroerde, R.J.A. Catalyst for the Selective Oxidation of Sulfur Compounds to Elemental Sulfur. U.S. Patent 6919296, 19 January 2005.
41. Borsboom, J.A.; Lagas, J.A.; Berben, P.H. The Superclaus Process Increases Sulfur Recovery. In *AT ACHEMA-88*; Frankfurt Am Main: Frankfurt, Germany, 1988.
42. Van Nisselrooy, P.F.M.T.; Lagas, J.A. Superclaus Reduced SO_2 Emission by the Use of a Selective Oxidation Catalyst. *Catal. Today* **1993**, *16*, 263–271. [CrossRef]
43. Besher, E.M.; Meisen, A. Low-Temperature Fluidized Bed Claus Reactor Performance. *Chem. Eng. Sci.* **1990**, *45*, 3035–3045. [CrossRef]
44. Santo, S.; Rameshni, M. The Challenges of Designing Grass Root Sulphur Recovery Units with a Wide Range of H_2S Concentration from Natural Gas. *J. Nat. Gas Sci. Eng.* **2014**, *18*, 137–148. [CrossRef]
45. Chekalov, L.V.; Sanaev, J.I. Electric Precipitator. R.U. Patent 2563481, 20 September 2015.
46. Bassani, A.; Pirola, C.; Maggio, E.; Pettinau, A.; Frau, C.; Bozzano, G.; Pierucci, S.; Ranzi, E.; Manenti, F. Acid Gas to Syngas (AG2S^{TM}) Technology Applied to Solid Fuel Gasification: Cutting H_2S and CO_2 Emissions by Improving Syngas Production. *Appl. Energy* **2016**, *184*, 1284–1291. [CrossRef]
47. Pirola, C.; Ranzi, E.; Manenti, F. Technical Feasibility of AG2S^{TM} Process Revamping. Computer. *Aided Chem. Eng.* **2017**, *40*, 385–390.
48. Ismagilov, Z.R.; Khairulin, S.R.; Ismagilov, F.R. Russian Refiner Tests New one-stage H_2S Removal process. *Oil Gas J.* **1994**, *7*, 81–82.
49. Ismagilov, Z.R.; Khairulin, S.R.; Barannik, G.B.; Kerzhentsev, M.A.; Nemkov, V.V.; Parmon, V.N. A Method for Cleaning the Blow-Off Gases of Wells from Hydrogen Sulfide. USSR A. S. 1608109, 9 February 1988.
50. Clark, P.D. Partial Oxidation of Hydrogen Sulfide in the Manufacture of Hydrogen, Sulfur, Ethylene and Propylene. In *SULPHUR*; Alberta Sulphur Research Ltd.: Caglary, AB, Canada, 2003.
51. Gamson, B.W.; Elkins, R.H. Sulfur from Hydrogen Sulfide. *Chem. Eng. Progress.* **1953**, *49*, 203.
52. Zagoruiko, A.N. Development of the Process of Obtaining Elementary Sulfur Method of Claus in Non-Stationary Mode. Ph.D. Thesis, Boreskov Institute of Catalysis, Novosibirsk, Russia, 1991.
53. Klein, J.; Henning, K.-D. Catalytic Oxidation of Hydrogen Sulphide on Activated Carbons. *Fuel* **1984**, *63*, 1064–1067. [CrossRef]
54. Pan, Z.; Shan, W.H.; Feng, H.; Smith, J.M. Kinetics of the Self-Fouling Oxidation of Hydrogen Sulfide on Activated Carbon. *AIChE J.* **1984**, *30*, 1021–1025.
55. Prokopenko, V.S.; Zemlyansky, N.N.; Vasilenko, V.A.; Artyushenko, G.V. Purification of Natural Gas from Hydrogen Sulfide by the Method of Direct Oxidation. *Gas Ind.* **1977**, *9*, 44–46.
56. Meeyoo, V.; Trimm, D.L.; Cant, N.W.J. Adsorption-Reaction Processes for the Removal of Hydrogen Sulphide from Gas Streams. *Chem. Technol. Biotechnol.* **1997**, *68*, 411–416. [CrossRef]
57. Sun, Y.; He, J.; Wang, Y.; Yang, G.; Sun, G.; Sage, V. Experimental and CFD study of H_2S oxidation by activated carbon prepared from cotton pulp black liquor. *Process Saf. Environ. Prot.* **2020**, *134*, 131–139. [CrossRef]
58. Adib, F.; Bagreev, A.; Bandosz, T.J. Effect of pH and Surface Chemistry on the Mechanism of H_2S Removal by Activated Carbons. *J. Colloid Interface Sci.* **1999**, *216*, 360–369. [CrossRef]

59. Liu, Y.; Song, C.; Wang, Y.; Cao, W.; Lei, Y.; Feng, Q.; Chen, Z.; Liang, S.; Xu, L.; Jiang, L. Rational designed Co@N-doped carbon catalyst for high-efficient H₂S selective oxidation by regulating electronic structures. *Chem. Eng. J.* **2020**, *401*, 1–10. [CrossRef]

60. Sun, M.; Wang, X.; Pan, X.; Liu, L.; Li, Y.; Zhao, Z.; Qiu, J. Nitrogen-rich hierarchical porous carbon nanofibers for selective oxidation of hydrogen sulfide. *Fuel Process. Technol.* **2019**, *191*, 121 128. [CrossRef]

61. Wu, X.X.; Schwartz, V.; Overbury, S.H.; Armstrong, T.R. Desulfurization of Gaseous Fuels Using Activated Carbons as Catalysts for the Selective Oxidation of Hydrogen Sulfide. *Energy Fuels* **2005**, *19*, 1774–1782. [CrossRef]

62. Fang, H.B.; Zhao, J.T.; Fang, Y.T.; Huang, J.J.; Wang, Y. Selective Oxidation of Hydrogen Sulfide to Sulfur Over Activated Carbon-Supported Metal Oxides. *Fuel* **2013**, *108*, 143–148. [CrossRef]

63. Tsai, J.H.; Jeng, F.T.; Chiang, H.L. Removal of H₂S from Exhaust Gas by Use of Alkaline Activated Carbon. *Adsorpt. J. Int. Adsorpt. Soc.* **2001**, *7*, 357–366. [CrossRef]

64. Xiao, Y.; Wang, S.; Wu, D.; Yuan, Q. Catalytic Oxidation of Hydrogen Sulfide Over Unmodified and Impregnated Activated Carbon. *Sep. Purif. Technol.* **2008**, *59*, 326–332. [CrossRef]

65. Abatzoglou, N.; Boivin, S. A Review of Biogas Purification Processes. *Biofuels Bioprod. Bioref.* **2009**, *3*, 42–71. [CrossRef]

66. Seredych, M.; Bandosz, T.J. Adsorption of Hydrogen Sulfide on Graphite Derived Materials Modified by Incorporation of Nitrogen. *Mater. Chem. Phys.* **2009**, *113*, 946–952. [CrossRef]

67. Bashkova, S.; Armstrong, T.R.; Schwartz, V. Selective Catalytic Oxidation of Hydrogen Sulfide on Activated Carbons Impregnated with Sodium Hydroxide. *Energy Fuels* **2009**, *23*, 1674–1682. [CrossRef]

68. Bashkova, S.; Baker, F.S.; Wu, X.; Armstrong, T.R.; Schwartz, V. Activated Carbon Catalyst for Selective Oxidation of Hydrogen Sulphide: On the Influence of Pore Structure, Surface Character-Istics, and Catalytically-Active Nitrogen. *Carbon* **2007**, *45*, 1354–1363. [CrossRef]

69. Primavera, A.; Trovarelli, A.; Andreussi, P.; Dolcetti, G. The Effect of Water in the Low-Temperature Catalytic Oxidation of Hydrogen Sulfide to Sulfur Over Activated Carbon. *Appl. Catal. A* **1998**, *173*, 185–192. [CrossRef]

70. Bashkova, S.; Bagreev, A.; Locke, D.C.; Bandosz, T.J. Adsorption of SO₂ on Sewage Sludge-Derived Materials. *Environ. Sci. Technol.* **2001**, *35*, 3263–3269. [CrossRef]

71. Chiang, H.L.; Tsai, J.H.; Tsai, C.L.; Hsu, Y.C. Adsorption Characteristics of Alkaline Activated Carbon Exemplified by Water Vapor, H₂S, and CH₃SH Gas. *Sep. Sci. Technol.* **2000**, *35*, 903–918. [CrossRef]

72. Harry, W. Regeneration of Zeolites Used for Sulfur Removal. *Oil Gas J.* **1961**, *167*, 99–116.

73. Bandosz, T.J. (Ed.) Desulfurization on Activated. In *Carbons in Activated Carbon Surfaces in Environmental Remediation*; Elsevier Ltd.: New York, NY, USA, 2006.

74. Cal, M.P.; Strickler, B.W.; Lizzio, A.A.; Gangwal, S.K. High Temperature Hydrogen Sulfide Adsorption on Activated Carbon: II. Effects of Gas Temperature, Gas Pressure and Sorbent Regeneration. *Carbon* **2000**, *38*, 1767–1774. [CrossRef]

75. Feng, W.; Kwon, S.; Borguet, E.; Vidic, R. Adsorption of Hydrogen Sulfide onto Activated Carbon Fibers: Effect of Pore Structure and Surface Chemistry. *Environ. Sci. Technol.* **2005**, *39*, 9744–9749. [CrossRef]

76. Bagreev, A.; Bandosz, T.J. A Role of Sodium Hydroxide in the Process of Hydrogen Sulfide Adsorption/Oxidation on Caustic-Impregnated Activated Carbons. *Ind. Eng. Chem. Res.* **2002**, *41*, 672–679. [CrossRef]

77. Mikhalovsky, S.V.; Zaitsev, Y.P. Catalytic Properties of Activated Carbons I. Gas-Phase Oxidation of Hydrogen Sulphide. *Carbon* **1997**, *35*, 1367–1374. [CrossRef]

78. Gardner, T.H.; Berry, D.A.; David Lyons, K.; Beer, S.K.; Freed, A.D. Fuel Processor Integrated H₂S Catalytic Partial Oxidation Technology for Sulfur Removal in Fuel Cell Power Plants. *Fuel* **2002**, *81*, 2157–2166. [CrossRef]

79. Wu, X.; Kercher, A.K.; Schwartz, V.; Overbury, S.H.; Armstrong, T.R. Activated Carbons for Selective Catalytic Oxidation of Hydrogen Sulfide to Sulfur. *Carbon* **2005**, *43*, 1087–1090. [CrossRef]

80. Bandosz, T.J.; Bagreev, A.; Adib, F.; Turk, A. Unmodified versus Caustics-Impregnated Carbons for Control of Hydrogen Sulfide Emissions from Sewage Treatment Plants. *Environ. Sci. Technol.* **2000**, *34*, 1069–1074. [CrossRef]

81. Bagreev, A.; Bandosz, T.J. On the Mechanism of Hydrogen Sulfide Removal from Moist Air on Catalytic Carbonaceous Adsorbents. *Ind. Eng. Chem. Res.* **2005**, *44*, 530–538. [CrossRef]

82. Rhodes, C.; Riddel, S.A.; West, J.; Williams, B.P.; Hutchings, G.J. The Low-Temperature Hydrolysis of Carbonyl Sulfide and Carbon Disulfide: A Review. *Catal. Today* **2000**, *59*, 443–464. [CrossRef]

83. Sun, F.; Liu, J.; Chen, H.; Zhang, Z.; Qiao, W.; Long, D.; Ling, L. Nitrogen-Rich Mesoporous Carbons: Highly Efficient, Regenerable Metal-Free Catalysts for Low-Temperature Oxidation of H₂S. *ACS Catal.* **2013**, *3*, 862–870. [CrossRef]

84. Ledoux, M.J.; Vieira, R.; Pham-Huu, C.; Keller, N. New Catalytic Phenomena on Nanostructured (Fibers and Tubes) Catalysts. *J. Catal.* **2003**, *216*, 333–342. [CrossRef]

85. Ismagilov, Z.R.; Shalagina, A.E.; Podyacheva, O.Y.; Kvon, R.I.; Ismagilov, I.Z.; Kerzhentsev, M.A.; Barnakov, C.N.; Kozlov, A.P. Synthesis of Nitrogen-Containing Carbon Materials for Solid Polymer Fuel Cell Cathodes. *Kinet. Catal.* **2007**, *48*, 581–588. [CrossRef]

86. Ismagilov, Z.R.; Shalagina, A.E.; Podyacheva, O.Y.; Ischenko, A.V.; Kibis, L.S.; Boronin, A.I.; Chesalov, Y.A.; Kochubey, D.I.; Romanenko, A.I.; Anikeeva, O.B.; et al. Structure and Electrical Conductivity of Nitrogen-Doped Carbon Nanofibers. *Carbon* **2009**, *47*, 1922–1929. [CrossRef]

87. Nhut, J.-M.; Nguyen, P.; Pham-Huu, C.; Keller, N.; Ledoux, M.-J. Carbon Nanotubes as Nanosized Reactor for the Selective Oxidation of H₂S Into Elemental Sulfur. *Catal. Today* **2004**, *91*, 91–97. [CrossRef]

88. Nhut, J.M.; Pesant, L.; Tessonnier, J.P.; Winé, G.; Guille, J.; Pham Huu, C.; Ledoux, M.J. Mesoporous Carbon Nanotubes for Use as Support in Catalysis and as Nanosized Reactors for One-Dimensional Inorganic Material Synthesis. *Appl. Catal. A* **2003**, *254*, 345–363. [CrossRef]

89. Chen, Q.; Wang, J.; Liu, X.; Zhao, X.; Qiao, W.; Long, D.; Ling, L. Alkaline Carbon Nanotubes As Effective Catalysts for H2S Oxidation. *Carbon* **2011**, *49*, 3773–3780. [CrossRef]

90. Mohamadalizadeh, A.; Towfighi, J.; Adinehnia, M.; Bozorgzadeh, H.R. H_2S Oxidation by Multi-Wall Carbon Nanotubes Decorated with Tungsten Sulfide. *Korean J. Chem. Eng.* **2013**, *30*, 871–877. [CrossRef]

91. Ba, H.; Duong-Viet, C.; Liu, Y.; Nhut, J.-M.; Granger, P.; Ledoux, M.J.; Pham-Huu, C. Nitrogen-Doped Carbon Nanotube Spheres As Metal-Free Catalysts for the Partial Oxidation of H_2S. *Comptes Rendus Chim.* **2016**, *19*, 1303–1309. [CrossRef]

92. Chizari, K.; Deneuve, A.; Ersen, O.; Florea, I.; Liu, Y.; Edouard, D.; Janowska, I.; Begin, D.; Pham-Huu, C. Nitrogen-Doped Carbon Nanotubes as a Highly Active Metal-Free Catalyst for Selective Oxidation. *Chem. Sus. Chem.* **2012**, *5*, 102–108. [CrossRef]

93. De Jong, K.P.; Geus, J.W. Carbon Nanofibers: Catalytic Synthesis and Applications. *Catal. Rev.-Sci. Eng.* **2000**, *42*, 481–510. [CrossRef]

94. Kuvshinov, G.G.; Shinkarev, V.V.; Glushenkov, A.M.; Boyko, M.N.; Kuvshinov, D.G. Catalytic Properties of Nanofibrous Carbon in Selective Oxidation of Hydrogen Sulfide. *China Particuol.* **2006**, *4*, 70–72. [CrossRef]

95. Shinkarev, V.V.; Glushenkov, A.M.; Kuvshinov, D.G.; Kuvshinov, G.G. New Effective Catalysts Based on Mesoporous Nanofibrous Carbon for Selective Oxidation of Hydrogen Sulfide. *Appl. Catal. B* **2009**, *85*, 180–191. [CrossRef]

96. Shinkarev, V.V.; Kuvshinov, G.G.; Zagoruiko, A.N. Kinetics of H2S Selective Oxidation by Oxygen at the Carbon Nanofibrous Catalyst. *Reac. Kinet. Mech. Cat.* **2018**, *123*, 625–639. [CrossRef]

97. Chen, Q.; Wang, Z.; Long, D.; Liu, X.; Zhan, L.; Liang, X.; Qiao, W.; Ling, L. Role of Pore Structure of Activated Carbon Fibers in the Catalytic Oxidation of H_2S. *Ind. Eng. Chem. Res.* **2010**, *49*, 3152–3159. [CrossRef]

98. Coelho, N.M.A.; da Cruz, G.M.; Vieira, R. Effect of Temperature and Water on the Selective Oxidation of H_2S to Elemental Sulfur on a Macroscopic Carbon Nanofiber Based Catalyst. *Catal. Lett.* **2012**, *142*, 108–111. [CrossRef]

99. Liu, B.-T.; Ke, Y.-X. Enhanced Selective Catalytic Oxidation of H_2S Over Ce-Fe/AC Catalysts at Ambient Temperature. *J. Taiwan Inst. Chem. Eng.* **2020**, *110*, 28–33. [CrossRef]

100. Puri, B.R. Studies of Catalytic Reactions on Activated Carbons. *Carbon* **1982**, *20*, 139. [CrossRef]

101. Bouzaza, A.; Laplanche, A.; Marsteau, S. Adsorption-Oxidation of Hydrogen Sulfide on Activated Carbon Fibers: Effect of the Composition and the Relative Humidity of the Gas Phase. *Chemosphere* **2004**, *54*, 481–488. [CrossRef] [PubMed]

102. Ziółek, M.; Dudzik, Z. Structural Changes of the NaX Zeolite During $H_2S + O_2$ Reaction. *React. Kinet. Catal. Lett.* **1980**, *14*, 213–217. [CrossRef]

103. Ziółek, M.; Dudzik, Z. Catalytically Active Centres in the $H_2S + O_2$ Reaction on Faujasites. *Zeolites* **1981**, *1*, 117–122. [CrossRef]

104. Lee, J.D.; Park, N.-K.; Han, K.B.; Ryu, S.O.; Lee, T.J. Influence of Reducing Power on Selective Oxidation of H_2S Over V_2O_5 Catalyst in IGCC System. *Stud. Surf. Sci. Catal.* **2006**, *159*, 425–428.

105. Nguyen, P.; Edouard, D.; Nhut, J.-M.; Ledoux, M.J.; Pham, C.; Pham-Huu, C. High Thermal Conductive β-SiC for Selective Oxidation of H_2S: A New Support for Exothermal Reactions. *Appl. Catal. B Environ.* **2007**, *76*, 300–310. [CrossRef]

106. Keller, N.; Pham-Huu, C.; Estournes, C.; Ledoux, M.J. Low Temperature Use of SiC-Supported NiS_2-Based Catalysts for Selective H_2S Oxidation Role of SiC Surface Heterogeneity and Nature of the Active Phase. *Appl. Catal. A Gen.* **2002**, *234*, 191–205. [CrossRef]

107. Keller, N.; Pham-Huu, C.; Estournes, C.; Ledoux, M.J. Continuous Process for Selective Oxidation of H_2S Over SiC-Supported Iron Catalysts into Elemental Sulfur Above Its Dewpoint. *Appl. Catal. A Gen.* **2001**, *217*, 205–217. [CrossRef]

108. Zhang, X.; Tang, Y.; Qu, S.; Da, J.; Hao, Z. H_2S-Selective Catalytic Oxidation: Catalysts and Processes. *ACS Catal.* **2015**, *5*, 1053–1067. [CrossRef]

109. Alkhazov, T.G.; Bagirov, R.A.; Dovlatova, S.M.; Filatova, O.E.; Kuliev, A.M.; Vartanov, A.A. Catalyst for Oxidizing Hydrogen Sulphide in Gaseous Phase to Elementary Sulphur. USSR Certiificate of Authorship 665939, 5 June 1979.

110. Alkhazov, T.G.; Bagdasaryan, B.V.; Mamedova, R.I.; Vartanov, A.A. Catalyst for Oxidizing Hydrogen Sulphide to Sulphur. USSR Certiificate of Authorship 882589, 23 November 1981.

111. Davydov, A.A.; Marshneva, V.I.; Shepotko, M.L. Metal Oxides in Hydrogen Sulfide Oxidation by Oxygen and Sulfur Dioxide I. the Comparison Study of the Catalytic Activity. Mechanism of the Interactions Between H_2S and SO_2 on Some Oxides. *Appl. Catal. A Gen.* **2003**, *244*, 93–100. [CrossRef]

112. Alkhazov, T.G.H.; Vartanov, A.A. Prjamoe gheteroghenno-katalytytcheskoe okyslenye serovodoroda v ehlementarnuju seru. *Yzvestyja VUZov. Neftj Y Ghaz.* **1981**, *3*, 45–49.

113. Batygina, M.V.; Dobrynkin, N.M.; Kirichenko, O.A.; Khairulin, S.R.; Ismagilov, Z.R. Studies of Supported Oxide Catalysts in the Direct Selective Oxidation of Hydrogen Sulfide. *React. Kinet. Catal. Lett.* **1992**, *48*, 55–63. [CrossRef]

114. Zheng, X.; Li, Y.; Zhang, L.; Shen, L.; Xiao, Y.; Zhang, Y.; Au, C.; Jiang, L. Insight Into the Effect of Morphology on Catalytic Performance of Porous CeO2 Nanocrystals for H_2S Selective Oxidation. *Appl. Catal. B Environ.* **2019**, *252*, 98–110. [CrossRef]

115. Ismagilov, Z.R.; Shkrabina, R.A.; Barannik, G.B.; Kerzhentsev, M.A. New Catalysts and Processes for Environment Protection. *React. Kinet. Catal. Lett.* **1995**, *55*, 489–499. [CrossRef]

116. Alkhazov, T.G.; Balyberdina, I.T.; Filatova, O.E.; Khendro, M.; Korotaev, Y.P.; Vartanov, A.A. Method of Producing Elemental sulphur. USSR A.s. 856974, 23 August 1981.

117. Amyrghuljan, N.S. *Materyalyh Pjatoj Respublykanskoj Konferentsyy Po Okyslyteljnomu Gheteroghennomu Katalyzu*; Sociology and Law of Azerbaijan National Academy of Sciences: Baku, Azerbaijan, 1981.
118. Geus, J.W. Preparation and Properties of Iron Oxide and Metallic Iron Catalysts. *Appl. Catal.* **1986**, *25*, 313–333. [CrossRef]
119. Terorde, R.J.A.M.; van den Brink, P.J.; Visser, L.M.; van Dillen, A.J.; Geus, J.W. Selective oxidation of hydrogen sulfide to elemental sulfur using iron oxide catalysts on various supports. *Catal. Today* **1993**, *17*, 217–224. [CrossRef]
120. Terorde, R.J.A.M.; de Jong, M.C.; Crombag, M.J.D.; van den Brink, P.J.; van Dillen, A.J.; Geus, J.W. Selective oxidation of hydrogen sulfide on a sodium promoted iron oxide on silica catalyst V. *New Dev. Sel. Oxid. II* **1994**, 861–868.
121. Khairulin, S.R.; Parmon, V.N.; Kuznetsov, V.V.; Batuev, R.A.; Tryasunov, B.G.; Teryaev, T.N.; Mazgarov, A.M.; Vildanov, A.F.; Golovanov, A.N.; Garaiev, A.M.; et al. Methods of purification of coke gas from hydrogen sulfide. H_2S recycling processes. Direct catalytic oxidation. Developments of the Institute of Catalysis SB RAS (review). *Altern. Energy Ecol.* **2014**, *19*, 86–106.
122. Ismagilov, Z.R.; Khairulin, S.R.; Ismagilov, F.R. The Technology of Catalytic Purification of Hydrogen Sulfide Containing Gases Over Honeycomb Monolith Catalysts. *Proc. First Int. Semin. Monolith. Honeycomb Supports Catal.* **1995**, *2*, 206–207.
123. Barannik, G.B.; Dobrynkin, N.M.; Ismagilov, F.R.; Ismagilov, Z.R.; Khajrulin, S.R.; Navalikhin, P.G.; Podshivalin, A.V. Method of Cleaning Gases from Sulfurous Compounds. R.U. Patent 2144495, 20 January 2000.
124. Khairulin, S.R.; Ismagilov, Z.R.; Kerzhentsev, M.A. Direct Selective Oxidation of Hydrogen Sulfide to Elementary Sulfur. In *Process for Geothermal Steam Purification 2nd World Congress on Environmental Catalysis*; Extended Abstracts: Miami, FL, USA, 1998.
125. Palma, V.; Barba, D.; Gerardi, V. Honeycomb-Structured Catalysts for the Selective Partial Oxidation of H_2S. *J. Clean. Prod.* **2016**, *111*, 69–75. [CrossRef]
126. Eom, H.; Jangb, Y.; Choib, S.Y.; Leea, S.M.; Kim, S.S. Application and Regeneration of Honeycomb-Type Catalysts for the Selective Catalytic Oxidation of H_2S to Sulfur from Landfill Gas. *Appl. Catal. A Gen.* **2020**, *590*, 117365. [CrossRef]
127. Broecker, F.J.; Gettert, H.A.; Kaempfer, K. Desulfurization of H2S-Containing Gases. U.S. Patent 4507274, 26 March 1985.
128. Miles, J.R.; Zhou, T. Forward Models for Gamma Ray Measurement Analysis of Subterranean Formations. International Patent Application 201002791, 7 January 2011.
129. Jategaonkar, S.; Kay, B.; Braga, T.; Srinivas, G.; Gebhard, S. SulfaTreat-DO: Direct Oxidation for Hydrogen Sulfide Removal. In Proceedings of the 84th Annual GPA Convention, San Antonio, TX, USA, 13–15 March 2005.
130. Royan, T.; Wichert, E. Options for Small-Scale Sulfur Recovery. *SPE Prod. Fac.* **1997**, *12*, 267–272. [CrossRef]
131. Kettner, R.; Liermann, N. New Claus Tail-Gas Process Proved in German Operation. *Oil Gas J.* **1982**, *11*, 63–66.
132. Borsboom, J.; Lagas, J.A.; Berben, P.H. *Sulphur-88 Conference Proceedings*; British Sulphur Corp.: London, UK, 1988.
133. Lagas, J.A.; Borsboom, J.; Johannes, B.; Geus, P.H.; John, W. Process for Recovering Sulfur from Sulfur-Containing Gases. U.S. Patent 4988494, 13 April 1987.
134. Julian, R.H. Chapter 10-Catalysis in the Production of Energy Carriers from Oil. Ross Contemporary Catalysis. *Fundam. Curr. Appl.* **2019**, 233–249.
135. Avdzhiev, G.R.; Dobrynkin, N.M.; Ismagilov, Z.R.; Khajrulin, S.R.; Koryabkina, N.A.; Krasilnikova, V.A.; Lygalova, A.S.; Ryabchenko, P.V.; Shkrabina, R.A. Method of Catalyst Preparing for Sulfur Production from Hydrogen Sulfide. R.U. Patent 1829182, 20 November 1990.
136. Dobrynkin, N.M.; Ismagilov, Z.R.; Khajrulin, S.R.; Koryabkina, N.A.; Shcherbilin, V.B.; Shkrabina, R.A. Method for Preparation of Catalyst for Production of Sulfur from Hydrogen Sulfide. R.U. Patent 2035221, 20 May 1992.
137. Ismagilov, Z.R. Monolithic Catalyst Design, Engineering and Prospects of Application for Environmental Protection in Russia Reaction. *Kinet. Catal. Lett.* **1997**, *60*, 215–218. [CrossRef]
138. Barannik, G.B.; Dobrynin, G.F.; Dobrynkin, N.M.; Ismagilov, F.R.; Ismagilov, Z.R.; Khajrulin, S.R.; Kulikovskaya, N.A. Method of Preparing Catalyst. R.U. Patent 2069586, 27 November 1996.
139. Ismagilov, Z.R.; Yashnik, S.A.; Shikina, N.V.; Kuznetsov, V.V.; Babich, I.V.; Moulijn, J.A. Development of the monolithic catalyst for deep recovery of elemental sulfur from technological off-gases of metallurgical coke plants and chemical refineries of crude oil II. Synthesis, characterization and testing of monolithic impregnated and wash coated catalysts for direct H2S oxidation. In Proceedings of the 1st Nordic Symposium on Catalysis, Oulu, Finland, 16–18 August 2004.
140. Khairulin, S.R. The Study of Reaction of Direct Catalytic Hydrogen Sulfide Oxidation and Development the Technologies of Gas Purification from Hydrogen Sulfide. Ph.D. Thesis, Institute of Technical Chemistry, Perm, Russia, 1998.
141. Ismagilov, Z.R.; Khairulin, S.R.; Kerzhentsev, M.A.; Mazgarov, A.M.; Vildanov, A.F. Development of Catalytic Technologies of Gases Purification from Hydrogen Sulfide Based on Direct Selective catalytic Oxidation of H_2S to Elemental Sulfur. Euro-Asian. *J. Chem. Technol.* **1999**, *1*, 49–56. [CrossRef]
142. Ismagilov, Z.R.; Kuznetsov, V.V.; Arendarskii, D.A.; Khairulin, S.R.; Kerzhentsev, M.A. Investigation of the Reaction of Hydrogen Sulphide Oxidation by Optical and Kinetic Methods. In Proceedings of the 11th ICC, Baltimore, MD, USA, 30 June–5 July 1996.
143. Yashnik, S.A.; Kuznetsov, V.V.; Ismagilov, Z.R.; Babich, I.V.; Moulijn, J.A. Development of the monolithic catalyst for deep recovery of elemental sulfur from technological off-gases of metallurgical coke plants and chemical refineries of crude oil I. FTIR study of surface acidity of aluminas and their activity in H2S oxidation. In Proceedings of the 11th Nordic Symposium on Catalysis, Oulu, Finland, 16–18 August 2004.
144. Ismagilov, Z.R.; Kerzhentsev, M.; Kuznetsov, V.; Golovanov, A.; Garitullin, R.; Zaklev, F.; Takhautdinov, S. The process of H_2S selective catalytic oxidation for on-site purification of hydrocarbon gaseous feedstock; technology demonstration. In Proceedings of the 10th Natural Gas Conversion Symposium, Doha, Qatar, 2–7 March 2013.

145. Ismagilov, Z.R. Catalytic Fuel Combustion-A Way of Reducing Emissions of Nitrogen Oxides. *Catal. Rev. Sci. Eng.* **1990**, *32*, 51–103. [CrossRef]
146. Ismagilov, Z.R. *Catalysis for Energy Production Chemistry for 21st Century Monograph*; Thomas, J.M., Zamaraev, K.I., Eds.; Blackwell Scientific Publication: Oxford, UK, 1992; pp. 337–357.
147. Ismagilov, Z.R.; Kerzhentsev, M.A.; Shikina, N.V.; Khairulin, S.R. Catalytic processes for the treatment of mixed organic waste containing radionuclides. *Chem. Sustain. Dev.* **2021**, in press. [CrossRef]
148. Ismagilov, Z.R.; Kerzhentsev, M.A. Fluidized Bed Catalytic Combustion. *Catal. Today* **1999**, *47*, 339–346. [CrossRef]
149. Alkhazov, T.G.; Barannik, G.B.; Ismagilov, F.R.; Ismagilov, Z.R.; Ivanov, A.A.; Kerzhentsev, M.A.; Khairulin, S.R.; Nemkov, V.V.; Parmon, V.N.; Zamaraev, K.I. Method for the Purification of Hydrogen Sulfide-Containing Gases. U.S. Patent 4886649, 12 December 1989.
150. Ismagilov, Z.R.; Khairulin, S.R.; Ismagilov, F.R.; Kerzhetsev, M.A. Direct oxidation of hydrogen sulphide. Hydrocarbon Technology International. *Winter* **1995**, *1994*, 59–64.
151. Khairulin, S.R.; Ismagilov, Z.R.; Kerzhentsev, M.A. Direct Heterogeneous-Catalytic H_2S Oxidation to Elemental Sulfur. *Khim. Prom.* **1996**, *4*, 265–268.
152. Ismagilov, Z.R.; Kerzhentsev, M.A.; Khairulin, S.R.; Kuznetsov, V.V. One-Stage Catalytic Methods of Acid Gases Purification from Hydrogen Sulfide. *Chem. Sustain. Dev.* **1999**, *7*, 375–396.
153. Ismagilov, Z.R.; Kerzhentsev, M.A.; Khairulin, S.R. Catalytic Purification of Geothermal Steam from Hydrogen Sulfide. *Chem. Sustain. Dev.* **1999**, *7*, 443–449.
154. Ismagilov, Z.R.; Yashnik, S.A.; Moulijn, J.A.; Startsev, A.N.; Boronin, A.I.; Stadnichenko, A.I.; Kriventsov, V.V.; Kasztelan, S.; Guillaume, D.; Makkee, M. Deep Desulfurization of Diesel Fuel on Bifunctional Monolithic Nanostructured Pt-Zeolite Catalysts. *Catal. Today* **2009**, *144*, 235–250. [CrossRef]
155. Ismagilov, Z.R.; Khairulin, S.R.; Parmon, V.N.; Sadykov, A.F.; Golovanov, A.A.; Yarullin, R.S.; Gibadukov, M.M.; Mazgarov, A.M.; Takhautdinov, S.F.; Zakiev, F.A.; et al. Direct Catalytic Oxidation of Hydrogen Sulfide as a Process for the Purification of Oil-Associated Gases. Experience of the Operation of the First Industrial Installation. *Gazokhimiya* **2011**, *3*, 57–60.
156. Ismagilov, Z.; Khairulin, S.; Kuznetsov, V.; Golovanov, A.; Garifullin, R.; Zakiev, F.; Takhautdinov, S. Field testing of the process of H_2S selective oxidation for purification of oil-associated gases. In Proceedings of the 7th International Conference on Environmental Catalysis (ICEC 2012), Lyon, France, 2–6 September 2012.
157. Ismagilov, Z.; Parmon, V.; Yarullin, R.; Mazgarov, A.; Khairulin, S.; Kerzhentsev, M.; Golovanov, A.; Vildanov, A.; Garifullin, R. The Process of H_2S Selective Catalytic Oxidation for on-site. Purification of Hydrocarbon Gaseous Feedstock-Technology. Demonstration at Bavly Oil Field in Republic of Tatarstan. In Proceedings of the XII European Congress on Catalysis EuropaCat XII, Kazan, Russia, 30 August–4 September 2015.
158. Vildanov, A.F.; Golovanov, A.A.; Ismagilov, Z.R.; Kerzhentsev, M.A.; Mazgarov, A.M.; Filippov, A.G.; Khairullin, S.R. Installation for the Processing of Hydrogen Sulfide-Containing Gases. R.U. Patent 149826, 20 January 2015.
159. Khairulin, S.R.; Ismagilov, Z.R.; Kerzhentsev, M.A.; Salnikov, A.V.; Loginov, R.I.; Philippov, A.G.; Vildanov, A.F.; Mazgarov, A.M. Carbon Materials for Gas Purification from Hydrogen Sulphide and Prospects of Their Use in Base Technologies for Associated Petroleum Gas Treatment. *Chem. Sustain. Dev.* **2018**, *26*, 679–689.
160. Ismagilov, Z.R.; Khairulin, S.R.; Filippov, A.G.; Mazgarov, A.M.; Vildanov, A.F. Direct Heterogeneous Catalytic Oxidation of Hydrogen Sulphide for Associated Petroleum Gas Treatment. *Chem. Sustain. Dev.* **2017**, *25*, 535–543.
161. Khairulin, S.R.; Ismagilov, Z.R.; Shabalin, O.N.; Komarov, F.F. Direct heterogeneous-catalytic oxidation of hydrogen sulfide for the purification of associated petroleum gases. In Proceedings of the III Russian Congress on Catalysis "Roskataliz", Nizhny Novgorod, Russia, 22–26 May 2017.
162. Burov, V.V.; Golovanov, A.A.; Ismagilov, Z.R.; Khajrulin, S.R.; Komarov, F.F.; Lotfullin, N.N.; Parmon, V.N.; Shabalin, O.N. Plant for Hydrogen-Sulphide-Containing Gas Cleaning Process. R.U. Patent 2639912, 11 November 2016.
163. Golubeva, I.A.; Khairullina, G.R.; Starynin, A.Y.; Karatun, O.N. The Production Analysis of Sulfur Using the Claus Method at Oil and Gas Industry of Russia, Unsolved Problems. *Oil Gas Chem.* **2017**, *3*, 5–12.
164. Khairulin, S.R.; Kerzhentsev, M.A.; Salnikov, A.V.; Ismagilov, Z.R. Basic technologies of direct catalytic oxidation of H_2S to sulfur. In Proceedings of the XXI Mendeleev Congress on General and Applied Chemistry, St. Petersburg, Russia, 9–13 September 2019.

catalysts

Editorial

Catalysts and Processes for H$_2$S Conversion to Sulfur

Daniela Barba

Department of Industrial Engineering, University of Salerno, Giovanni Paolo II, 132, 84084 Fisciano, Italy; dbarba@unisa.it

The hydrogen sulfide (H$_2$S) is one of the main byproducts in natural gas plants, refineries, heavy oil upgraders, and metallurgical processes. It is a toxic gas and is classified as hazardous industrial waste. The exploitation of hydrogen sulfide as fuel using conventional combustion technologies is forbidden and criminalized by the more stringent environmental policies due to its deleterious effect like the SO$_2$ formation, which is the main responsible for acidic precipitation. There are different technologies for the removal of hydrogen sulfide but are characterized by high costs and limited H$_2$S conversion efficiency. Hydrogen Sulfide is usually removed by the well-known Claus process, which is mainly used in refineries, natural gas processing plants for the treatment of rich-H$_2$S gas streams, but it is not economically profitable because the hydrogen is lost as water. An interesting alternative could be to produce simultaneously sulfur and hydrogen by thermal catalytic decomposition of H$_2$S, even if the amount of energy requested to achieve extremely high temperatures, a low hydrogen yield and the need for subsequent separation stages represent the main drawbacks to an industrial application.

Therefore, the challenge is to realize the H$_2$S abatement in a one-reaction step in the presence of an active catalyst and very selective to sulfur already at low temperature. The choice of the catalyst plays a fundamental role in assuring a high grade of H$_2$S removal with a lower selectivity to SO$_2$.

Consequently, the development of innovative processes, or also the optimization of the most common technologies employing new catalysts for the H$_2$S abatement, is welcomed to the Special Issue "Catalysts and Processes for H$_2$S Conversion to Sulfur".

The Special Issue is particularly devoted to the preparation of novel powdered/structured supported catalysts and the physical-chemical characterization, to the study of the aspects concerning the stability, reusability as well as of phenomena that could underlie the deactivation of the catalyst. The Special Issue covers also the kinetic modelling of the reaction system, by the identification of the main reactions to provide information about the reaction mechanisms, allowing so to optimize the reactor design, maximizing the activity of the catalyst.

This special issue contains 7 articles and 1 communication regarding the desulfurization of sour gases and fuel oil, the synthesis of novel adsorbents and catalysts for the H$_2$S abatement. In the following, a brief description of the papers included in this issue is provided to serve as an outline to encourage further reading.

Chen et al. have investigated porous carbonaceous materials for the reduction of H$_2$S emission during swine manure agitation. Two biochars, highly alkaline and porous made from corn stover and red oak were tested. The authors have verified the possibility of using surficial biochar treatment for short-term mitigation of H$_2$S emissions during and shortly after manure agitation [1].

Bao et al. have used the waste solid as a wet absorbent to purify the H$_2$S and phosphine from industrial tail gas. The reaction mechanism of simultaneous removal of H$_2$S and phosphine by manganese slag slurry was investigated. Best efficiency removal of both H$_2$S and phosphine was obtained by the modified manganese slag slurry [2].

The desulfurization of sour gases was studied by Duong-Viet et al., over carbon-based nanomaterials in the form of N-doped networks by the coating of a ceramic SiC. The chem-

check for
updates

Citation: Barba, D. Catalysts and Processes for H$_2$S Conversion to Sulfur. *Catalysts* **2021**, *11*, 903. https://doi.org/10.3390/catal11080903

Received: 13 July 2021
Accepted: 24 July 2021
Published: 26 July 2021

Publisher's Note: MDPI stays neutral with regard to jurisdictional claims in published maps and institutional affiliations.

ical and morphological properties of the nano-doped carbon phase/SiC-based composite were controlled to get more effective and robust catalysts able to remove H_2S from sour gases under severe desulfurization conditions such as high GHSVs and concentrations of aromatics as sour gas stream contaminants [3].

Li et al., have carried out the oxidative desulfurization of fuel oil for the removal of dibenzothiophene by using imidazole-based polyoxometalate dicationic ionic liquids. Three kinds of catalyst were synthesized and tested under different conditions [4]. The catalytic performance of the catalysts was studied under different conditions by removing the dibenzothiophene from model oil. The authors have identified a catalyst with an excellent DBT removal efficiency under optimal operating conditions.

The H_2S and SO_2 removal at low temperature was investigated by Ahmad et al., over eco-friendly sorbents from the raw and calcined eggshells. They have studied the effect of relative humidity and reaction temperatures. The best adsorption capacity for H_2S and SO_2 were obtained at a high calcination temperature of eggshell [5].

Zulkefli et al. have prepared a zinc acetate supported over the commercial activated carbon for the H_2S capture by adsorption. The optimization conditions for the adsorbent synthesis were carried out using RSM and the Box–Behnken experimental design. Several factors and levels were evaluated, including the zinc acetate molarity, soaked period, and soaked temperature, along with the response of the H_2S adsorption capacity and the surface area [6].

Vanadium-sulfide-based catalysts supported on ceria were used for the direct and selective oxidation of H_2S to sulfur and water at a lower temperature. Barba et al. have performed a screening of catalysts with different vanadium loading in order to study the catalytic performance in terms of H_2S conversion and SO_2 selectivity. The effect of the temperature, contact time, and H_2S inlet concentration was studied over the catalyst that has exhibited the highest H_2S removal efficiency and the lowest SO_2 selectivity [7].

The H_2S adsorption was studied over a novel kind of hydrochar adsorbent derived from chitosan or starch and modified by CuO-ZnO. Zang et al., have investigated the formation of CuO-ZnO on hydrochar, the effect of the hydrochar species, the adsorption temperature and the adsorption mechanism [8].

As Guest Editor, I would like to thank all the authors who contributed to this Special Issue. Their contributions represent interesting and innovative examples of the current research trends in the field of H_2S removal from liquid and gas streams.

I also wish to thank the editorial staff of Catalysts for their help to organize this issue.

I hope that the topics presented in this issue will inspire readers to further investigate new materials and solutions to reduce significantly the presence of pollutants such as H_2S, SO_2 and other sulfur-based compounds, and so pursuing the objective of "zero emissions" in the atmosphere.

Funding: This research received no external funding.

Conflicts of Interest: The author declares no conflict of interest.

References

1. Chen, B.; Koziel, J.A.; Białowiec, A.; Lee, M.; Ma, H.; Li, P.; Meiirkhanuly, Z.; Brown, R.C. The Impact of Surficial Biochar Treatment on Acute H_2S Emissions during Swine Manure Agitation before Pump-Out: Proof-of-the-Concept. *Catalysts* **2020**, *10*, 940. [CrossRef]
2. Bao, J.; Wang, X.; Li, K.; Wang, F.; Wang, C.; Song, X.; Sun, X.; Ning, P. Reaction Mechanism of Simultaneous Removal of H_2S and PH_3 Using Modified Manganese Slag Slurry. *Catalysts* **2020**, *10*, 1384. [CrossRef]
3. Duong-Viet, C.; Nhut, J.-M.; Truong-Huu, T.; Tuci, G.; Nguyen-Dinh, L.; Pham, C.; Giambastiani, G.; Pham-Huu, C. Tailoring Properties of Metal-Free Catalysts for the Highly Efficient Desulfurization of Sour Gases under Harsh Conditions. *Catalysts* **2021**, *11*, 226. [CrossRef]
4. Li, J.; Guo, Y.; Tan, J.; Hu, B. Polyoxometalate Dicationic Ionic Liquids as Catalyst forExtractive Coupled Catalytic Oxidative Desulfurization. *Catalysts* **2021**, *11*, 356. [CrossRef]
5. Ahmad, W.; Sethupathi, S.; Munusamy, Y.; Kanthasamy, R. Valorization of Raw and Calcined Chicken Eggshell for Sulfur Dioxide and Hydrogen Sulfide Removal at Low Temperature. *Catalysts* **2021**, *11*, 295. [CrossRef]

6. Zulkefli, N.N.; Masdar, M.S.; Wan Isahak, W.N.R.; Abu Bakar, S.N.H.; Abu Hasan, H.; Mohd Sofian, N. Application of Response Surface Methodology for Preparationof $ZnAC_2$/CAC Adsorbents for Hydrogen Sulfide (H_2S) Capture. *Catalysts* **2021**, *11*, 545. [CrossRef]
7. Barba, D.; Vaiano, V.; Palma, V. Selective Catalytic Oxidation of Lean-H_2S Gas Stream to Elemental Sulfur at Lower Temperature. *Catalysts* **2021**, *11*, 746. [CrossRef]
8. Zang, L.; Zhou, C.; Dong, L.; Wang, L.; Mao, J.; Lu, X.; Xue, R.; Ma, Y. One-Pot Synthesis of Nano CuO-ZnO Modified Hydrochar Derived from Chitosan and Starch for the H_2S Conversion. *Catalysts* **2021**, *11*, 767. [CrossRef]

catalysts

Article

One-Pot Synthesis of Nano CuO-ZnO Modified Hydrochar Derived from Chitosan and Starch for the H₂S Conversion

Lihua Zang [1,2], Chengxuan Zhou [2], Liming Dong [1], Leilei Wang [3], Jiaming Mao [4], Xiaomin Lu [5], Rong Xue [2] and Yunqian Ma [1,2,*]

1   Key Laboratory of Cleaner Production and Integrated Resource Utilization of China National Light Industry, Beijing Technology and Business University, Beijing 100048, China; zlh@qlu.edu.cn (L.Z.); donglm@btbu.edu.cn (L.D.)
2   College of Environmental Science and Engineering, Qilu University of Technology (Shandong Academy of Science), Jinan 250353, China; zhouchengxuan@ipe.ac.cn (C.Z.); xr@qlu.edu.cn (R.X.)
3   Ecology Institute of Shandong Academy of Sciences, Qilu University of Technology (Shandong Academy of Sciences), Jinan 250014, China; wangll@qlu.edu.cn
4   College of Environmental Science and Engineering, Beijing Forestry University, Beijing 100083, China; mjm2020@bjfu.edu.cn
5   Department of Forest Biomaterials, North Carolina State University, Raleigh, NC 27695, USA; xlu13@ncsu.edu
*   Correspondence: mayq@qlu.edu.cn

Abstract: A novel kind of hydrochar adsorbent, modified by CuO-ZnO and derived from chitosan or starch, was synthesized for H₂S adsorption. The prepared adsorbent was characterized by BET, XRD, EDX, SEM, and XPS. The results showed that the modified hydrochar contained many amino groups as functional groups, and the nanometer metal oxide particles had good dispersion on the surface of the hydrochar. The maximum sulfur capacity reached 28.06 mg/g-adsorbent under the optimized conditions. The amine group significantly reduced the activation energy between H₂S and CuO-ZnO conducive to the rapid diffusion of H₂S among the lattices. Simultaneously, cationic polyacrylamide as a steric stabilizer could change the formation process of CuO and ZnO nanoparticles, which made the particle size smaller, enabling them to react with H₂S sufficiently easily. This modified hydrochar derived from both chitosan and starch could be a promising adsorbent for H₂S removal.

Keywords: hydrochar; adsorbent; mixed metal oxides; H₂S conversion

Citation: Zang, L.; Zhou, C.; Dong, L.; Wang, L.; Mao, J.; Lu, X.; Xue, R.; Ma, Y. One-Pot Synthesis of Nano CuO-ZnO Modified Hydrochar Derived from Chitosan and Starch for the H₂S Conversion. *Catalysts* **2021**, *11*, 767. https://10.3390/catal11070767

Academic Editor: Daniela Barba

Received: 18 May 2021
Accepted: 15 June 2021
Published: 24 June 2021

Publisher's Note: MDPI stays neutral with regard to jurisdictional claims in published maps and institutional affiliations.

1. Introduction

Hydrogen sulfide (H₂S), a poisonous, odorous, and corrosive gas, commonly exists in industrial gases such as coal gasification gas, natural gas, and biogas. H₂S is harmful to humans and livestock, and it not only brings corrosion to metal pipes and reaction devices in the industrial production process, but also causes catalyst poisoning, which affects product quality [1,2]. H₂S is easy to burn, generating SO₂ as a combustion product. Whether through combustion or direct emissions, it can exert a severe impact on the atmospheric environment. Therefore, H₂S should be fixed on some materials or removed from the production process and environment.

At present, there are many industrial methods to remove H₂S. According to production conditions and desulfurization costs, the methods of H₂S removal in industrial processes can be classified into wet flue gas desulfurization (WFGD) and dry flue gas desulfurization (DFGD) [3,4]. The desulfurizer of WFGD is a liquid that absorbs and separates H₂S with large processing capacity and mature technology. WFGD has some disadvantages—for example, high energy consumption, secondary pollution, and high regeneration cost. DFGD has mainly been used to remove H₂S at low concentrations, and it has the advantages of high H₂S removal efficiency and low cost [5]. The commonly used dry desulfurization methods in industrial processes are the zinc oxide method, iron oxide method, manganese desulfurization method, Claus method, etc. [6–10].

Biochar is defined as a solid, carbon-rich product obtained from biomass through various thermochemical technologies [11–14]. Pyrochar and hydrochar are two kinds of biochar prepared from the pyrolysis and hydrothermal carbonization (HTC) of biomass [15], respectively. The biomass includes straw, sawdust, the dung of herbivores, etc. [16]. Biochar is a promising alternative adsorbent for toxic gas and wastewater treatment [17,18]. Most researchers have focused on pyrochar, but there are a few reports on the application of hydrochar in H_2S capture. HTC is an auspicious approach for the use of waste biomass. Compared with pyrochar, hydrochar is suitable for dealing with wet biomass directly, with lower energy consumption [19–21]. Hydrochar also has higher yield and cation exchange capacity, and, during the production process, no PAHs are released. On the surface of hydrochar, it has more oxygen-containing functional groups, which is favorable for H_2S capture and oxidation [22,23]. Although hydrochar has such abundant advantages, it has been rarely used to treat gaseous pollutants, especially H_2S.

Chitosan, insoluble in water and organic solvents, is a natural macromolecular aminopolysaccharide with a yield second only to cellulose [24]. Moreover, cornstarch is the world's largest source of starch, accounting for approximately 65% of the total amount worldwide [25]. In this work, chitosan and cornstarch were used to synthesize hydrochar, which was modified by CuO-ZnO in a one-pot process. The aim of this work was to explore the formation of CuO-ZnO on hydrochar, the physical and chemical properties of this new material, the desulfurization products, and the mechanisms. This work provides new insights into the development and application of hydrochar products.

2. Results and Discussion

2.1. Basic Physical and Chemical Properties of Hydrochars

The results of the EDX analyses of the hydrochars are shown in Figure 1. The mass yield (the ratio of product to the original raw biomass), ultimate analysis by EDX (C, O, N, Zn, Cu), specific surface area, pore volume, and average pore diameter are reported in Table 1. It was found that the type of precursor had a significant effect on the yield of hydrochar (Table 1). The mass yields of the solids recovered changed depending on the content and type of the precursor. With the increase in chitosan content, the yield of hydrochar and nitrogen content increased [26], because the hydrochar yield is related to the solubility of the precursors in water, and the solubility of starch is much higher than that of chitosan [24,25]. Simultaneously, the nitrogen in the hydrochar products mainly came from chitosan, and a small amount of nitrogen came from polyacrylamide; therefore, with the increase in the chitosan content, the nitrogen content increased.

Under the same synthetic conditions, when the ratio of chitosan to starch was 1:1, the specific surface area of hydrochar reached 30.102 m^2/g. It was found that, although the added amount of $ZnCl_2$ and $CuCl_2$ in the precursor was the same, the detected content of Zn in the product was much lower than that of copper. One possible reason is that the pH value of the filtrate during hydrothermal carbonization kept decreasing, and ZnO could not remain stable under the slightly acidic conditions, while CuO was stable under the acidic conditions.

Table 1. The yields, specific surface area, average pore diameter, and elemental composition of the synthesized hydrochars.

Hydrochar	Yield (%)	SSA	PV	APD	Elemental Composition (%)				
					C	O	Zn	Cu	N
S5C5	14.83	30.102	0.071	8.9264	48.49	38.73	0.34	6.68	2.58
S10C0	11.01	12.939	0.041	1.2464	36.42	54.34	0.37	5.57	0.05
S3C7	29.89	16.456	0.069	1.456	38.85	42.07	0.36	6.97	5.93
S7C3	14.44	17.211	0.073	1.6234	43.20	45.37	0.66	5.10	1.18
S0C10	32.37	14.653	0.046	1.093	43.33	39.64	0.41	6.35	7.13
S5C5N	9.7	8.162	0.042	0.891	49.38	36.97	0.42	5.59	2.05

Note: Specific surface area (SSA), Pore volume (PV), Average pore diameter (APD).

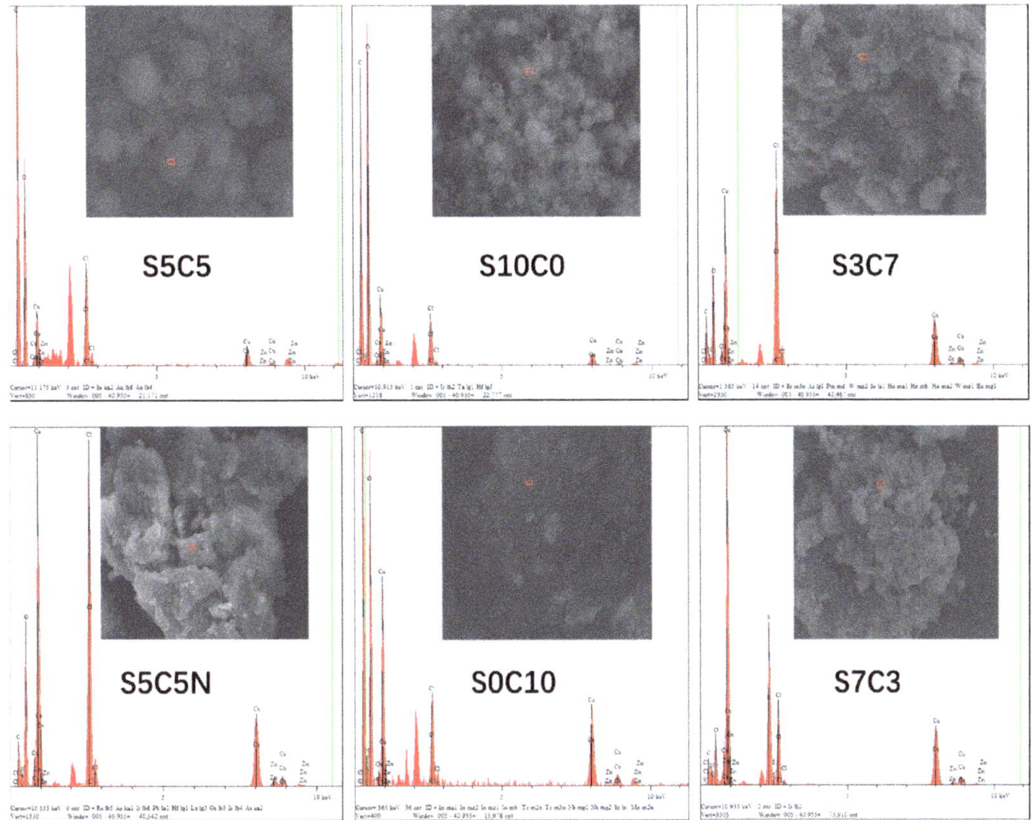

Figure 1. EDX images of the hydrochars.

FT-IR spectra of the hydrochars were used to determine the functional groups contained in the sample, and these are shown in Figure 2. The C=C in aromatic groups showed an adsorption peak between 1613 cm^{-1} and 1718 cm^{-1}. For carbonyl in -COOH and CO-NH, the regions from 1400 cm^{-1} to 1500 cm^{-1} were ascribed to C-N and C-O groups' stretching vibration. The peak at 1033 cm^{-1} was assigned to C-O stretching or O-H bending vibrations. The broad band at 3400 cm^{-1} can be assigned to the existence of the N-H structure. Finally, the peaks at 1033 cm^{-1} and 1403 cm^{-1} were suggested to be C-N and C-O, respectively. According to the elemental analysis and FT-IR analysis, it was proven that amine groups and oxygen-containing groups on the surface of the hydrochar were abundant.

The surface morphology of several hydrochars is shown in Figure 3. It was found that with the change in the starch and chitosan content in the precursors, the hydrochars presented different microstructures. Among all the hydrochars without polyacrylamide, the precursor starch mainly showed carbon particles and carbon spheres with diameters from 10 μm to 100 μm, while the precursor chitosan mainly had a porous cellular structure. However, the hydrochar with polyacrylamide was dense, with no regular shape, and consisted of some carbon particles, because both chitosan and cationic polyacrylamide contained positive charges, which could make the distribution of the system more uniform. Furthermore, metal oxide clusters were not observed in any of the photomicrographs. The hydrochar samples with different starch and chitosan content were analyzed by XRD (in Figure 4), and this showed that the crystallinity of metal oxides in the hydrochar was low. There may be ZnO in S5C5, S5C5N, S10C0, and free Cu in S0C10 due to the reducibility of chitosan [27]. It indicated that the distribution of active metal sites in the hydrochar was relatively uniform or that metal oxides were embedded in carbon spheres or carbon particles [28].

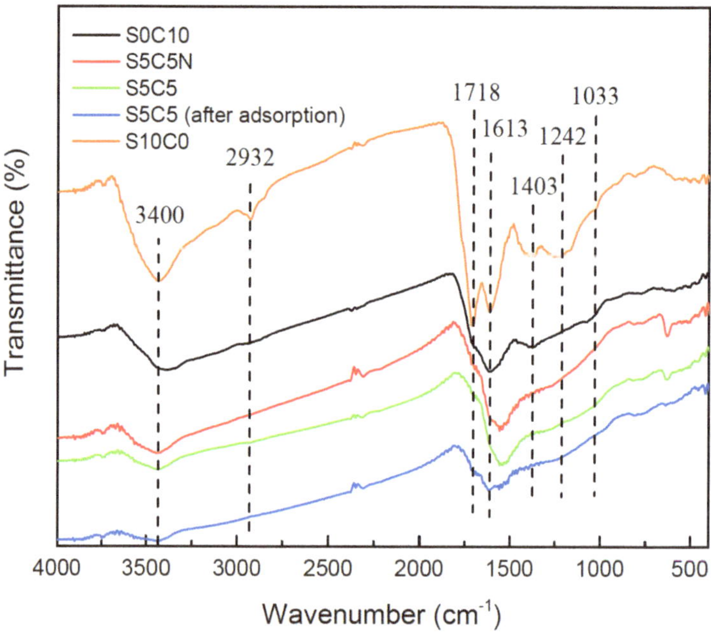

Figure 2. FT-IR spectra of the hydrochars.

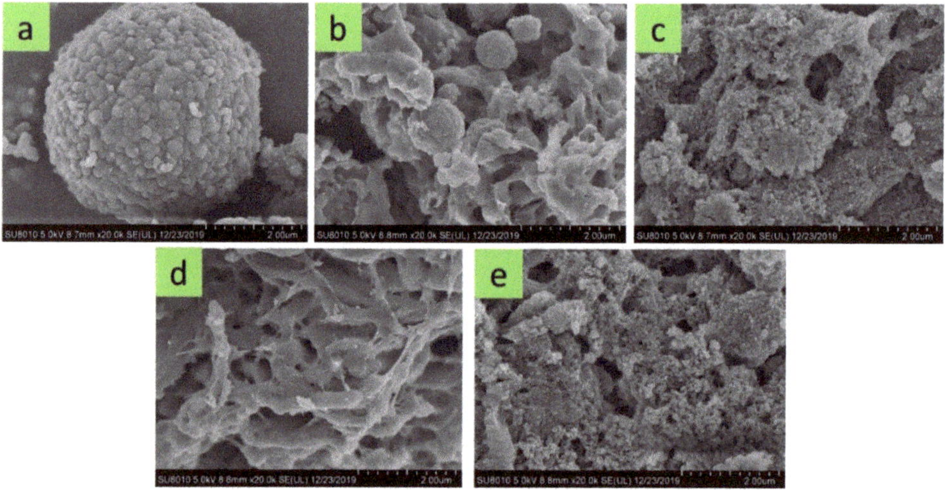

Figure 3. SEM images of different hydrochars. (**a**) S10C0; (**b**) S5C5; (**c**) S5C5N and (**d**) S0C10 before adsorption; (**e**) S5C5 after adsorption.

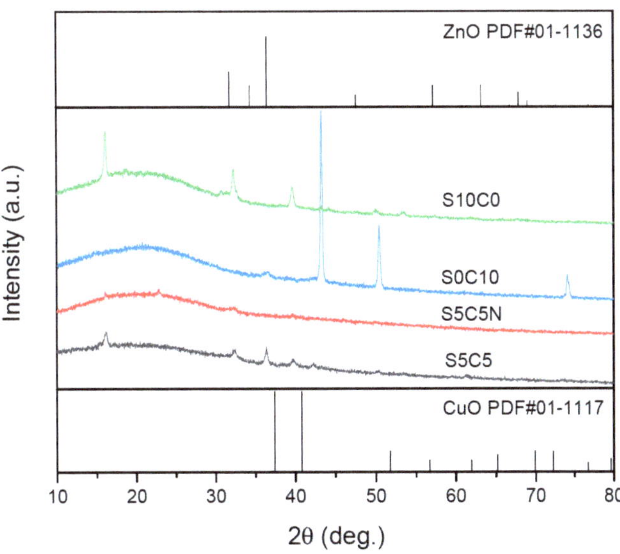

Figure 4. X-Ray diffraction patterns of hydrochar samples.

Notably, no metal oxide aggregates were observed in any of the samples. Based on previous results [29,30], we believe that both the molecular weight and concentration of PAM significantly affected the morphology of the end-products. For the synthesis of nanoporous materials at a large scale, the approach was facile and had many potential applications. In addition, it was also applicative for the synthesis of other materials with a high surface area and nanoporous structures. It has been suggested that the addition of polyacrylamide can affect the morphology of the hydrochar. As an important capping agent, PAM has been widely used to synthesize materials with various nanostructures (nanorods, nanowires, nanoplates, nanocubes, etc.). The exact function of PAM on the shape selectivity is not yet fully understood; however, we believe that the selective adsorption of PAM on various crystallographic planes (newly formed CuO, ZnO, or hydrochar particles) suppressed their intrinsic anisotropic growth [30]. With an N-C=O group, PAM was easily attached to the surfaces of these materials and limited the growth of the crystal faces. Selective interactions between PAM and the different surface planes of the CuO or ZnO may greatly influence the growth direction and rate and ultimately result in particles with different shapes [30]. For an oxidation catalyst, its effectiveness can be mainly attributed to the adsorption and desorption of gas molecules from its surface.

2.2. H_2S Adsorption Performance

2.2.1. Effect of Hydrochar Species

In this section, a series of single-factor experiments were carried out to determine the effect of the adsorbent in the desulfurization system. The H_2S removal efficiency (%) and breakthrough sulfur capacity were selected as the evaluation index. The ratio of chitosan to starch had a significant influence on H_2S removal. The sulfur capacities of hydrochars with different molar ratios of chitosan to starch under 180 °C were measured and are shown in Figure 5. It can be observed that the sulfur capacity of S5C5 was higher than that of other hydrochars and the addition of polyacrylamide had a great impact on H_2S adsorption.

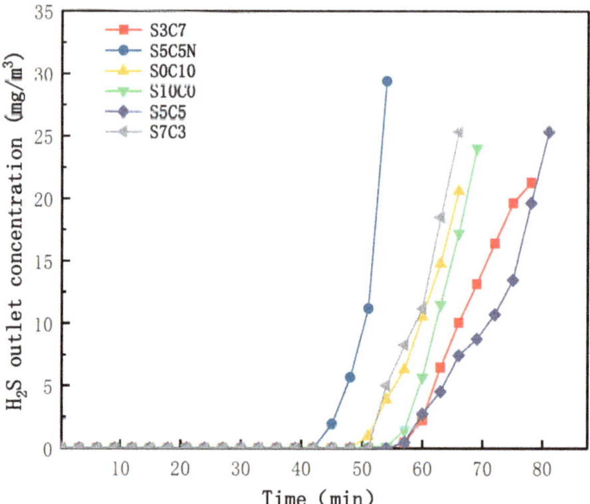

Figure 5. The breakthrough curves of hydrochar S5C5 with different molar ratios of chitosan to starch (T, 230 °C; auxiliary agent, cationic PAM; cationic PAM concentration, 2.0 g/L).

2.2.2. Effect of Auxiliary Agents on H_2S Removal

Cationic PAM, polyvinylpyrrolidone, and neutral PAM were used as the auxiliary agents in the synthesis of the hydrochar. According to the experimental results, it was found that the addition of polyacrylamide and its concentration can affect the sulfur capacity. The effect of different auxiliary agents on H_2S removal by hydrochar S5C5 is shown in Figure 6. Among the three auxiliary agents, the cationic PAM-synthesized hydrochar showed the best performance for H_2S removal. In order to explore the best composition of precursors, the PAM concentrations were also optimized, as shown in Figure 7. With the increasing of the PAM concentration from 0.5 g/L to 3.0 g/L, the sulfur capacity decreased. Cationic PAM with a positive charge can attract to and interact with chitosan of a negative charge, leading to a good combination of cationic PAM in hydrochar, but the best amount of cationic PAM depended on the amount of chitosan.

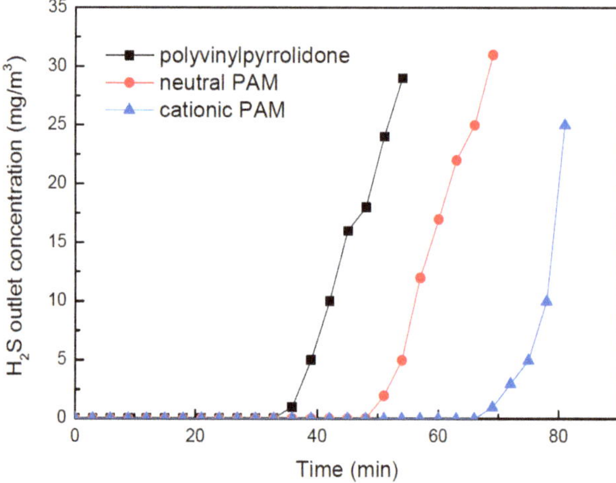

Figure 6. The breakthrough curves of hydrochar S5C5 with different auxiliary agents (T, 230 °C; auxiliary agent concentration, 2.0 g/L).

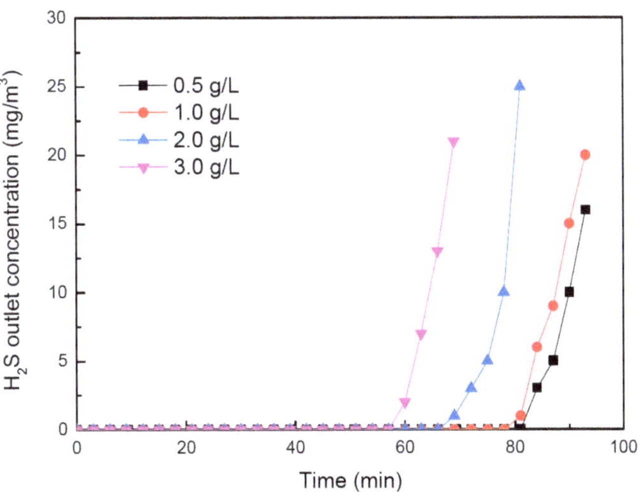

Figure 7. The breakthough curves of hydrochar S5C5 with cationic PAM of different concentrations (T, 230 °C; auxiliary agent, cationic PAM).

2.2.3. Effect of Adsorption Temperature on H$_2$S Removal

The effect of adsorption temperature on H$_2$S removal by hydrochar S5C5 is shown in Figure 8. With the desulfurization temperature increasing, the desulfurization ability of hydrochar S5C5 was clearly improved. This result indicated that the desulfurization ability of S5C5 modified by metal oxide was lower than the sorbent derived from the molecular sieve (SBA-15 or MCM-41) with modification or adsorbents with high metal content; however, it was far higher than other types of common active carbon [31–33]. As is known, a low temperature is beneficial for H$_2$S adsorption. Raising the temperature could enhance the molecular mobility and interaction between each reactant to promote H$_2$S adsorption by increasing the reaction rate; however, a higher temperature would be an obstruction to H$_2$S adsorption in an exothermic reaction.

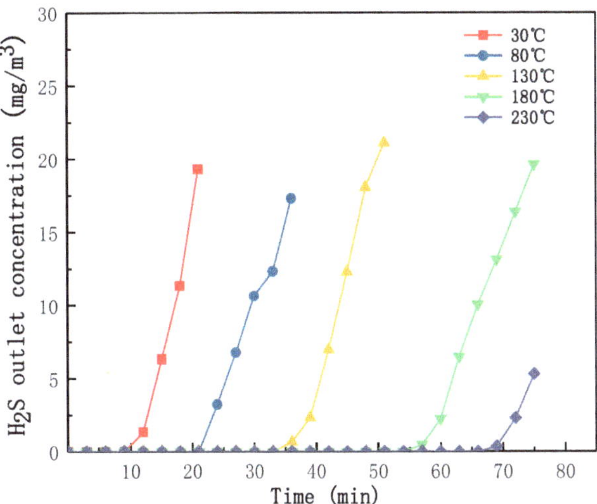

Figure 8. The breakthough curves of hydrochar S5C5 under different adsorption temperatures (auxiliary agent, cationic PAM; cationic PAM concentration, 0.5 g/L).

In the adsorption and oxidation process of H_2S, the change in the oxygen functional groups on the surface of the hydrochar plays an important role. The quinone and carbonyl groups on the surface of the hydrochar can react with molecular H_2S. The C=O bond and C=C bond were broken and combined with H_2S to form the S-O bond. This process was endothermic and favored the rising temperature. When the ratio of chitosan to starch is 1:1, the cationic PAM concentration is 0.5 g/L, and the temperature is 230 °C, the maximum sulfur capacity of the hydrochar S5C5 is 28.06 mg/g-adsorbent.

2.3. Adsorption Mechanism

The adsorption of H_2S on the hydrochar consisted of three parts: the first part is the reaction of H_2S with the metal active sites, such as CuO and ZnO [34]; the second part is the reaction of H_2S with the oxygen-containing functional groups and carbon on the surface of the hydrochar to form C-S bonds and O-S bonds; the third part is the physical adsorption process of H_2S on the surface of the hydrochar [33].

To further investigate the reaction mechanism, the chemical valence states of the element in the whole process were analyzed by XPS. The XPS spectra of S, Cu, and Zn are shown in Figures 9–11.

The XPS spectrum of S in hydrochar S5C5 after H_2S adsorption is shown in Figure 9. The valence state of S was confirmed within the binding energy in the range of 162-172 eV. This showed that S 2p3/2 and S^{2-} 2p2/3 appeared at 167.5 eV and 161.7 eV, respectively. The peak at 163.4 eV may vary due to the existence of the structure of C-S [33]. The Cu^+ was confirmed by the Cu 2p3/2 binding energy in the range of 930 eV to 937 eV, and it showed that Cu^+ 2p3/2 appeared at 932.56 eV, and the Cu^{2+} was assigned to the binding energy of the XPS contribution from 928 to 937 eV with a satellite contribution in the range of 937–947 eV, and it appeared at 934.61 eV. The Zn^{2+} was confirmed by the Zn 2p3/2 that appeared at 1021.77 eV. Therefore, the existing sulfur, zinc, and copper in the hydrochar S5C5 were CuS, ZnS, Cu_2S, sulfur, and a C-S bond.

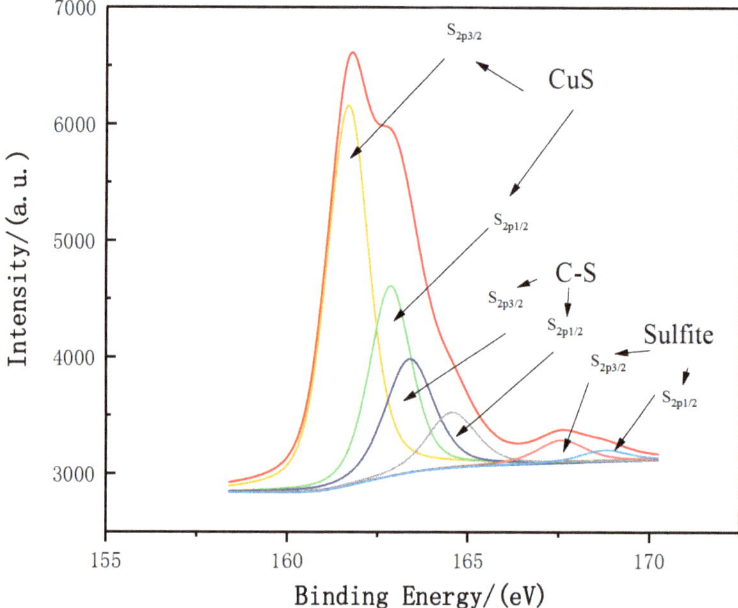

Figure 9. The XPS spectrum of S in hydrochar S5C5 after H_2S adsorption.

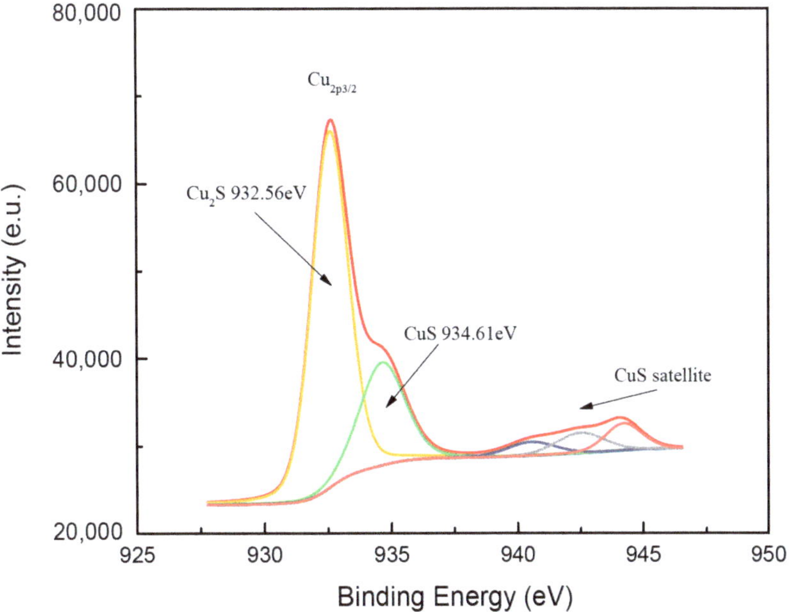

Figure 10. The XPS spectrum of Cu in hydrochar S5C5 after H_2S adsorption.

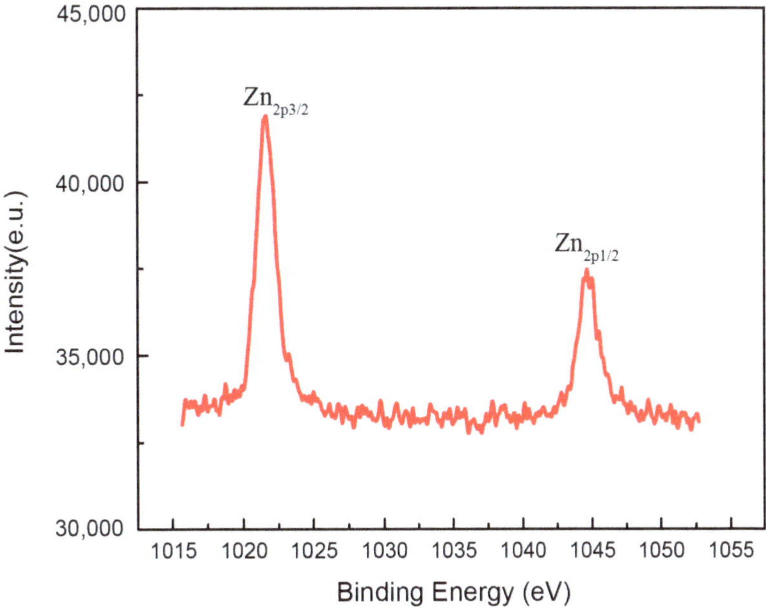

Figure 11. The XPS spectrum of Zn in hydrochar S5C5 after H_2S adsorption.

During the hydrothermal reaction, $CuCl_2$ and $ZnCl_2$ reacted to form CuO and ZnO. The reaction equation can be described as follows:

$$CuCl_2 + H_2O = Cu(OH)_2 + HCl \tag{1}$$

$$ZnCl_2 + H_2O = Zn(OH)_2 + HCl \tag{2}$$

$$Cu(OH)_2 = CuO + H_2O \tag{3}$$

$$Zn(OH)_2 = ZnO + H_2O \tag{4}$$

At the same time, copper (II) oxide has a certain oxidation capacity; H_2S can be oxidized partly to elemental sulfur and a fraction of the sulfide ions that have not been oxidized can form Cu_2S [35]. Hydrochar, rich in oxygen-containing functional groups, can combine with H_2S to form a C-S bond and S-O bond [33,36]. Therefore, the form of sulfur after adsorption can be confirmed to be sulfur and a C-S bond. H_2S also can react with oxygen-containing functional groups to form sulfates in the absence of oxygen [33]. There was no oxygen gas to participate in this adsorption, so sulfate did not exist in the product.

3. Materials and Methods

3.1. Materials

The reagents, all of analytical grade, used in this experiment were purchased directly without further purification. Copper chloride, zinc chloride, cationic polyacrylamides (CPAM), and chitosan (low viscosity, deacetylation >90%) were purchased from Shanghai Macklin Biochemical Co., Ltd (Shanghai, China). Cornstarch (Pharmaceutical grade) was purchased from Shanghai Aladding Biochemical Technology Co., Ltd (Shanghai, China). H_2S standard gas of 1% and N_2 of 99.999% were provided by Jinan Deyang Special Gas Co., Ltd (Jinan, China). The solution was prepared using laboratory-made deionized water ($18.3 \ M\Omega \cdot cm^{-1}$).

3.2. Preparation of Adsorbent

There were six types of hydrochar synthesized in the experiment. All of them were composed of chitosan and cornstarch, with a cationic polyacrylamide solution (an auxiliary agent) with different dosages. The abbreviation and composition of the synthesized hydrochar samples are shown in Table 2. Taking C5S5 as an example, chitosan (3.60 g), cornstarch (3.60 g), $ZnCl_2$ (0.84 g), and $CuCl_2$ (0.84 g) were placed in a mortar, ground evenly with force, and then moved to a glass beaker, following by the addition of 45 mL of cationic polyacrylamide solution (0.5, 1.0, 2.0, and 3.0 g/L). While being treated with ultrasound, the precursors were stirred vigorously until a light blue color appeared. The sample was placed in a hydrothermal reactor and heated up to 230 °C for 4 h. The product was washed repeatedly using deionized water until the pH of the rinsed water stabilized. Other hydrochars were synthesized using the same method with different molar ratios of the precursors.

Table 2. Abbreviations and compositions of six kinds of hydrochar.

Sample	Carbon Precursor (Dosage)	Auxiliary Agent	Metal Oxide Precursor (Dosage)
S5C5	Starch (3.60 g) + Chitosan (3.60 g)		
S10C0	Starch (7.20 g)	Cationic polyacrylamide solution	$ZnCl_2$ (0.84 g) + $CuCl_2$ (0.84 g)
S3C7	Starch (2.16 g) + Chitosan (5.04 g)		
S7C3	Starch (5.04 g) + Chitosan (2.16 g)		
S0C10	Chitosan (7.20 g)		
S5C5N	Starch (3.60 g) + Chitosan (3.60 g)	None	

3.3. Characterization of Hydrochar

The characterization of materials was investigated using a Fourier-transform infrared (FTIR) spectrophotometer (IRAffinity-1s, Shimadzu, Kyoto, Japan). X-ray diffraction (XRD) patterns of hydrochar samples were recorded on an X-ray diffractometer (SmartLab, Rigaku, Tokyo, Japan) and carried out in the 2θ range from 10° to 80°. The surface morphologies of materials were observed by scanning electron microscope (SEM) apparatus (Regulus 8220, Hitachi, Tokyo, Japan). The element composition and valence state of materials were explored by X-ray photoelectron spectroscopy (XPS) with a multifunctional imaging electron spectrometer (ESCALAB 250XI, Thermo Fisher, Waltham, MA, America). The specific surface areas of materials were measured using the Brunauer–Emmett–Teller (BET) method, and the pore size distribution was calculated using the Barrett–Joymer–Halenda (BJH) method from the isotherm of the adsorption branch with an automatic specific surface area and porosity analyzer (TriStar II 3020, Micromeritics, Norcross, GA, USA).

3.4. Batch H_2S Adsorption Experiments

The mixed gas was prepared by blending H_2S standard gas with N_2, both quantified by flow indicators (D08-1F), which were purchased from Beijing Sevenstar Electronics Co., Ltd. (Beijing, China); the concentration of H_2S was measured by the gas analyzer (TH-990S) from Wuhan Tianhong Instrument Group. After adsorption, the sulfur capability was calculated by Equation (1):

$$H_2S \ removal \ efficiency(\%) = \frac{C_{in} - C_{out}}{C_{in}} \times 100\% \tag{5}$$

C_{in} and C_{out} (mg·m^{-3}) were the inlet and outlet concentration of H_2S in the gas mixture, respectively. A diagram of the test devices for the evaluation of desulfurization performance is shown in Figure 12.

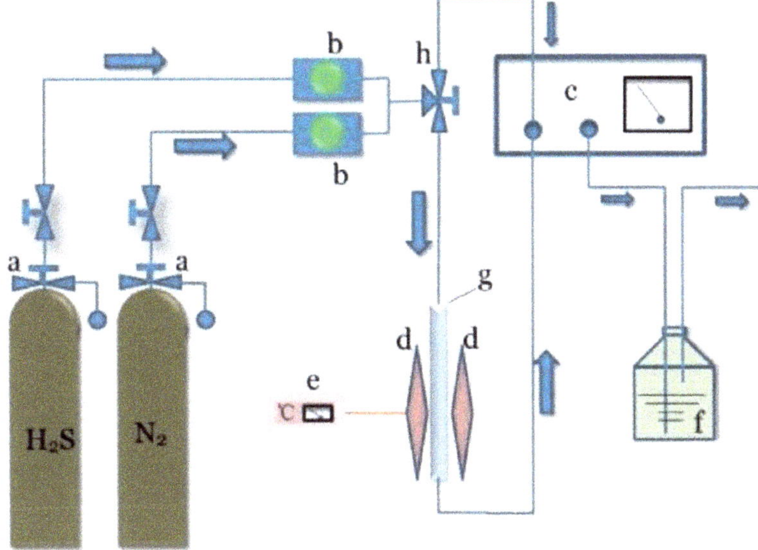

Figure 12. The diagram of test devices for the evaluation of desulfurization performance. (**a**) Pressure reducing valve; (**b**) mass flow controller; (**c**) portable hydrogen sulfide concentration analyzer; (**d**) tube furnace; (**e**) temperature controller; (**f**) concentrated lye; (**g**) quartz tube; (**h**) three-way valve.

To test sulfur capacity, a quartz tube was used and its diameter and height were 6 mm and 100 mm, respectively. The adsorption temperature was controlled by a tube furnace. In the tests, a gas mixture containing 3000 ppm (4617 mg/m^3) of H_2S (nitrogen as balance gas) was passed through the quartz tube filled with adsorbent of 0.5 g, under a gas flow rate of 100 mL/min. The outlet H_2S gas was absorbed by KOH solution. The breakthrough sulfur capacity (S_b, mg/g) was calculated in the stage from the beginning to when the H_2S outlet concentration was higher than 20 mg/m^3 by Equation (2).

$$S_b = \frac{M_S}{M_{H_2S}} \times \frac{Q_{H_2S}}{m} \left[\int_0^t (C_{in} - C_{out})dt \right] \times 10^{-6} \tag{6}$$

where S_b represents the breakthrough sulfur capacity of sorbents (mg/g), M_S and M_{H_2S} are the molar weight of sulfur (32.06 g/mol) and H_2S (34.06 g/mol), respectively; m is the weight of sorbents; Q_{H_2S} is the H_2S gas flow rate; t is the reaction time for desulfurization (min), and C_{in} and C_{out} are the inlet and outlet concentration of H_2S (mg/m^3), respectively. When t is the saturation adsorption time, Equation (2) was also used to calculate the max sulfur capacity (C_m).

4. Conclusions

For H_2S adsorption, this study provides a method for the synthesis of hydrochar, obtained by the hydrothermal reaction of chitosan, starch, cationic polyacrylamide aqueous solution, $ZnCl_2$, and $CuCl_2$. The experimental results showed that the hydrochar contained many amino groups as functional groups, and the nano-scaled metal oxide particles had good dispersion on the surface of the hydrochar. The amine group significantly reduced the activation energy of H_2S and CuO-ZnO, which was conducive to the rapid diffusion of H_2S among the lattices. At the same time, cationic polyacrylamide as a steric stabilizer can change the formation process of CuO and ZnO nanoparticles, making the particle size smaller and allowing it to react more easily with H_2S sufficiently. When the ratio of chitosan to starch is 1:1, the temperature is 230 °C, and the cationic PAM concentration is 0.5 g/L, the maximum sulfur capacity of the hydrochar S5C5 is 28.06 mg/g-adsorbent. Therefore, modified hydrochar may be a promising adsorbent for H_2S removal.

Author Contributions: Conceptualization, Y.M. and L.Z.; methodology, J.M. and C.Z.; software, R.X.; validation, Y.M. and L.Z.; formal analysis, L.W.; resources, L.D. and L.Z.; data curation, J.M. and Y.M.; writing—original draft preparation, J.M., C.Z. and Y.M.; writing—review and editing, X.L.; funding acquisition, L.D. and L.Z. All authors have read and agreed to the published version of the manuscript.

Funding: This work was funded by the Open Research Fund Program of Key Laboratory of Cleaner Production and Integrated Resource Utilization of China National Light Industry, grant number CP-2020-YB8, Natural Science Foundation of Shandong Province, grant number ZR2020QB199, and Qilu University of Technology (Shandong Academy of Sciences) Youth Doctor Cooperation Fund, grant number 2019BSHZ0028.

Conflicts of Interest: The authors declare no conflict of interest.

References

1. Coenen, K.; Gallucci, F.; Hensen, E.; Annaland, M.V.S. Adsorption behavior and kinetics of H_2S on a potassium-promoted hydrotalcite. *Int. J. Hydrog. Energy* **2018**, *43*, 20758–20771. [CrossRef]
2. Garcia-Arriaga, V.; Alvarez-Ramirez, J.; Amaya, M.; Sosa, E. H_2S and O_2 influence on the corrosion of carbon steel immersed in a solution containing 3M diethanolamine. *Corros. Eng.* **2010**, *52*, 2268–2279. [CrossRef]
3. Xiao, Y.H.; Wang, S.D.; Wu, D.Y.; Yuan, Q. Experimental and simulation study of hydrogen sulfide adsorption on impregnated activated carbon under anaerobic conditions. *J. Hazard. Mater.* **2008**, *153*, 1193–1200. [CrossRef] [PubMed]
4. Eow, J.S. Recovery of sulfur from sour acid gas: A review of the technology. *Environ. Prog. Sustain. Energy* **2010**, *21*, 143–162. [CrossRef]
5. Zhao, T.; Yao, Y.; Li, D.; Wu, F.; Zhang, C.; Gao, B. Facile low-temperature one-step synthesis of pomelo peel biochar under air atmosphere and its adsorption behaviors for Ag(I) and Pb(II). *Sci. Total Environ.* **2018**, *640–641*, 73–79. [CrossRef]
6. Zhang, X.; Dou, G.Y.; Wang, Z.; Li, L.; Wang, Y.F.; Wang, H.L.; Hao, Z.P. Selective catalytic oxidation of H_2S over iron oxide supported on alumina-intercalated Laponite clay catalysts. *J. Hazard. Mater.* **2013**, *260*, 104–111. [CrossRef] [PubMed]
7. Yasyerli, S. Cerium–manganese mixed oxides for high temperature H_2S removal and activity comparisons with V–Mn, Zn–Mn, Fe–Mn sorbents. *Chem. Eng. Process. Process Intensif.* **2008**, *47*, 577–584. [CrossRef]

8. Khudenko, B.M.; Gitman, G.M.; Wechsler, T.E.P. Oxygen Based Claus Process for Recovery of Sulfur from H_2S Gases. *J. Environ. Eng.* **1993**, *119*, 1233–1251. [CrossRef]

9. Garces, H.F.; Galindo, H.M.; Garces, L.J.; Hunt, J.; Morey, A.; Suib, S.L. Low temperature H_2S dry-desulfurization with zinc oxide. *Microporous Mesoporous Mater.* **2010**, *127*, 190–197. [CrossRef]

10. Zhang, X.; Tang, Y.Y.; Qu, S.Q.; Da, J.W.; Hao, Z.P. H_2S-Selective Catalytic Oxidation: Catalysts and Processes. *ACS Catal.* **2015**, *5*, 1053–1067. [CrossRef]

11. Liu, Z.; Wang, Z.; Chen, H.; Cai, T.; Liu, Z. Hydrochar and pyrochar for sorption of pollutants in wastewater and exhaust gas: A critical review. *Environ. Pollut.* **2021**, *268*, 115910. [CrossRef]

12. Bridgwater, A.V.; Meier, D.; Radlein, D. An Overview of Fast Pyrolysis of Biomass. *Org. Geochem.* **1999**, *30*, 1479–1493. [CrossRef]

13. Liang, J.; Shan, G.C.; Sun, Y.F. Catalytic fast pyrolysis of lignocellulosic biomass: Critical role of zeolite catalysts. *Renew. Sustain. Energy Rev.* **2021**, *139*, 110707. [CrossRef]

14. Hardy, B.; Leifeld, J.; Knicker, H.; Dufey, J.E.; Deforce, K.; Cornélis, J.-T. Long term change in chemical properties of preindustrial charcoal particles aged in forest and agricultural temperate soil. *Org. Geochem.* **2017**, *107*, 33–45. [CrossRef]

15. Luz, F.C.; Volpe, M.; Fiori, L.; Manni, A.; Cordiner, S.; Mulone, V.; Rocco, V. Spent coffee enhanced biomethane potential via an integrated hydrothermal carbonizationanaerobic digestion process. *Bioresour. Technol.* **2018**, *256*, 102–109.

16. Liu, Y.X.; Yao, S.; Wang, Y.Y.; Lu, H.H.; Brar, S.K.; Yang, S.M. Bio- and hydrochars from rice straw and pig manure: Inter-comparison. *Bioresour. Technol.* **2017**, *235*, 332–337. [CrossRef]

17. Xiao, X.; Chen, B.L.; Chen, Z.M.; Zhu, L.Z.; Schnoor, J.L. Insight into Multiple and Multi-level Structures of Biochars and Their Potential Environmental Applications: A Critical Review. *Environ. Eng. Technol.* **2018**, *52*, 5027–5047.

18. Sun, P.Z.; Li, Y.X.; Meng, T.; Zhang, R.C.; Song, M.; Ren, J. Removal of sulfonamide antibiotics and human metabolite by biochar and biochar/H_2O_2 in synthetic urine. *Water Res.* **2018**, *147*, 91–100. [CrossRef] [PubMed]

19. Liu, Z.G.; Balasubramanian, R. Upgrading of waste biomass by hydrothermal carbonization (HTC) and low temperature pyrolysis (LTP): A comparative evaluation. *Appl. Energy* **2014**, *114*, 857–864. [CrossRef]

20. Titirici, M.M.; White, R.J.; Falco, C.; Sevilla, M. Black perspectives for a green future: Hydrothermal carbons for environment protection and energy storage. *Energy Environ. Sci.* **2012**, *5*, 6796. [CrossRef]

21. Brunner, G. Near critical and supercritical water. Part I. Hydrolytic and hydrothermal processes. *J. Supercrit. Fluids* **2009**, *47*, 373–381. [CrossRef]

22. Xia, Y.; Yang, T.X.; Zhu, N.M.; Li, D.; Chen, Z.L.; Lang, Q.Q.; Liu, Z.G.; Jiao, W.T. Enhanced adsorption of Pb(II) onto modified hydrochar: Modeling and mechanism analysis. *Bioresour. Technol.* **2019**, *288*, 121593. [CrossRef]

23. Dawood, S.; Sen, T.K.; Phan, C. Synthesis and characterization of slow pyrolysis pine cone bio-char in the removal of organic and inorganic pollutants from aqueous solution by adsorption: Kinetic, equilibrium, mechanism and thermodynamic. *Bioresour. Technol.* **2017**, *246*, 76–81. [CrossRef] [PubMed]

24. Simsir, H.; Eltugral, N.; Karagoz, S. Hydrothermal carbonization for the preparation of hydrochars from glucose, cellulose, chitin, chitosan and wood chips via low-temperature and their characterization. *Bioresour. Technol.* **2017**, *246*, 82–87. [CrossRef] [PubMed]

25. Wang, B.; Gao, W.; Kang, X.M.; Dong, Y.Q.; Liu, P.F.; Yan, S.X.; Yu, B.; Guo, L.; Cui, B.; El-Aty, A.M.A. Structural changes in corn starch granules treated at different temperatures. *Food Hydrocoll.* **2021**, *118*, 106760. [CrossRef]

26. Tekin, K.; Karagöz, S.; Bektaş, S. A review of hydrothermal biomass processing. *Renew. Sustain. Energy Rev.* **2014**, *40*, 673–687. [CrossRef]

27. Almughamisi, M.S.; Khan, Z.A.; Alshitari, W.; Elwakeel, K.Z. Recovery of chromium(VI) oxyanions from aqueous solution using $Cu(OH)_2$ and CuO embedded chitosan adsorbents. *J. Polym. Environ.* **2020**, *28*, 47–60. [CrossRef]

28. Xu, X.W.; Huang, G.Q.; Qi, S. Properties of AC and 13X zeolite modified with $CuCl_2$ and $Cu(NO_3)_2$ in phosphine removal and the adsorptive mechanisms. *Chem. Eng. J.* **2017**, *316*, 563–572. [CrossRef]

29. Peng, Y.; Liu, Z.Y.; Yang, Z.H. Polymer-Controlled Growth of CuO Nanodiscs in the Mild Aqueous Solution. *Chin. J. Chem.* **2009**, *27*, 1086–1092. [CrossRef]

30. Zhou, M.; Gao, Y.; Wang, B.; Rozynek, Z.; Fossum, J.O. Carbonate-Assisted Hydrothermal Synthesis of Nanoporous CuO Microstructures and Their Application in Catalysis. *Eur. J. Inorg. Chem.* **2010**, *5*, 729–734. [CrossRef]

31. Mureddu, M.; Ferino, I.; Rombi, E.; Cutrufello, M.G.; Deiana, P.; Ardu, A.; Musinu, A.; Piccaluga, G.; Cannas, C. ZnO/SBA-15 composites for mid-temperature removal of H_2S: Synthesis, performance and regeneration studies. *Fuel* **2012**, *102*, 691–700. [CrossRef]

32. Dhage, P.; Samokhvalov, A.; Repala, D.; Duin, E.C.; Bowman, M.; Tatarchuk, B.J. Copper-Promoted ZnO/SiO_2 Regenerable Sorbents for the Room Temperature Removal of H_2S from Reformate Gas Streams. *Ind. Eng. Chem. Res.* **2010**, *49*, 8388–8396. [CrossRef]

33. Li, Y.R.; Lin, Y.T.; Xu, Z.C.; Wang, B.; Zhu, T.Y. Oxidation mechanisms of H_2S by oxygen and oxygen-containing functional groups on activated carbon. *Fuel Process. Technol.* **2019**, *189*, 110–119. [CrossRef]

34. Falco, D.G.; Montagnaro, F.; Balsamo, M.; Erto, A.; Deorsola, F.A.; Lisi, L.; Cimino, S. Synergic effect of Zn and Cu oxides dispersed on activated carbon during reactive adsorption of H_2S at room temperature. *Microporous Mesoporous Mater.* **2018**, *257*, 135–146. [CrossRef]
35. Kim, S.-J.; Na, C.W.; Hwang, I.-S.; Lee, J.-H. One-pot hydrothermal synthesis of CuO ZnO composite hollow spheres for selective H_2S detection. *Sens. Actuators B Chem.* **2012**, *168*, 83–89. [CrossRef]
36. Boutillara, Y.; Tombeur, J.L.; De Weireld, G.; Lodewyckx, P. In-situ copper impregnation by chemical activation with $CuCl_2$ and its application to SO_2 and H_2S capture by activated carbons. *Chem. Eng. J.* **2019**, *372*, 631–637. [CrossRef]

catalysts

Article

Selective Catalytic Oxidation of Lean-H$_2$S Gas Stream to Elemental Sulfur at Lower Temperature

Daniela Barba *, **Vincenzo Vaiano** and **Vincenzo Palma**

Department of Industrial Engineering, University of Salerno, Via Giovanni Paolo II, 132, 84084 Fisciano, Italy; vvaiano@unisa.it (V.V.); vpalma@unisa.it (V.P.)
* Correspondence: dbarba@unisa.it

Abstract: Ceria-supported vanadium catalysts were studied for H$_2$S removal via partial and selective oxidation reactions at low temperature. The catalysts were characterized by N$_2$ adsorption at 77 K, Raman spectroscopy, X-ray diffraction techniques, and X-ray fluorescence analysis. X-ray diffraction and Raman analysis showed a good dispersion of the V-species on the support. A preliminary screening of these samples was performed at fixed temperature (T = 327 °C) and H$_2$S inlet concentration (10 vol%) in order to study the catalytic performance in terms of H$_2$S conversion and SO$_2$ selectivity. For the catalyst that exhibited the higher removal efficiency of H$_2$S (92%) together with a lower SO$_2$ selectivity (4%), the influence of temperature (307–370 °C), contact time (0.6–1 s), and H$_2$S inlet concentration (6–15 vol%) was investigated.

Keywords: hydrogen sulfide; H$_2$S selective partial oxidation; sulfur; sulfur dioxide; vanadium-based catalysts

Citation: Barba, D.; Vaiano, V.; Palma, V. Selective Catalytic Oxidation of Lean-H$_2$S Gas Stream to Elemental Sulfur at Lower Temperature. *Catalysts* **2021**, *11*, 746. https://doi.org/10.3390/catal11060746

Academic Editor: Stefano Cimino

Received: 28 May 2021
Accepted: 16 June 2021
Published: 18 June 2021

Publisher's Note: MDPI stays neutral with regard to jurisdictional claims in published maps and institutional affiliations.

1. Introduction

Hydrogen sulfide (H$_2$S) is a common gas pollutant, which is harmful to human health with deleterious effects on many industrial catalysts, and represents the main source of acid rains when it is oxidized to sulfur dioxide (SO$_2$) [1]. Many attempts have been focused on H$_2$S removal from gaseous streams due to the worldwide increase in restrictive emission standards. Today, H$_2$S-removal-based processes include wet scrubbing [2], biological methods [3], adsorption [4], and selective catalytic oxidation [5]. Among these purification processes, selective catalytic oxidation seems to be very promising for lean-H$_2$S gas streams, where the concentration of hydrogen sulfide is in the range 0.1–10 vol%.

Typically, lean-H$_2$S gas is characteristic of tail gas treating (<5 wt% H$_2$S), crude petroleum (0.3–0.8 wt% H$_2$S), and natural gas streams (0.03–0.3 wt% H$_2$S), although in this last case the H$_2$S can also reach 30 wt% [6].

The selective catalytic oxidation of H$_2$S into elemental sulfur is one of the treatment methods employed for the removal of H$_2$S from the Claus process tail gas [7,8]. This reaction can be performed above or below the sulfur dew point (180 °C) and the processes used are super-Claus, doxosulfreen (Elf-Lurgi), and the mobil direct oxidation process (MODOP) [9]. The super-Claus process, developed in 1985, is continuously being improved and allows achievement of a desulfurization efficiency of ~99.5% at 240 °C in the presence of iron- and chromium-based catalysts supported on alumina or silica [10]. In MODOP, the direct oxidation of H$_2$S into elemental sulfur occurs on a TiO$_2$-based catalyst that deactivates in the presence of water [11]. In the super-Claus process, H$_2$S is oxidized without removing water from the tail gas. Metal-oxide-based catalysts, such as Al$_2$O$_3$, TiO$_2$, V$_2$O$_5$, Mn$_2$O$_3$, Fe$_2$O$_3$, and CuO are the most used and investigated for H$_2$S-selective catalytic oxidation [12]. Indeed, vanadium oxides have been investigated as active phases for H$_2$S selective oxidation and are used as bulk V$_2$O$_5$ [13], mixed with other metals [14], or supported over commercial [15] and mesoporous materials [16].

In our previous works, vanadium-based catalysts supported on different metal oxides (CeO_2, TiO_2, and $CuFe_2O_4$) were investigated for H_2S removal from biogas by partial and selective oxidation reactions in the temperature range 50–250 °C [17]. The optimization of the V_2O_5 loading (2.55–50 wt%) was performed on the CeO_2 support at the temperature of 150 °C [18]. The 20 wt% V_2O_5/CeO_2 catalyst showed the best catalytic performance in terms of H_2S conversion (99%) and sulfur selectivity (99%) at 150 °C, by feeding a very diluted stream containing only 500 ppm of H_2S [19]. Structured catalysts starting from a cordierite carrier in the form of a monolith honeycomb were also prepared, characterized, and tested at low temperature and evidenced high activity and very low SO_2 selectivity [20,21].

Based on these obtained promising results, in this study, vanadium-based catalysts supported on ceria were prepared, characterized, and tested in the presence of a lean-H_2S gas stream containing a H_2S concentration higher than 5 vol%, which is a typical concentration of the Claus process tail gas. A preliminary screening of the catalysts with different vanadium loadings was carried out at 327 °C, in order to identify the catalyst formulation able to maximize the H_2S conversion and depress the SO_2 formation in the presence of 10 vol% of H_2S. The effect of the main operating parameters, such as temperature, contact time, and H_2S inlet concentration, was also investigated.

2. Results and Discussion

2.1. Catalytic Activity Test

First of all, the reaction system was studied in the presence of 10 vol% of H_2S at the temperature of 327 °C without the catalyst (Figure 1).

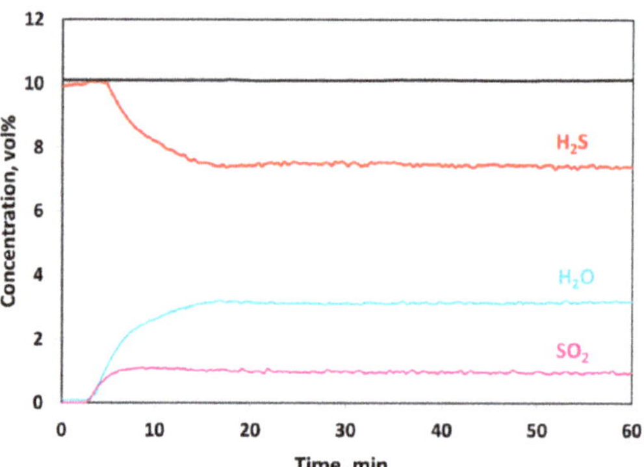

Figure 1. Activity test without catalyst (T = 327 °C, H_2S = 10 vol%, residence time = 0.6 s).

Figure 1 shows the behavior of H_2S, H_2O, and SO_2 during 1 h of time on stream.

After the first 5 min, the feed stream was sent to the reactor and the formation of SO_2 and water could be observed. The sulfur formation was not detectable because of the removal by the gaseous stream in the sulfur trap. The final H_2S conversion was 26%, while the SO_2 selectivity was high enough (~39%). The SO_3 formation (m/z = 80) was not observed either for the test in the absence of a catalyst or for all the catalytic tests. In Table 1, the values obtained by the test carried out without the catalyst were compared with the ones expected by the thermodynamic equilibrium and with the experimental data achieved with 20 V-CeO_2 catalyst.

Table 1. Comparison between non-catalytic system, catalytic system, and equilibrium (T = 327 °C, H_2S = 10 vol%).

	No Catalyst	20 V-CeO$_2$	Equilibrium
H$_2$S Conversion, %	26 (±1.5)	92 (±1.5)	90
SO$_2$ Selectivity, %	38.5 (±2)	4 (±2)	6
SO$_2$, vol%	1	0.4	0.5

It is evident that the reaction system without the catalyst is very far from the equilibrium conditions; in fact, the expected H_2S conversion and SO_2 concentration would be, respectively, 90% and 0.5 vol%. Conversely, the catalytic performance of the 20 V-CeO$_2$ sample is very close to that expected from the equilibrium, confirming the key role of the catalyst for maximizing the H_2S conversion and inhibiting the SO_2 formation.

The screening of the vanadium-based catalysts was performed at 327 °C and the catalytic activity of the V-CeO$_2$ samples was also compared with the support (CeO$_2$) and with the bulk V$_2$O$_5$. For each sample, the catalytic performance under steady-state conditions is reported in Figure 2.

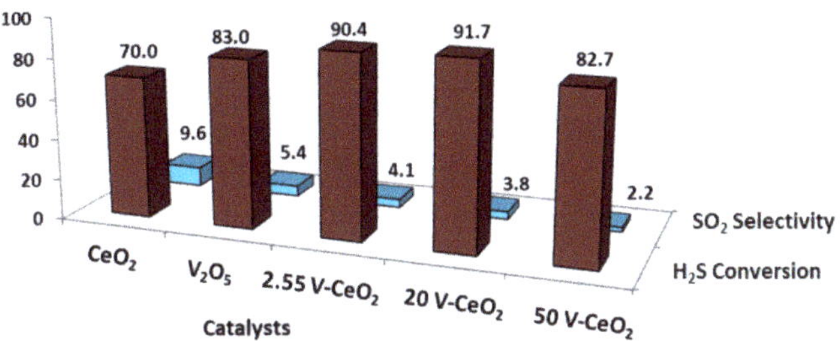

Figure 2. Catalytic performance of the different catalysts under steady-state conditions (T = 327 °C, H_2S = 10 vol%, contact time = 0.6 s).

The best catalytic performance can be observed for the catalysts having a V$_2$O$_5$ loading of 2.55 and 20 wt%, for which the H_2S conversion and SO_2 selectivity values are very similar. Although the lowest SO_2 selectivity (2.2%) was observed for the 50 V-CeO$_2$ sample, it unfortunately showed the lowest H_2S conversion (83%).

The influence of the temperature was investigated for the 20 V-CeO$_2$ catalyst and the experimental data for H_2S conversion and SO_2 selectivity were compared with the equilibrium data (red and blue lines, respectively) (Figure 3).

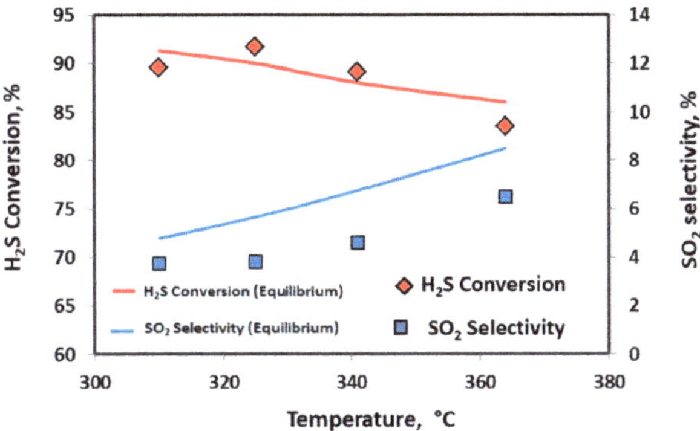

Figure 3. Temperature effect over 20 V-CeO$_2$ catalyst on the H_2S conversion and SO_2 selectivity (H_2S = 10 vol%, contact time = 0.6 s).

As it is possible to observe from Figure 3, the H_2S conversion is very close to the equilibrium values (red line) while the SO_2 selectivity is, in the overall investigated temperature range, slightly below the equilibrium calculation (blue line), evidencing that the catalyst is able to inhibit the SO_2 formation. The effect of the H_2S concentration, in the range 6–15 vol%, was then evaluated at the temperature of 327 °C (Figure 4)

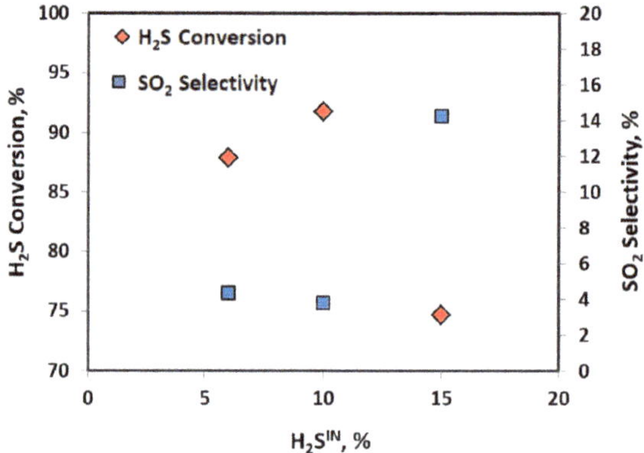

Figure 4. Influence of the H_2S inlet concentration over 20 V-CeO$_2$ catalyst on the H_2S conversion and SO_2 selectivity (O_2/H_2S = 0.5, T = 327 °C, contact time = 0.6 s).

The highest value of H_2S conversion and the lowest SO_2 selectivity were observed when the H_2S inlet concentration was 10 vol%. In the presence of a feed stream more concentrated in H_2S (15 vol%), the conversion was drastically reduced to 75% and the SO_2 concentration was about 1.5 vol%; in this case, the selectivity increase is of one magnitude order (14%) with respect to the other obtained values. In Table 2, the equilibrium data are compared with those obtained experimentally at different H_2S inlet concentrations.

Table 2. Equilibrium and experimental data by varying the H_2S inlet concentration (T = 327 °C, contact time = 0.6 s).

H_2S^{IN}, vol%	xH_2S, %	xH_2S Eq., %	S SO_2, %	sSO$_2$ Eq., %
6	88 (±1.5)	89	4.4 (±2)	4
10	92 (±1.5)	90	4 (±2)	6
15	75 (±1.5)	90	14 (±2)	9

Based on the data listed in Table 2, it is possible to see that the reaction system deviates from the equilibrium values especially in presence of 15 vol% of H_2S.

The influence of the contact time on the catalytic performance is reported in Figure 5. For comparison, the equilibrium data for both H_2S conversion and SO_2 selectivity at the temperature of 327 °C are also shown.

The catalytic performance resulted in little affected from the variation of the contact time. In particular, it is noteworthy to evidence that the H_2S conversion is quite close to the equilibrium values, while the SO_2 selectivity is in all cases below the equilibrium value.

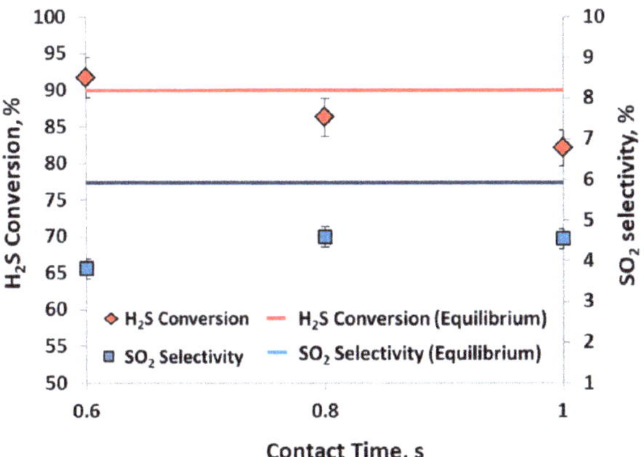

Figure 5. Effect of the contact time over 20 V-CeO$_2$ catalyst on the H$_2$S conversion and SO$_2$ selectivity (T = 327 °C, H$_2$S = 10 vol%).

2.2. Catalyst Characterization

The nominal and measured vanadium oxide content of the catalysts before the activation step is reported in Table 3.

Table 3. Theoretical and measured vanadium content of the catalysts before the sulfuration.

Sample	V$_2$O$_5$ Nominal wt%	% V$_2$O$_5$ Measured wt%
2.55 V-CeO$_2$	2.55	2.7
20 V-CeO$_2$	20	22
50 V-CeO$_2$	50	51

The results reported evidence that the nominal V$_2$O$_5$ loading is very close to the measured loading.

The specific surface areas of the fresh and used catalysts are reported in Table 4.

Table 4. Specific surface area (SSA, m^2/g) of the fresh and used catalysts.

Sample	CeO$_2$	V$_2$O$_5$	2.55 V-CeO$_2$	20 V-CeO$_2$	50 V-CeO$_2$
Fresh	29	8	25	22	20
Used	17	2	17	4	14

The lowest SSA was observed for the sample that was not supported (V$_2$O$_5$). In particular, the value of bulk V$_2$O$_5$ (8 m^2/g) decreased more than 50% after the catalytic test. For the fresh V-CeO$_2$ catalysts, the values of SSA were slightly lower than the CeO$_2$ support (~30 m^2/g). After the catalytic activity tests, the SSA decrease was likely due to the sulfur deposition on the catalyst surface. This aspect was more evident for the 20 V-CeO$_2$ sample (SSA = 4 m^2/g) and was confirmed by XRD and Raman characterizations.

Raman spectra of the support and fresh catalysts are shown in Figure 6. The Raman spectrum for pure CeO$_2$ shows the main band at 460 cm^{-1}, ascribable to ceria in the typical cubic crystal structure of fluorite-type cerium oxide [22,23]. The 2.55 V-CeO$_2$ sample shows that such Raman band slightly shifted to 465 cm^{-1}, while in the case of the catalysts with the highest V loading this band shifted up to 454 cm^{-1}. A more detailed discussion of these results is reported in the Supplementary Materials (Figure S1).

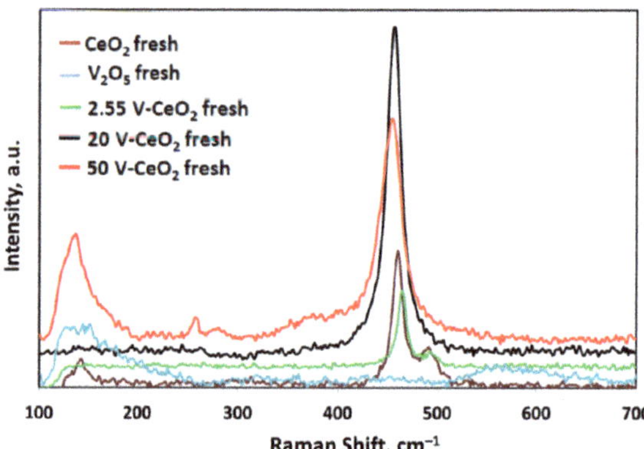

Figure 6. Raman spectra of CeO$_2$, V$_2$O$_5$, and 2.55, 20, 50 V-CeO$_2$ fresh catalysts.

The XRD spectra of CeO$_2$ and the fresh catalysts are shown in Figure 7. All the catalysts exhibit the characteristic peaks of CeO$_2$ at 28.3°, 32.8°, 47.3°, 56.1°, 58°, and 69°, corresponding to diffraction planes indexed as (1 1 1), (2 0 0), (2 2 0), (3 1 1), (2 2 2), and (4 0 0), respectively [24]. These patterns are ascribable to the typical cubic crystal structure of fluorite-type cerium oxide [25]. No additional reflections attributable to V$_2$O$_5$ are detectable, evidencing that the sulfuration of the catalysts completely occurred [26]. Furthermore, there were no peaks detected that related to typical vanadium sulfides (VS$_2$, VS$_4$, V$_2$S$_3$, V$_3$S$_4$) that might have formed following the sulfuration treatment [27].

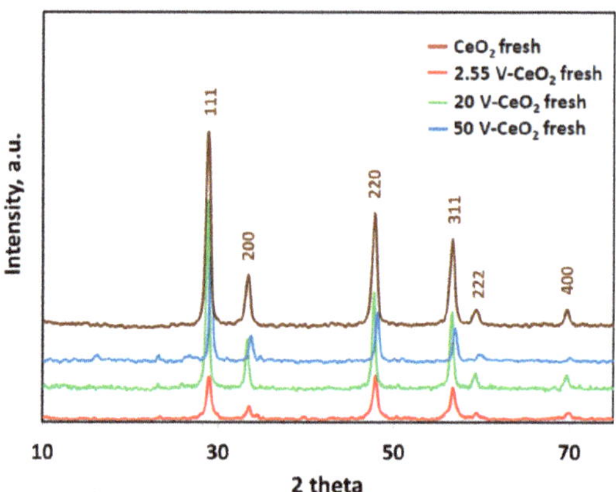

Figure 7. XRD spectra of CeO$_2$ and 2.55, 20, 50 V-CeO$_2$ fresh catalysts.

In Figure 8, the Raman spectra of the fresh samples (CeO$_2$ and V-CeO$_2$ catalysts) are compared with the used catalysts. The used CeO$_2$, equally to the fresh one, has the characteristic Raman peak perfectly centered at 460 cm^{-1} (Figure 8a) [22,23]. A slight shift of this Raman band up to 465 cm^{-1}, 457 cm^{-1}, and 454 cm^{-1} (Figure 8b–d) is detectable for 2.55 V-CeO$_2$, 20 V-CeO$_2$, and 50 V-CeO$_2$ fresh catalysts, respectively, as already previously observed (Figure 6). A detailed discussion of the Raman results is reported in the Supplementary Materials (Figure S2).

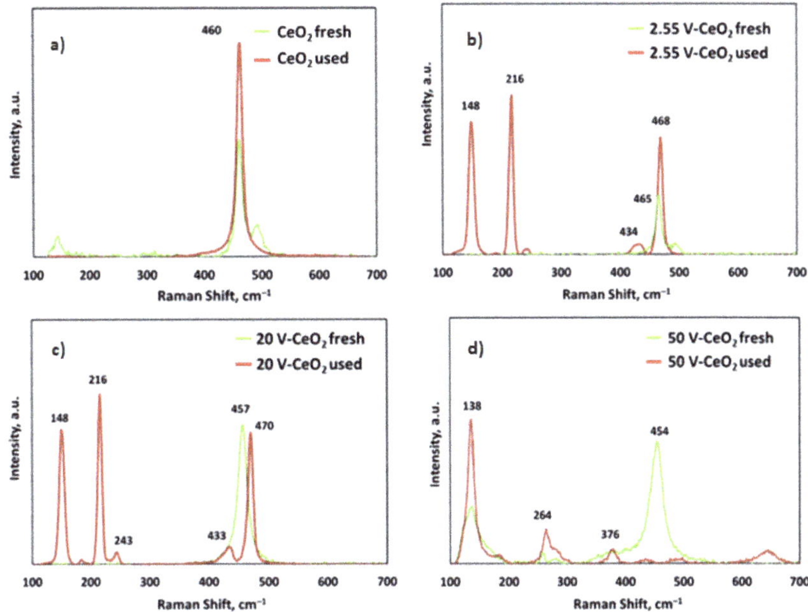

Figure 8. Raman spectra of fresh and used catalysts: CeO_2 (**a**), 2.55 V-CeO_2 (**b**), 20 V-CeO_2 (**c**), 50 V-CeO_2 (**d**).

Furthermore, the absence of any characteristic bands of the vibrational modes of crystalline V_2O_5 [28] and V = O stretching vibration ascribable to monovanadate species (VO_4^{3-}) denotes that the sulfuration of the catalysts occurred completely [16]. The Raman spectrum of the bulk V_2O_5 after the catalytic test is reported in Figure 9.

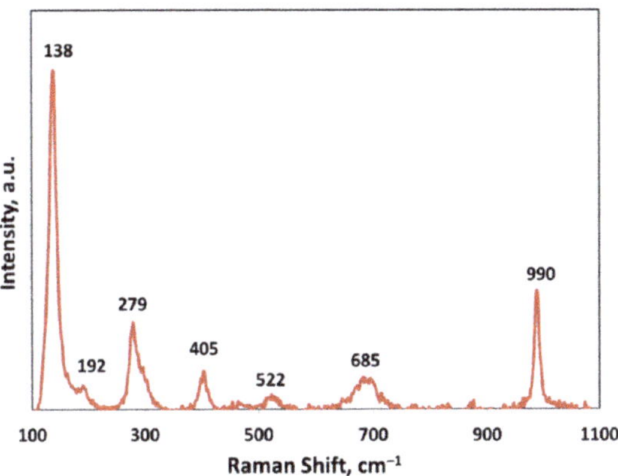

Figure 9. Raman spectra of V_2O_5 used.

The Raman bands at 140, 192, 282, 405, 688, and 993 cm^{-1} are characteristic of the vanadium sulfide in VS_2 form, as reported in the literature [29]. In particular, all the signals correspond to the rocking combination and stretching vibrations of V–S bonds or their combination [30]. Moreover, no bands related to the formation of vanadyl sulfate (984 cm^{-1} and 1060 cm^{-1}) were observed [31].

In Figure 10, the XRD patterns of the fresh samples are compared with the used ones. There are no differences between the XRD spectra of the fresh/used bulk CeO_2; for the used sample less intensity of the peaks is observed, which is likely due to the sulfur deposition (Figure 10a). For the used 2.55 V-CeO$_2$ catalyst, in addition to the characteristic peaks of the CeO$_2$ fresh sample, a signal is visible at 2θ = 23° due to the sulfur formation [32], as also confirmed from Raman analysis (Figure 10b). The spectra of the used 20 V CeO$_2$ catalyst (Figure 10c) are different, where other peaks attributable to the sulfur are observable at 2θ = 23°, 24°, 26°, 27°, and 28° [32]. For the 50 V-CeO$_2$ catalyst, the XRD spectrum of the fresh sample is perfectly stackable with that of the used sample (Figure 10d) because all the peaks are ascribable only to the CeO$_2$ support.

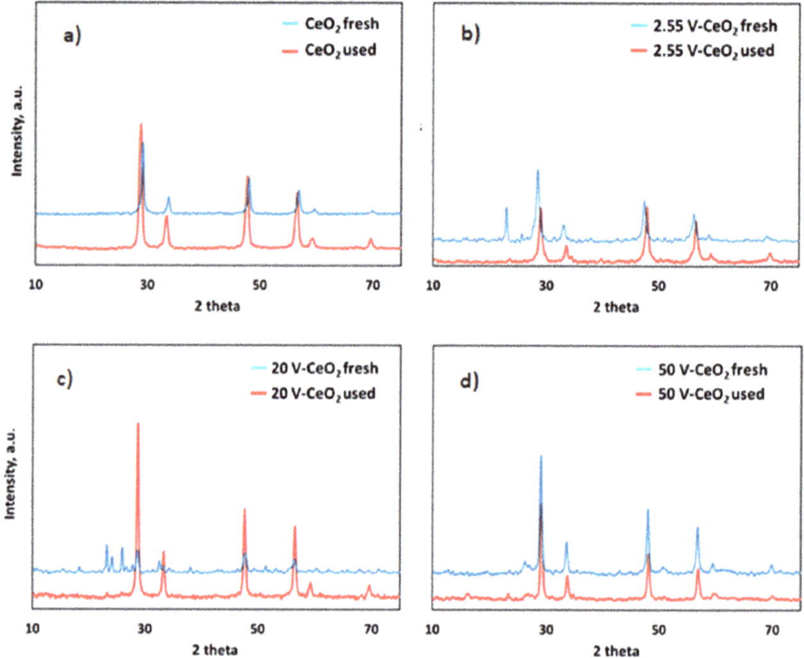

Figure 10. XRD spectra of fresh and used catalysts: CeO$_2$ (**a**), 2.55 V-CeO$_2$ (**b**), 20 V-CeO$_2$ (**c**), 50 V-CeO$_2$ (**d**).

The average crystallite size of ceria for the different catalysts, calculated with the Scherrer equation, are listed in Table 5.

Table 5. Average crystallite size (<L>, nm) of the fresh and used catalysts.

Sample	Fresh	Used
CeO$_2$	13	21
2.55 V-CeO$_2$	16	19
20 V-CeO$_2$	18	19
50 V-CeO$_2$	24	25

As it is possible to observe from Table 5, the increase of the V-loading for the different catalysts has involved an increase in the crystallite size of the CeO$_2$, as reported in the literature for supported vanadium catalysts [25]. The average crystallite size of bulk CeO$_2$ before the catalytic tests was 13 nm; it increased to 24 nm for the catalyst having the highest V-content (50 V-CeO$_2$). Relatively to the catalysts, there is a negligible variation of the ceria average crystallite size between fresh and used samples.

The only significant variation between fresh and used samples was obtained for the support; the greater segregation of the CeO_2 after the catalytic activity tests involved the increase of the crystallite dimension (21 nm). The segregation of the CeO_2 crystallite may be due to the high SO_2 formation observed on the support in the absence of the active phase; in fact, among the catalysts, the highest value of SO_2 selectivity (~10%) was obtained for the CeO_2 at 327 °C as previously reported in Figure 2. The reaction temperature could favor the formation of sulfate species and also the oxygen in the ceria lattice could facilitate the CeO_2 sulfuration [33]; therefore, the reaction between CeO_2 and SO_2 could occur, leading to the formation of cerium sulfate $Ce(SO_4)_2$, which is stable at high temperature and decomposes between 722 and 843 °C to CeO_2 [34].

3. Materials and Methods

3.1. Catalyst Preparation and Characterization

The preparation of vanadium-based catalysts supported on ceria with different loading of active phase (2.55, 20, and 50 wt% V_2O_5 nominal loading) was described in detail in our previous work [19]. All the reactants were provided by Sigma Aldrich. After the calcination, the sulfuration procedure was carried out in a quartz reactor containing the catalyst to be sulfurized. In particular, the activation step was realized by feeding a gaseous stream containing N_2 and H_2S at 20 vol%, by increasing the temperature from ambient temperature up to 200 °C with a heating rate of 10 °C/min for 1 h. Finally, the catalysts were reduced to the size 38–180 μm. For simplicity, the catalysts are named in the paper as follows: 2.55 V-CeO$_2$, 20 V-CeO$_2$, 50 V-CeO$_2$, where "2.55" means the nominal V loading (wt%) expressed as V_2O_5. The sulfurized samples before the testing are named "fresh", while they are named "used" after the catalytic activity test.

The catalysts were characterized by nitrogen adsorption at 77 K, Raman spectroscopy and X-ray diffraction. The specific surface area was evaluated with a Costech Sorptometer 1040 (Costech International, Firenze, Italy) by using N_2 and He, respectively, as adsorptive and carrier gas. The powder catalysts were treated at 150 °C for 30 min in a He flow prior to testing. A BET method multipoint analysis based on N_2 adsorption/desorption isotherms at 77 K was used to evaluate the specific surface area of the fresh and used catalysts. X-ray diffraction (XRD) was performed using a Brucker D2 Phaser (Germany) using CuKa radiation (λ = 1.5401 A°). Laser Raman spectra of the catalysts were obtained in air with a Dispersive MicroRaman (Invia, Renishaw, Italy), equipped with a 514 nm diode-laser, in the range of 100–2000 cm^{-1} Raman shift. The V-content of the fresh catalysts (expressed as V_2O_5 wt%) was evaluated by X-ray fluorescence (XRF) spectra by using an ARL QUANT'X EDXRF spectrometer (ThermoFisher Scientific, Italy).

3.2. Experimental Apparatus

The catalytic activity tests were performed in the laboratory plant schematized in Figure 11.

The laboratory plant is made of three sections: feed, reaction, and analysis sections. The feed stream containing H_2S, O_2, and N_2 is sent by a three-way valve to the reactor, or in bypass position to the analyzer to verify the composition. All gases came from SOL S.p.A with a purity degree of 99.999% for N_2, O_2, and SO_2, and 99.5% for H_2S.

The reaction system comprises a furnace, a reactor, and a sulfur abatement trap. The quartz-made reactor, consisting of a tube of 300 mm length and an internal diameter of 19 mm, is housed in a vertical furnace heated with silicon carbide (SiC)-based resistances. At the bottom of the reactor are a reactant inlet and a thermocouple sheet concentric to the reactor. The catalytic bed is placed in the isothermal zone of the reactor and the temperature is measured continuously by a K-type thermocouple. In the head of the reactor is welded a trap for the sulfur abatement, which is made of an expansion vessel that allows the sulfur to liquefy, involving its separation by the gaseous stream. This trap is maintained at the temperature of 250 °C.

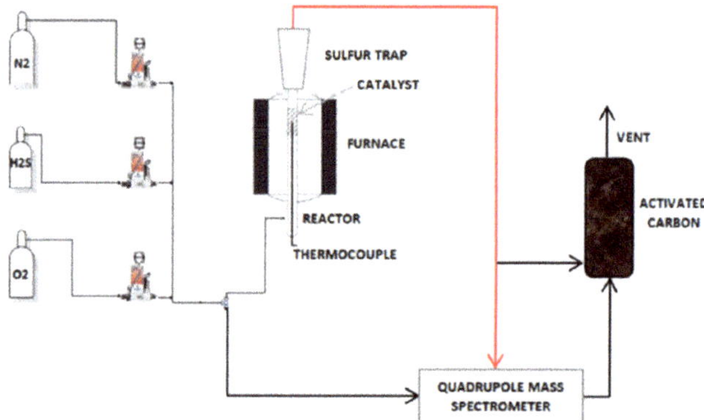

Figure 11. Scheme of the apparatus plant.

All the lines downstream of the reactor were heated at the temperature of 170 °C to avoid sulfur solidification and possible clogging of the mass spectrometer capillary and to maintain the water in the gas phase for the analysis. The analysis of the gaseous stream (H_2S, O_2, N_2, SO_2, SO_3) was performed with the mass spectrometer quadrupole (Hiden HPR-20) (Warrington, United Kingdom).

Finally, the abatement of unconverted H_2S was realized by adsorption on activated carbons loaded in a special vessel having a capacity of 10 Lt. Furthermore, the entire apparatus plant was housed under the hood and isolated from the external environment in order to avoid gas leakage.

The operating conditions of the catalytic activity tests are listed in Table 6.

Table 6. Operating conditions.

Operating Conditions	
Temperature	200–367 °C
Contact Time	0.6–1 s
Catalyst Volume	3 cm^3
Total Flow Rate	180–300 Ncc·min^{-1}
GHSV	3600–6000 h^{-1}
H_2S^{IN} concentration	6–15 vol%
O_2/H_2S	0.5

H_2S conversion (x H_2S) and the SO_2 selectivity (s SO_2) were calculated by using the following relationship (Equations (1) and (2)), by considering the gas phase volume change to be negligible:

$$x\ H_2S,\ \% = ((H_2S^{IN} - H_2S^{OUT})/H_2S^{IN})\cdot 100 \tag{1}$$

$$s\ SO_2,\ \% = (SO_2^{OUT}/(H_2S^{IN} - H_2S^{OUT}))\cdot 100 \tag{2}$$

For the equilibrium calculation, the *GasEq* program was used, software (0.7.0.9 version, Chris Morley) based on the minimization of Gibbs free energy, which is able to calculate the equilibrium product composition of an ideal gaseous mixture when there are a lot of simultaneous reactions. The thermodynamic analysis was carried out considering the following chemical species that could be present at equilibrium: H_2S, O_2, SO_2, S_2, S_6, S_8, H_2O, and nitrogen.

Calibration Procedure

The calibration procedure is required in order to measure the concentration of all the species that could be in the gas stream for analysis and, for this reason, it must be performed prior to carrying out experimental tests. However, it could be necessary to repeat the calibration every time the process conditions are changed (e.g., after the replacement of the capillary, the filaments, change of pressure chamber value) or when the signal seems to be affected by derivative effects. The measurements could be affected by interference due to the presence of ions of different molecules having the same m/z ratio. Each molecule has a matrix of interference, which defines the "weight" of the disturbance of other molecules on the partial pressure of the molecule in the phase of calibration. The partial pressure obtained, net of the relative interference, must be corrected by a response factor, thus returning the actual partial pressure of each molecule in the stream analyzed. At this point, it is possible to calculate the correct concentration of each component. The calibration procedure is characterized by different steps:

(1) Report in a table the partial pressure of the all-mass fragment for each concentration of the component to calibrate;
(2) Construct the matrix of interference and the response factors table relatively for the component you are calibrating;
(3) Calculate the concentrations of the component calibrated by considering the relative interference of other species on the component to calibrate and correcting the measure by its response factor.

After the calibration, which is carried out in a by-pass position, the feed stream can be sent to the reactor. The reactor, before each test, is purged with nitrogen to avoid humidity and/or impurity and is heated up to the reaction temperature at which the feed stream is sent.

Similarly, at the end of the activity test, the reactor is cooled down with nitrogen to room temperature.

4. Conclusions

The H_2S selective oxidation reaction to sulfur and water was investigated over vanadium-sulfide-based catalysts supported on CeO_2. The catalysts were prepared with different vanadium loading and were characterized before and after the catalytic tests with different techniques. X-ray diffraction and Raman analysis showed a good dispersion of the V-species on the support because the V-sulfide presence was not detected on the different catalysts. The only vanadium sulfide in VS_2 form was observed for the bulk V_2O_5 after the catalytic tests. Furthermore, the presence of the sulfur was observed especially over the used catalysts at lower V-loading by Raman and SSA analysis.

From the preliminary screening of the catalysts performed at 327 °C, the higher catalytic activity was observed over the 2.55 V-CeO_2 and 20 V-CeO_2 catalysts, with H_2S conversion, respectively, of 90% and 92%, and SO_2 selectivity of ~4%. No SO_3 formation and catalyst deactivation phenomena by the sulfur deposition were observed. The effect of the temperature, contact time, and H_2S inlet concentration was studied over 20 V-CeO_2 catalysts. By increasing the H_2S inlet concentration (up to 15 vol%), the conversion decreased from 86% to 75% with an SO_2 concentration of about 1.5 vol%. The effect of the contact time was almost negligible on the H_2S conversion and SO_2 selectivity, while the temperature had a significant influence. In the range of temperatures investigated (300–370 °C), the H_2S conversion was very close to the equilibrium values while the SO_2 selectivity was below the equilibrium calculation, evidencing that the catalyst is effectively able to inhibit SO_2 formation.

Based on the obtained results, the ceria-supported vanadium catalysts could be considered good candidates to carry out the selective oxidation of H_2S to sulfur by an H_2S-lean gas stream (e.g., natural gas, Claus process tail gas) at very low temperature.

Supplementary Materials: The following are available online at https://www.mdpi.com/article/10.3390/catal11060746/s1: Figure S1: Raman Spectra of CeO2, V2O5 and 2.55, 20, 50 V-CeO2 fresh catalysts; Figure S2: Raman Spectra of fresh and used catalysts CeO2 (a), 2.55 V-CeO2 (b), 20 V-CeO2 (c), 50 V-CeO2 (d).

Author Contributions: Conceptualization, D.B. and V.V.; methodology, D.B.; software, D.B.; validation, D.B. and V.V.; formal analysis, V.P.; investigation, D.B.; resources, V.V. and V.P.; data curation, V.P.; writing—original draft preparation, D.B.; writing—review and editing, V.V.; visualization, V.P. and V.V; supervision, V.P.; project administration, D.B.; funding acquisition, V.P. All authors have read and agreed to the published version of the manuscript.

Funding: This research received no external funding.

Conflicts of Interest: The authors declare no conflict of interest.

References

1. Forzatti, P.; Lietti, L. Catalyst deactivation. *Catal. Today* **1999**, *52*, 165–181. [CrossRef]
2. Wang, R. Investigation on a new liquid redox method for H_2S removal and sulfur recovery with heteropoly compound. *Sep. Purif. Technol.* **2003**, *31*, 111–121. [CrossRef]
3. Duan, H.; Yan, R.; Koe, L.C.; Wang, X. Combined effect of adsorption and biodegradation of biological activated carbon on H_2S biotrickling filtration. *Chemosphere* **2007**, *66*, 1684–1691. [CrossRef]
4. Seredych, M.; Bandosz, T.J. Sewage sludge as a single precursor for development of composite adsorbents/catalysts. *Chem. Eng. J.* **2007**, *128*, 59–67. [CrossRef]
5. Yasyerli, S.; Dogu, G.; Ar, I.; Dogu, T. Dynamic analysis of removal and selective oxidation of H_2S to elemental sulfur over Cu–V and Cu–V–Mo mixed oxides in a fixed bed reactor. *Chem. Eng. Sci.* **2004**, *59*, 4001–4009. [CrossRef]
6. Elmawgoud, H.A.; Elshiekh, M.; Abdelkreem, M.; Khalil, S.A.; Alsabagh, A.M. Optimization of petroleum crude oil treatment using hydrogen sulfide scavenger. *Egypt. J. Pet.* **2019**, *28*, 161–164. [CrossRef]
7. Pi, J.H.; Lee, D.H.; Lee, J.D.; Jun, J.H.; Park, N.K.; Ryu, S.O.; Lee, T.J. The study on the selective oxidation of H_2S over the mixture zeolite NaX-WO_3 catalysts. *Korean J. Chem. Eng.* **2004**, *21*, 126–131. [CrossRef]
8. Lee, J.D.; Han, G.B.; Park, N.K.; Ryu, S.O.; Lee, T.J. The selective oxidation of H_2S on V_2O_5/zeolite-X catalysts in an IGCC system. *J. Ind. Eng. Chem.* **2006**, *12*, 80–85.
9. Wiheeb, A.D.; Shamsudin, I.K.; Ahmad, M.A.; Murat, N.M.; Kim, J.; Othman, M.R. Present technologies for hydrogen sulfide removal from gaseous mixtures. *Rev. Chem. Eng.* **2013**, *29*, 449–470. [CrossRef]
10. Keller, N.; Pham-Huu, C.; Crouzet, C.; Ledoux, M.J.; Savin-Poncet, S.; Nougayrede, J.B.; Bousquet, J. Direct oxidation of H_2S into S: New catalysts and processes based on SiC support. *Catal. Today* **1999**, *53*, 535–542. [CrossRef]
11. Zhang, X.; Tang, Y.; Qu, S.; Da, J.; Hao, Z. H_2S-selective catalytic oxidation: Catalysts and processes. *ACS Catal.* **2015**, *5*, 1053–1067. [CrossRef]
12. Davydov, A.A.; Marshneva, V.I.; Shepotko, M.L. The comparison study of the catalytic activity. Mechanism of the interactions between H_2S and SO_2 on some oxides. *Appl. Catal. A* **2003**, *244*, 93–100. [CrossRef]
13. Li, K.T.; Hyang, M.Y.; Cheng, W.D. Vanadium-based mixed-oxide catalysts for selective oxidation of hydrogen sulfide to sulfur. *Ind. Eng. Chem. Res.* **1996**, *35*, 621–626. [CrossRef]
14. Park, D.W.; Byung, B.H.; Ju, W.D.; Kim, M.I.; Kim, K.H.; Woo, H.C. Selective oxidation of hydrogen sulfide containing excess water and ammonia over Bi-V-Sb-O catalysts. *Korean J. Chem. Eng.* **2005**, *22*, 190–195. [CrossRef]
15. Kalinkin, P.; Kovalenko, O.; Lapina, O.; Khabibulin, D.; Kundo, N. Kinetic peculiarities in the low-temperature oxidation of H_2S over vanadium catalysts. *J. Mol. Catal. A Chem.* **2002**, *178*, 173–180. [CrossRef]
16. Soriano, M.D.; Jimenez-Jimenez, J.; Concepcion, P.; Jimenez- Lopez, A.; Rodriguez-Castellon, E.; Lopez Nieto, J.M. Selective oxidation of H_2S to sulfur over vanadia supported on mesoporous zirconium phosphate heterostructure. *Appl. Catal. B Environ.* **2009**, *92*, 271–279. [CrossRef]
17. Palma, V.; Barba, D.; Ciambelli, P. Screening of catalysts for H_2S abatement from biogas to feed molten carbonate fuel cells. *Int. J. Hydrogen Energy* **2013**, *38*, 328–335. [CrossRef]
18. Palma, V.; Barba, D. Low temperature catalytic oxidation of H_2S over V_2O_5/CeO_2 catalysts. *Int. J. Hydrogen Energy* **2014**, *39*, 21524–21530. [CrossRef]
19. Palma, V.; Barba, D. Vanadium-ceria catalysts for H_2S abatement from biogas to feed to MCFC. *Int. J. Hydrogen Energy* **2017**, *42*, 1891–1898. [CrossRef]
20. Palma, V.; Barba, D.; Gerardi, V. Honeycomb-structured catalysts for the selective partial oxidation of H_2S. *J. Clean Prod.* **2016**, *111*, 69–75. [CrossRef]
21. Palma, V.; Barba, D. Honeycomb V_2O_5-CeO_2 catalysts for H_2S abatement from biogas by direct selective oxidation to sulfur at low temperature. *Chem. Eng. Trans.* **2015**, *43*, 1957–1962. [CrossRef]
22. Gu, X.; Ge, J. Structural, redox and acid–base properties of V_2O_5/CeO_2 catalyst. *Thermochim. Acta* **2006**, *451*, 84–93. [CrossRef]
23. Escribano, V.S.; López, E.F.; Panizza, M.; Resini, C.; Amores, J.M.G.; Busca, G. Characterization of cubic ceria–zirconia powders by X-ray diffraction and vibrational and electronic spectroscopy. *Solid State Sci.* **2003**, *5*, 1369–1376. [CrossRef]
24. Sun, C.; Li, H.; Zhang, H.; Wang, Z.; Chen, T. Controlled synthesis of CeO_2 nanorods by a solvothermal method. *Nanotechnology* **2005**, *16*, 1454–1463. [CrossRef]
25. Radhika, T.; Sugunan, S. Structural and catalytic investigation of vanadia supported on ceria promoted with high surface area rice husk silica. *J. Mol. Catal. A Chem.* **2006**, *250*, 169–176. [CrossRef]

26. Singh, B.; Gupta, M.K.; Mishra, S.K.; Mittal, R.; Sastry, P.U.; Rolsc, S.; Lal Chaplot, S. Anomalous lattice behavior of vanadium pentaoxide (V_2O_5): X-ray diffraction, inelastic neutron scattering and ab initio lattice dynamics. *Phys. Chem. Chem. Phys.* **2017**, *19*, 17967. [CrossRef]

27. Liu, Y.Y.; Xu, L.; Guo, X.T.; Lv, T.T.; Pang, H. Vanadium sulfide based materials: Synthesis, energy storage and conversion. *J. Mater. Chem.* **2020**, *8*, 20781–20802. [CrossRef]

28. Holgrado, J.P.; Soriano, M.D.; Jimenez, J. Operando XAS and Raman study on the structure of a supported vanadium 409 oxide catalyst during the oxidation of H2S to sulfur. *Appl. Catal. B* **2010**, *92*, 271–279, 410. [CrossRef]

29. Huang, L.; Zhu, W.; Zhang, W.; Chen, K.; Wang, J.; Wang, R.; Yang, Q.; Hu, N.; Suo, Y.; Wang, J. Layered vanadium (IV) 414 disulfide nanosheets as a peroxidase-like nanozyme for colorimetric detection of glucose. *Microchim. Acta* **2018**, *185*, 415. [CrossRef]

30. Qu, Y.; Shao, M.; Shao, Y.; Yang, M.; Xu, J.; Kwok, C.T.; Shi, X.; Lu, Z.; Pan, H. Ultra-high electrocatalytic activity of VS_2 nanoflowers for efficient hydrogen evolution reaction. *J. Mater. Chem. A* **2017**, *5*, 15080–15086. [CrossRef]

31. Evans, J.C. The vibrational spectra and structure of the vanadyl ion in aqueous solution. *Inorg. Chem.* **1963**, *2*, 372–375. [CrossRef]

32. Xu, J.; Su, D.; Zhang, W.; Bao, W.; Wang, G. A nitrogen–sulfur co-doped porous graphene matrix as a sulfur immobilizer for high performance lithium–sulfur batteries. *J. Mater. Chem. A* **2016**, *4*, 17381–17393. [CrossRef]

33. Smirnov, M.Y.; Kalinkin, A.V.; Pashis, A.V.; Sorokin, A.M. Interaction of Al_2O_3 and CeO_2 Surfaces with SO_2 and $SO_2 + O_2$ studied by X-ray photoelectron spectroscopy. *J. Phys. Chem. B* **2005**, *109*, 11712–11719. [CrossRef] [PubMed]

34. Sharma, I.B.; Singh, V.; Lakhanpal, M. Study of thermal decomposition of ammonium cerium sulphate. *J. Therm.* **1992**, *38*, 1345–1355. [CrossRef]

catalysts

MDPI

Article

Application of Response Surface Methodology for Preparation of ZnAC$_2$/CAC Adsorbents for Hydrogen Sulfide (H$_2$S) Capture

Nurul Noramelya Zulkefli [1], Mohd Shahbudin Masdar [1,2,*], Wan Nor Roslam Wan Isahak [1], Siti Nur Hatika Abu Bakar [1], Hassimi Abu Hasan [1,3] and Nabilah Mohd Sofian [2]

[1] Department of Chemical & Process Engineering, Faculty of Engineering & Built Environment, UKM Bangi, Selangor 43600, Malaysia; amelyaz@yahoo.com (N.N.Z.); wannorroslam@ukm.edu.my (W.N.R.W.I.); nurhatika_abu@yahoo.com.my (S.N.H.A.B.); hassimi@ukm.edu.my (H.A.H.)

[2] Fuel Cell Institute, UKM Bangi, Selangor 43600, Malaysia; nabilah@ukm.edu.my

[3] Research Centre for Sustainable Process Technology, UKM Bangi, Selangor 43600, Malaysia

* Correspondence: shahbud@ukm.edu.my

Abstract: Hydrogen sulfide (H$_2$S) should be removed in the early stage of biogas purification as it may affect biogas production and cause environmental and catalyst toxicity. The adsorption of H$_2$S gas by using activated carbon as a catalyst has been explored as a possible technology to remove H$_2$S in the biogas industry. In this study, we investigated the optimal catalytic preparation conditions of the H$_2$S adsorbent by using the RSM methodology and the Box–Behnken experimental design. The H$_2$S catalyst was synthesized by impregnating commercial activated carbon (CAC) with zinc acetate (ZnAc$_2$) with the factors and level for the Box–Behnken Design (BBD): molarity of 0.2–1.0 M ZnAc$_2$ solution, soaked temperature of 30–100 °C, and soaked time of 30–180 min. Two responses including the H$_2$S adsorption capacity and the BET surface area were assessed using two-factor interaction (2FI) models. The interactions were examined by using the analysis of variance (ANOVA). Hence, the optimum point of molarity was 0.22 M ZnAc$_2$ solution, the soaked period was 48.82 min, and the soaked temperature was 95.08 °C obtained from the optimum point with the highest H$_2$S adsorption capacity (2.37 mg H$_2$S/g) and the optimum BET surface area (620.55 m^2/g). Additionally, the comparison of the optimized and the non-optimized catalytic adsorbents showed an enhancement in the H$_2$S adsorption capacity of up to 33%.

Keywords: adsorption; adsorbent; purification; H$_2$S removal; response surface methodology (RSM)

Citation: Zulkefli, N.N.; Masdar, M.S.; Wan Isahak, W.N.R.; Abu Bakar, S.N.H.; Abu Hasan, H.; Mohd Sofian, N. Application of Response Surface Methodology for Preparation of ZnAC$_2$/CAC Adsorbents for Hydrogen Sulfide (H$_2$S) Capture. *Catalysts* **2021**, *11*, 545. https://doi.org/10.3390/catal11050545

Academic Editor: Daniela Barba

Received: 5 March 2021
Accepted: 19 April 2021
Published: 24 April 2021

Publisher's Note: MDPI stays neutral with regard to jurisdictional claims in published maps and institutional affiliations.

1. Introduction

Agricultural industries, livestock ranches, and fuel industries generally generate some natural wastewaters and wastewaters that have a tremendous effect on the debate and pollution of water [1]. The anaerobic digestion of natural wastewater and wastewater does not mitigate this degradation; instead, it creates biogas, fertilized solids, and filtered sewage for subsequent beneficial use [2–4]. For example, biogas can be efficiently used for heat and energy substitution for gasoline in transport applications [5]. The biogas composition typically consists of roughly 40–75% of methane (CH$_4$), 25–40% of carbon dioxide (CO$_2$), 0.5–2.5% of nitrogen (N$_2$), 10–30 ppm(v) of ammonia (NH$_3$), and 1000–3000 ppm(v) of hydrogen sulfide (H$_2$S) [6,7]. These compositions, however, depend on the differential sources of the organic substrates.

In practice, the elimination of hydrogen sulfide (H$_2$S) in the oil and gas or biogas processing industries remains one of the key obstacles to the sustainable growth of profitable technologies [8]. H$_2$S is toxic at low concentrations (<1 ppm(v)), impacting the production of biogas and has life-threatening effects at higher concentrations (500 ppm(v)) [9,10]; hence, it is imperative to eliminate H$_2$S in the early stages of the purification system [11]. Several methods have been implemented to eliminate H$_2$S, such as the Clauss technique, which is primarily used in the oil and gas industries [12] that typically produce high concentrations of H_2S (>10,000 ppm(v)). Several technologies that are commonly used and commercialized for H_2S removal include chemical absorption [13], physical adsorption [14], biological treatment [15–18], and membrane technology [19,20].

The H_2S capture via a biological treatment is efficient and cost-effective; however, it needs a large upfront investment as compared to the dry-based processes. Even though this method is an environmentally friendly system, the separation and purification of H_2S may be difficult to carry out. In contrast, the liquid-based and membrane techniques for H_2S removal are not economically or energetically viable technologies [13]. However, the adsorption technology is the best and superior for H_2S removal even at low concentrations and temperatures [21–23]. Adsorption is the most commonly used technique for both large-scale and small-scale applications. All of these technologies are summarized and compared with the most relevant and alternative technology for H_2S removal in Table 1.

Table 1. Summary of different H_2S removal technologies.

H_2S Removal Technologies	Strength	Weakness	Comments
Clauss process [10]	- Very well-known process. - Cost is minimal since existing units are available.	- Tail gas treatment is challenging. - Mainly used in refineries, natural gas processing plants, or syngas plants that require high H_2S concentration.	- Most matured process for H_2S purification.
Adsorption [14]	- Widely used mesoporous materials utilized as adsorbents are low-cost raw materials. - Materials have high surface area and porosity, which leads to optimum adsorption capacity. - Normally impregnated with selective chemicals to enhance the adsorbent capability.	- Only selective chemicals capable of capturing H_2S gas are used. - Sulfur bound on the mesoporous materials degraded the adsorption performance.	- Solid adsorbent should be highly selective and have high capacity of H_2S adsorption. - Widely used for low H_2S concentration.
Absorption (liquid solution) [24]	- Process based on alkanolamines is matured and has been perfected over the last several decades. - Depends on the interaction strength of the gas molecules and solvents.	- High regeneration cost required and inefficient. - Could cause secondary pollution.	- Developing H_2S selective alkanolamines is a big challenge to overcome the liquid absorbents. - Current study focuses on ionic liquid to improve the absorption performance.
Biological [15–18]	- Alternative to very costly industrial methods. - Environmentally friendly.	- Requires highly sensitive of biological process to operate effectively. - Process system requires a special procedure. - Purification and separation studies are challenging.	- New research study for H_2S purification with high improvement potential.
Membrane [19,20]	- Rarely found membrane technologies used for H_2S separation. - In term of CO_2 separation, the membrane technology is more economical.	- Presence of H_2S gas increases the separation cost. - Selective membranes exhibit no significant difference in permeability.	- Hybrid process involving membrane or chemical absorption or adsorption can lead towards an overall better economical process.

Adsorption techniques [25–27] to remove H_2S typically involve mesoporous materials (activated carbon, zeolites, and/or silica) that are also widely known as catalysts because of their surface chemistry, high degree of microporosity, and developed surface area (which can exceed 1000 m^2/g) [28]. On both the macro and nanoscales, these materials may have crystalline and/or amorphous structures [16], but they can be further changed to adjust their physicochemical properties, thus improving their adsorption ability against the target molecules. As commercial activated carbon (AC) is often impregnated to increase the capacity of the adsorbents to absorb the adsorbates, it is also subjected to surface modification. The improvement was primarily based on increasing the basic surface area and the porous structure of the mesoporous materials by using chemical activation methods.

Impregnated adsorbents such as catalytic adsorbents, widely applied several chemicals based on alkalis (NaOH, KOH, and KI) [29–33], carbonate compounds (Cu), transition metal oxide compounds (Zn, Fe, and Cu), or metal acetic acid compounds (Zn) [14,34] can be used as solid catalysts or be dispersed as small grains on the surface of a supporting material. Selecting the precursors of active components as well as any necessary promoters and stirring them in a solvent is the first step in producing a supported catalyst. In the end, the active metal or precursor from the solvents is dispersed on the adsorbents' surface. In contrast to raw activated carbon (AC), impregnated adsorbents with both of these chemicals have a higher specific surface area, smaller particle sizes, and increased H_2S capability [32]. Despite this, a metal-supported catalyst ($ZnAc_2$/CAC) demonstrates favorable associations between the adsorbent's capabilities in capturing the adsorbate and develops better surface area. The dispersion of $ZnAc_2$ on the CAC surface normally acts as active sites to capture the adsorbate particles efficiently. Moreover, $ZnAc_2$ leads to an increase in the specific surface area by decreasing the particle size, which results in an increase in the H_2S adsorption capacity, as reported on the basis of the ZnO impregnated performance [35]. Impregnation from both chemicals ($ZnAc_2$ and ZnO) enhanced the adsorbent's capabilities through surface area and adsorption capacities.

For example, a study on the optimization of the CAC performance evaluated the optimal response using certain factors (molarity, time, humidity, temperature, and pH) and responses (adsorption capacity, surface area, selectivity, and percentage utilization). All the information obtained from these factors and responses can be used in an interaction study to determine the proposed optimization. Normally, the interaction parameters (condition variables) can be analyzed using two types of methods, namely univariate and multivariate optimization. Univariates have the slightest remedial effect relative to multivariates because of the capacity of the univariates to rely on one optimization variable; thus, the multivariate approach requires a design that adjusts all levels of variables simultaneously. For expository systems, this phase is crucial and the optimal operating conditions are determined using complex test designs, the Doehlert lattice (DM), central composite design (CCD), and three-level designs such as the Box–Behnken design (BBD) [36–38]. The relationship between the explanatory variables and the response variables [39] can be evaluated graphically by using the empirical data sufficient for the optimal area, thereby allowing new models to be developed and identified and the current product designs to be updated [40].

Therefore, in this study, we applied the BBD by using response surface methodology (RSM) to assess the influence of factors with a minimal number of experiments by evaluating and controlling the Zn acetate CAC impregnation. The response to the selected factors determined the H_2S adsorption capacity and characterizes the surface morphologies via the BET surface area of the impregnated CAC on the basis of the BBD recommendation.

2. Materials and Methods

2.1. Adsorbent Preparation

Effigen Carbon Sdn. Bhd, Malaysia, supplied granular commercial coconut activated carbon (CAC), which was sieved to obtain a particle size in the range of 3–5 mm. The selected CAC impregnation compound was zinc acetate ($ZnC_4H_6O_4$), which was purchased from Friendemann Schmidt Chemicals (Malaysia) and used as obtained without prior purification.

The impregnated CAC surface was prepared with 600 mL of distilled water for a 0.2–1.0 M zinc acetate solvent at 30–100 °C. In brief, 350 g of CAC was soaked into the solvent for 30–180 min before the distribution of the zinc acetate compound on the surface. The wet CAC was drained and dried at 120 °C overnight before being used for H_2S adsorption testing and is indicated as $ZnAc_2$/CAC. Moreover, the design of experiments (DOE) recommendation was submitted on the basis of the chosen molarity, soaked time, and soaked temperature for the preparation of the adsorbents.

2.2. Characterization

The surface area and the pore structure were analyzed by a Brunauer–Emmett–Teller (BET) surface area analysis using Micrometric ASAP 2010 Version 4.0.0. The surface area was obtained from the measurement of the BET isotherm, while the pore volumes and the standard pore volumes were calculated at P/Po of 0.98 by using the N_2 adsorption isotherm. Meanwhile, the micropore volume was calculated using the t-plot method. After degassing for 4 h at 150 °C, the textural properties of the sorbents were determined by N_2 adsorption–desorption at 196 °C with Quantachrome Autosorb 1 °C. The exact surface was extracted from the estimation of the BET.

The surface morphology and the chemical structure characterization for the optimized and non-optimized adsorbents were analyzed using the CARL ZEISS EVO MA10 and energy dispersive X-Ray analysis (EDX) with EDAX APOLLO X model. This characterization method was used to visualize the details of the adsorbent properties in terms of the structural morphology and to identify the elemental composition of the materials present on the surfaces of the adsorbent under an accelerating voltage of 10 kV.

2.3. H_2S Adsorption Test

In this study, the H_2S adsorption test was implemented using a laboratory-scale set-up of a single stainless-steel column (height and diameter of 0.3 m and 0.06 m, respectively), as shown in Figure 1. In brief, 75 g of the impregnated adsorbent ($ZnAc_2$/CAC) was loaded into the adsorber column and fed in with a commercial mixed gas H_2S/N_2 (5000-ppm(v) H_2S with balanced N_2). The adsorption test operated at ambient temperature, the flow rate and pressure gauge were mounted at 5.5 L/min and 1 bar. Due to the tolerable range for the gas exposed to the atmosphere and fuel cell devices, the H_2S breakthrough gas concentration at the outlet stream was set at 5–10 ppm(v) [41–43]. The outlet H_2S gas was detected using a customized portable H_2S analyzer (model GC310), which directly imported the data into the computer program. Then, the adsorption capacity of H_2S for each DOE suggestion was calculated according to the equation reported by Zulkefli et al. [44].

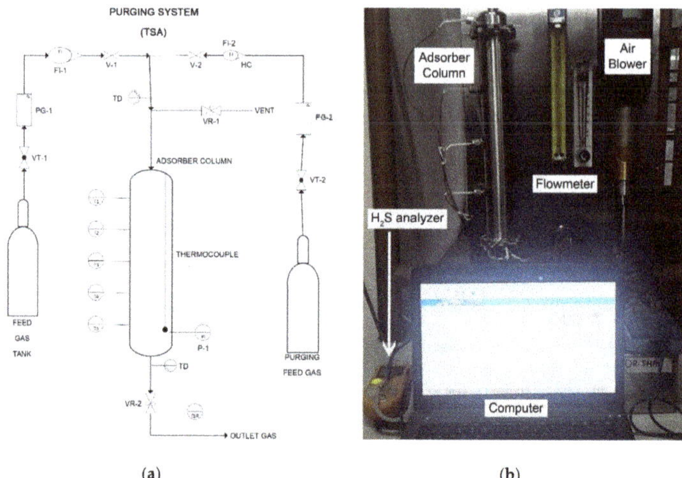

(**a**)   (**b**)

Figure 1. H$_2$S adsorption system [32]: (**a**) schematic diagram and (**b**) experimental H$_2$S adsorption test set-up.

2.4. Regeneration of Adsorbents

The desorption process for the adsorbents was followed by set-up in a previous study by Zulkefli et al. [32]. The spent adsorbents underwent a three-step purging process. In the first step, the spent adsorbents were run through an air blower for 30 min at 150 °C and a flow rate of 100 L/min. Secondly, the same operating parameters were applied to the column at ambient temperature for 30 min. In the final step, the N$_2$ gas was introduced into the stream; it was fed at 5.5 L/min for 30 min to purge out and stabilized the active site on the adsorbent surface before use in the next adsorption operation up to several cycles of adsorption–desorption.

2.5. Control Factors and Level Selection

It is possible to test the effect of quadratic interactions by using a BBD combined with response surface modeling and quadratic programming. This experimental approach used the regression design to show the result as a predictive function of variables with an impartial and limited variance. In this strategy, the graphical profile illustrates the summary of the response surface being examined [45]. The effects of three preparation factors were investigated: (A) ZnAc$_2$ solution molarity (M), (B) soaked time (min), and (C) soaked temperature (°C) on the CAC surface as well as the capture of the H$_2$S gas. Two responses were used, namely the H$_2$S adsorption capacity and the BET surface area, as a reference to the preparation factors.

The typical variables are coded separately as +1, −1, and 0 for the high, low, and center points; therefore, the units of the parameters are not relevant. Real variables (Xi) are coded by direct transformation as follows:

$$\chi_i = \frac{x_i - x_0}{\Delta x} \quad i = 1, 2, 3 \tag{1}$$

where χ_i is the encoded value of an independent variable, x_i is the actual value of an independent variable, x_0 is the actual value of a center point independent variable, and Δx is the phase shift value of an independent variable [46]. The process factors and factor levels of the adsorbent preparation state are described in Table 2.

Table 2. Process factors and factor levels of the adsorbent preparation state.

Factors		Level		
		−1	0	+1
A	Molarity (M)	0.2	0.6	1.0
B	Soaked period (min)	30	105	180
C	Soaked temperature (°C)	30	65	100

The data from the BBD were analyzed by multiple regression to fit the following quadratic polynomial model:

$$Y = \alpha_0 + \sum_{i=1}^{3} \alpha_i x_i + \sum_{i=1}^{3} \alpha_{ii} x_i^2 + \sum_{i=1}^{2} \sum_{j=2}^{3} \alpha_{ij} x_i x_j + \varepsilon \tag{2}$$

where Y is the response variables, α_0 is the model constant, α_i represents the linear coefficient, α_{ii} denotes the quadratic coefficient, α_{ij} is the interaction coefficient, and ε is the statistical error. The least-squares method is used to solve this set of Equation (2). BBD is a common experimental design for the technique of response surfaces in statistics and is a type of second-order rotatable or nearly rotatable design based on three-level incomplete factorial designs. Each design is a combination of a two-level (full or fractional) factorial design with an incomplete block design [47].

Both combinations for factorial design are placed through a certain number of factors in each block, while the other factors are held at the central values. The BBD is a good design for this technique because (1) it enables the calculation of the parameters of the quadratic model, (2) there are no runs where all the variables are at either +1 or −1 levels, and (3) the number of experiments (N) needed for the BBD to evolve is defined as follows [48]:

$$N = 2k(k-1) + C_N \tag{3}$$

where the number of variables is k and the number of center points is C_N. On the basis of Equation (3), the runs will be reduced to 17 with 3 major variables and 5 times the repetition in the center point to reduce the magnitude of error ($k = 3$ and $C_N = 5$). Three conditions were investigated, namely the molarity state, soaked time, and the soaked temperature; hence, 17 runs were executed.

2.6. Steps for Process Parameter Optimization

The steps followed for process optimization are shown in the flow chart in Figure 2. In this optimization, the molarity, soaked period, and the soaked temperature of the adsorbents were entered as the explanatory variables and the optimal adsorption capacity with the BET surface area of the adsorbents as the response variable.

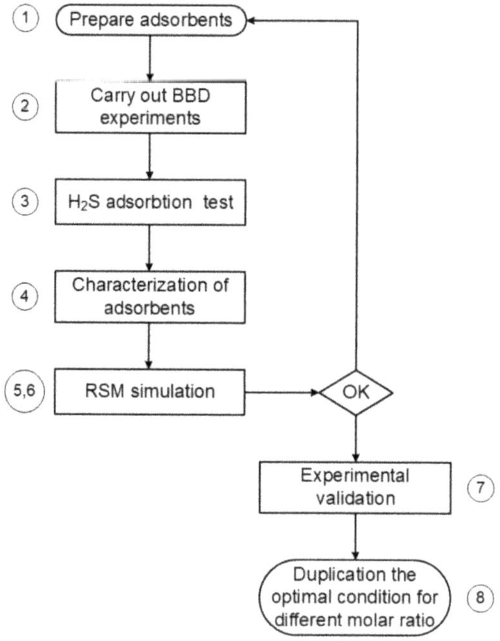

Figure 2. Process optimization.

Step 1. CAC impregnated with $ZnAc_2$ as suggested by the design tools.

Step 2. There were 17 run trials conducted, with the BBD matrix consisting of 12 different level combinations of the independent variables as well as three center point runs used to fit a second-order response surface and provide a measure of process stability and inherent variability [49]. The BBD design matrix along with the experimental values of the responses are shown in Table 2 (in terms of the coded factors).

Step 3. The adsorption capacity was calculated using an H_2S adsorption test and was determined for each of the 17 runs of the adsorbents.

Step 4. The BET surface was determined for each of the 17 runs.

Step 5. RSM simulation, including second-order regression and analysis of variance (ANOVA), was conducted.

Step 6. The optimal conditions for the different molarities were traced on the basis of the contour and the surface plots of the RSM simulation.

Step 7. The simulation and experimental results were verified.

Step 8. The standard parameter conditions were duplicated for different molar ratios and impregnated materials.

3. Results and Discussion

3.1. Box–Behnken Model Evaluation

Based on Equation (3), with three main factors and five replications at the center point to reduce the magnitude of error ($k = 3$ and $C_o = 5$), the runs were limited to 17, as detailed in Table 3. The obtained breakthrough curves for the three soaked periods, i.e., 30, 105, and 180 min, are presented in Figure 3a–c, respectively. The breakthrough curves are shown in three figures because of the large number of runs and for an easier understanding and interpretation of the obtained results. Thus, differences between the relative concentrations of the H_2S curves of the runs could be understood more easily. To decide about the adequacy of the model for the H_2S adsorption capacity, three different tests, namely the sequential model sum of squares, lack of fit test, and model summary statistics, were carried out in the present study. The data of the H_2S adsorption capacity in this research were subjected to a regression analysis to estimate the effect of the process variables.

Table 3. Box–Behnken and experimental data of responses' adsorption capacity and BET surface area.

No. of Run	ZnAc$_2$ Molarity, M (A)	Soaked Period, min (B)	Soaked Temperature, °C (C)	Adsorption Capacity, mg H$_2$S/g (Y1)	BET Surface Area, m^2/g (Y2)
1	0.20	105.00	30.00	1.81	842.74
2	1.00	105.00	30.00	2.03	584.01
3	0.60	105.00	65.00	0.47	544.52
4	0.60	30.00	30.00	0.67	765.23
5	1.00	105.00	100.00	0.56	757.80
6	0.20	30.00	65.00	1.75	692.65
7	0.60	105.00	65.00	1.84	737.64
8	1.00	180.00	65.00	1.47	698.38
9	1.00	30.00	65.00	0.58	825.52
10	0.60	30.00	100.0	2.37	198.31
11	0.60	180.00	30.00	1.36	694.21
12	0.60	105.00	65.00	1.48	726.12
13	0.20	180.00	65.00	2.11	612.05
14	0.60	105.00	65.00	2.23	473.72
15	0.60	180.00	100.00	1.70	485.28
16	0.20	105.00	100.00	2.14	555.00
17	0.60	105.00	65.00	0.89	731.88

The results shown in Table 3 can be compared with those of a previous study by Zulkefli et al. [32]. On the basis of [32], the adsorption capacity and the BET surface area were obtained at 1.83 mg H_2S/g and 656.75 m^2/g, respectively, at 0.2 M of ZnAc$_2$, soaked temperature of 65 °C, and soaked period of 30 min. Under similar conditions, the obtained values in this study were slightly different at 1.75 mg H_2S/g and 692.65 m^2/g for the adsorption capacity and the BET surface area, respectively, which was probably because of the differences in the preparation process. The decrease in the BET surface area in the adsorbent was caused by the blocking of some micropores with the chemical compound of ZnAc$_2$. This characterization of the pore structure influenced the adsorption profiles [50,51]. In contrast, the data in Table 3 show similar trends to those reported by Nakamura et al. [52]. The BET surface area decreased with an increase in the ZnAc$_2$ molarity, even though the BET surface area was slightly different for both cases because of the differences in the preparation conditions.

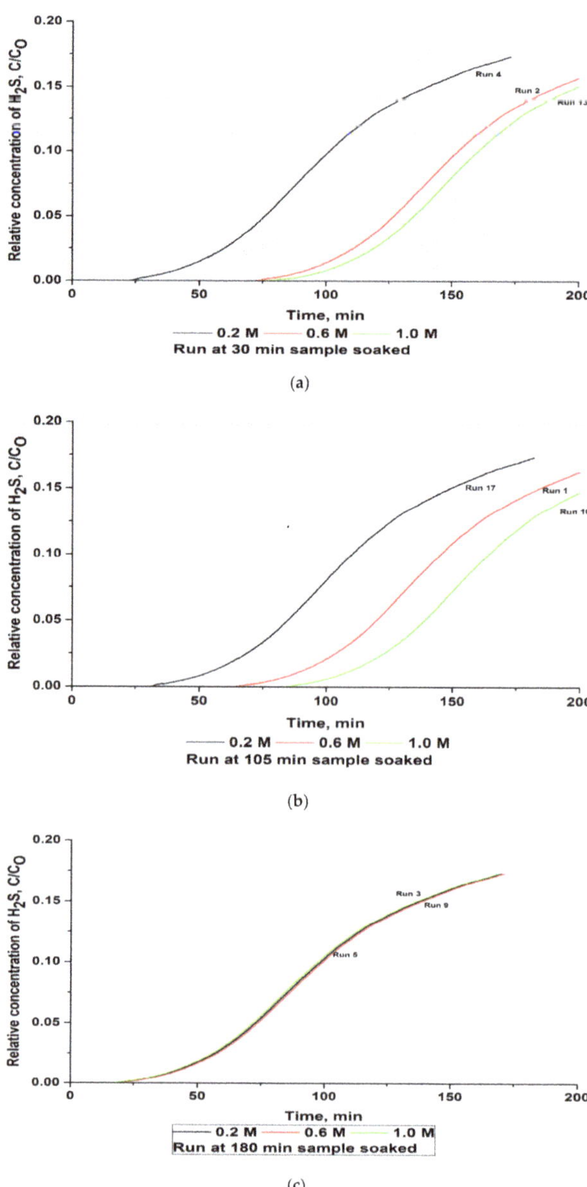

Figure 3. Breakthrough curve of H_2S versus $ZnAc_2$ molarity for a constant soaked time: (**a**) 30 min, (**b**) 105 min, and (**c**) 180 min.

3.2. ANOVA

The results were analyzed using the analysis of variance (ANOVA), a regression model, coefficient of determination (R^2), adjusted R^2, coefficient of variation (CV), and statistical–diagnostic and response plots. The analysis of variance (ANOVA) test is a robust and common statistical method in different applications. The ANOVA provides a statistical procedure that determines whether the means of several groups are equal or not. The Fisher's variance ratio, F-value, is used to test the significance of the model, individual variables, and their interactions [53,54]. Mean square (MSS) is the sum of squares divided by the degrees of freedom, for each source. The F-value is defined as $MSS_{variable}/MSS_{residual}$ and shows the relative contribution of the sample variance to the residual variance [55]. If the ratio deviates increasingly from 1, the samples are not from the same population, with more confidence.

The results of the ANOVA based on experimental data are shown in Figure 4 and Table 4. The model summary statistics showed that the excluding cubic model was aliased and the 2FI model was found to have the maximum adjusted R^2 values. Therefore, the 2FI model was chosen for further analysis.

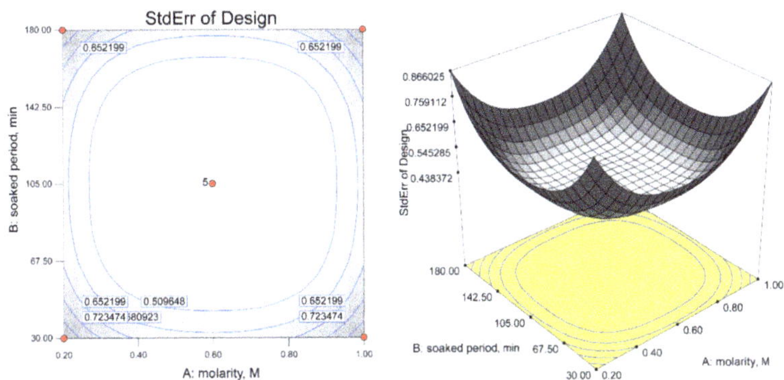

Figure 4. Design matrix evaluation for response surface of 2FI model.

Table 4. Adequacy of model for adsorption capacity response.

Source	Sum of Squares	DF	Mean Square	F Value	Prob > F	
Mean	38.14	1	38.14			Suggested
Linear	1.56	3	0.52	1.36	0.30	
2FI	1.35	3	0.45	1.25	0.34	Suggested
Quadratic	0.15	3	0.05	0.10	0.96	
Cubic	1.47	3	0.49	0.98	0.49	Aliased
Residual	2.00	4	0.50			
Total	44.66	17	2.63			
lack of fit						
Linear	2.97	9	0.33	0.66	0.72	
2FI	1.62	6	0.267	0.54	0.76	Suggested
Quadratic	1.47	3	0.49	0.98	0.49	
Cubic	0	0				Aliased
Pure error	2.00	4	0.50			

Source	Standard Deviation	R^2	Adjusted R^2	Predicted R^2	PRESS	
Linear	0.62	0.24	0.06	−0.29	8.41	
2FI	0.60	0.45	0.11	−0.58	10.29	Suggested
Quadratic	0.70	0.47	−0.21	−3.08	26.64	
Cubic	0.71	0.69	−0.22		+	Aliased

Next, the ANOVA of the adsorption capacity of H_2S is summarized in Table 5. If the calculated value of F is greater than that in the F table at a specified probability level, a statistically significant factor or interaction is obtained [56,57]. The F is defined as F = MSF/MSE, where MSF and MSE are the mean square of factors (interactions) and the mean square of errors, respectively. The ANOVA test revealed that the factors A, B, and C, and the interactions A × C and B × C proved to have a statistically significant effect on the H_2S adsorption capacity. The F value is an indication of the level of significance. A higher F denotes a more significant effect on the response.

Table 5. Analysis of variance (ANOVA) of adsorption capacity response.

Source	Sum of Squares	DF	Mean Square	F-Value	F-Value from Table ($p = 0.05$)	Prob > F	
Model	2.91	6	0.49	1.34	3.22	0.32	Not significant
A	1.26	1	1.26	3.48	4.96	0.09	
B	0.20	1	0.20	0.57	4.96	0.47	
C	0.099	1	0.099	0.27	4.96	0.61	
AB	0.070	1	0.070	0.19	4.96	0.67	
AC	0.82	1	0.82	2.26	4.96	0.16	
BC	0.46	1	0.46	1.29	4.96	0.28	
Residual	3.61	10	0.36	0.54	2.98	0.76	
Lack of fit	1.62	6	0.27				Not significant
Pure error	2.00	4	0.50				
Cor total	6.53	16					

We can compare the F-value from the calculations with the F-value obtained from the F-distribution table with the degree of freedom (DF) from the model and the error to discern the significance and the adequacy of the model [48]. An effect is statistically significant if the calculated F-value for the effect is greater than the F-value extracted from the table at the desired probability level. On the basis of the calculated p-value (prob > F), all the three factors, namely molarity, soaked period, and soaked temperature, and their interaction effects were found to be significant (Table 4). The regression equation obtained after the variance analysis yielded the level of the H_2S adsorption capacity. It included a linear relationship between all the main effects and the response. The final quadratic polynomial equations in terms of the coded and the actual variables are presented as follows:

$$\text{Adsorption capacity, } Q_{coded} = 1.50 - 0.40x_A + 0.16x_B + 0.11x_C + 0.13x_Ax_B - 0.45x_Ax_C - 0.34x_Bx_C \tag{4}$$

$$\text{Adsorption capacity, } Q_{actual} = -0.21 + 0.65x_A + 0.008x_B + 0.04x_C + 0.004x_Ax_B - 0.03x_Ax_C - 0.00013x_Bx_C \tag{5}$$

As seen in statistical studies, the values of prob > F below 0.05 signify that the model terms are significant. In this case, as shown in Table 5, models B and AC were significant with the value of prob > F of 0.05 and 0.03, respectively. The Fisher's F-value and the probability value of the regression model were found to be 1.34 and 0.32, respectively. This implied that the terms in the model had a significant effect on the response. The tabular F-value with the degree of freedom, $DF_{model} = 6$ and $DF_{error} = 10$, respectively, at the significance level of 0.05 ($F_{0.05,(6,10)} = 3.22$) was higher than the calculated F-value ($F_{0.05,(6,10)} = 1.34$), implied that most of the variation in the response could not be explained by the regression equation.

Then, the coefficient of determination, R^2, indicated the overall predictive capability of the model. From Table 6, the R^2 value of the model was determined to be 0.45. Therefore, we assumed that 45% of the total variations in the response can be explained by the model. However, this value of R^2 did not necessarily imply that the regression model was a suitable one. A negative prediction R^2 was defined as a better predictor of the H_2S adsorption capacity response for the current model. In this case, an adequate R^2 value of 4.40 was more than 4 as the ratio desirability, which indicated that the model navigated the design space. As was observed, the adjusted R^2 was close to R^2, emphasizing the high significance of the model. Another method to describe the variation of a model is to calculate the coefficient of variation (CV). While the values presented in Table 5 are not logically significant for the H_2S adsorption capacity, the low value of the coefficient of variation (C.V.% = 40.14) might reflect the fact that this model could have high reliability and good fitness.

Table 6. Model reliability analysis of adsorption capacity response.

Source	Result
Standard deviation	0.60
Mean	1.50
Coefficient variation (%), C.V	40.14
PRESS	10.29
R^2	0.45
Adjusted R^2	0.11
Prediction R^2	−0.58
Adequate precision	4.40

The response to the BET surface area also suggested the use of the 2FI model for further analysis through the ANOVA study based on the highest value obtained for the adjusted R^2 (0.35). Meanwhile, the F value obtained was an indication of the model significance level. As presented in Table 7, the 2FI model had the highest F value (2.65) of the considered models. Moreover, the highest values of the adjusted R^2 and the predicted R^2 could be a reason for the suggestion of the use of the 2FI model for further analysis.

Table 7. Adequacy of the model for BET surface area response.

Source	Sum of Squares	DF	Mean Square	F-Value	Prob > F	
Mean	7,813,331	1	7,813,331			
Linear	57,607.09	3	19,202.36	1.62	0.23	
2FI	68,436.96	3	22,812.32	2.65	0.11	Suggested
Quadratic	10,032.71	3	3344.236	0.31	0.82	
Cubic	13,827.22	3	4609.073	0.30	0.83	Aliased
Residual	62,118.94	4	15,529.74			
Total	8,025,354	17	472,079.7			
Lack-of-Fit Tests						
Source	Squares	DF	Square	Value	Prob > F	
Linear	92,296.88	9	10,255.21	0.66036	0.7227	
2FI	23,859.93	6	3976.654	0.256067	0.9323	Suggested
Quadratic	13,827.22	3	4609.073	0.29679	0.8270	
Cubic	0	0				Aliased
Pure error	62,118.94	4	15,529.74			

Source	Standard Deviation	R^2	Adjusted R^2	Predicted R^2	PRESS	
Linear	108.99	0.27	0.10	−0.21	257,363.7	
2FI	92.72	0.59	0.35	0.21	167,922.1	Suggested
Quadratic	104.16	0.64	0.18	−0.50	318,296.4	
Cubic	124.62	0.71	-0.17		+	Aliased

Based on the calculated *p*-value (prob > F), all the three factors and their interaction effects were found to be significant, as presented in Table 8. The regression equation obtained after the variance analysis provided the level for the BET surface area response. It also included a linear relationship between all the main effects and the response. The factors A, B, and C, and the interactions A × C and B × C proved to have statistically significant effects on the BET surface area. The final quadratic polynomial equations of the coded and the actual variables are presented in the equation below:

$$\text{BET surface area (coded)} = 677.94 + 20.41x_A - 73.97x_B - 36.22x_C - 11.63x_Ax_B + 115.38x_Ax_C - 60.50x_Bx_C \tag{6}$$

$$\text{BET surface area (actual)} = 957.86 - 443.97x_A + 0.74x_B - 3.56x_C - 0.39x_Ax_B + 8.24x_Ax_C - 0.02x_Bx_C \tag{7}$$

Table 8. Analysis of variance (ANOVA) of BET surface area responses.

Source	Sum of Squares	DF	Mean Square	F-Value	F-Value from Table ($p = 0.05$)	Prob > F	
Model	1.3×10^5	6	2.1×10^4	2.44	3.22	0.10	Not significant
A	3.3×10^3	1	3.3×10^3	0.39	4.96	0.55	
B	4.3×10^4	1	4.4×10^4	5.09	4.96	0.05	
C	1.0×10^4	1	1.0×10^4	1.22	4.96	0.30	
AB	541.25	1	541.25	0.06	4.96	0.81	
AC	5.3×10^4	1	5.3×10^4	6.19	4.96	0.03	
BC	1.5×10^4	1	1.5×10^4	1.70	4.96	0.22	
Residual	8.6×10^4	10	8.6×10^3		2.98		
Lack of fit	2.4×10^4	6	4.0×10^3	0.26		0.93	Not significant
Pure error	6.2×10^4	4	1.6×10^4				
Cor total	2.1×10^5	16					

The Fisher's F-value and the very low probability value of the regression model were found to be 2.44 and 0.10, respectively. This implied that the terms in the model had a significant effect on the response. The tabular F-value with a degree of freedom, DF_{model} = 6 and DF_{error} = 10, respectively, at the significance level of 0.05 ($F_{0.05,(6,10)}$ = 3.22) was higher than the calculated F-value ($F_{0.05,(6,10)}$ = 2.44), indicating that the variation in the response was not significant.

As shown in Table 9, the R^2 value obtained was 0.59, which could be assumed to be 59% of the total variation in the BET surface area response. The coefficient of variation (CV) indicated a lower value than that of the H_2S adsorption capacity response, which is 13.68% and had the highest chance for reliability and good fit of the model. The value of prediction R^2 = 0.21 was in reasonable agreement with the adjusted R^2 = 0.35. The adequate precision normally measures the signal-to-noise ratio. As shown in Table 9, the adequate precision marked at 5.10 and the ratios were more than 4, which indicated that the model was adequate for navigating the design space.

Table 9. Model reliability analysis of adsorption capacity response.

Source	Result
Standard deviation	92.72
Mean	677.94
Coefficient variation (%)	13.68
PRESS	16800
R^2	0.59
Adjusted R^2	0.35
Prediction R^2	0.21
Adequate precision	5.10

3.3. Contour Plots for H_2S Adsorption Capacity and BET Surface Area Responses

Response surface plots and contour plots are useful for the model equation image and perceiving the nature of the response surface. These plots are also useful in the study of the effect of process variables on the H_2S adsorption capacity and the BET surface area in a wider range of preparation conditions of the adsorbents. Furthermore, they can be used for designing the optimum conditions for adsorbent synthesis. Equations (5) and (7) were used to construct the contour plots for the H_2S adsorption capacity and the BET surface area against the molarity, soaked period, and soaked temperature, as shown in Figures 5 and 6. They depict the interaction of three main factors by keeping the other at its central level for two types of responses based on the refitted Equations (4) and (5) with the experimental data.

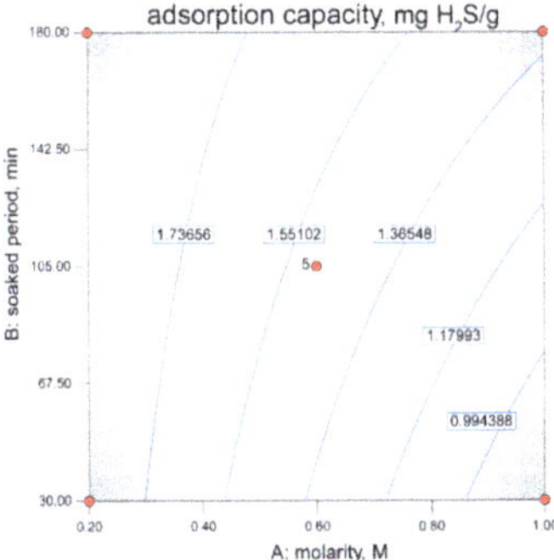 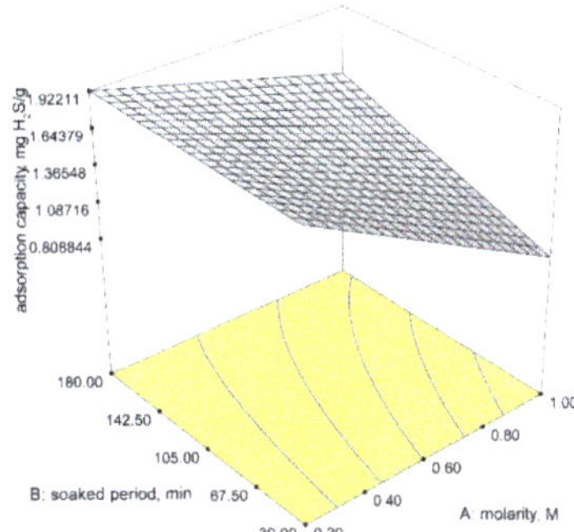

Figure 5. Contour plot describing the adsorption capacity response in soaked temperature function of ZnAc$_2$ molarity and soaked period (Soaked temperature: 65 °C).

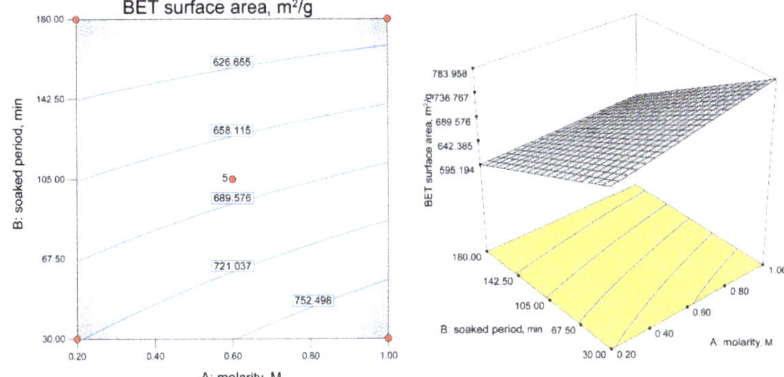

Figure 6. Contour plot describing the BET surface area response in soaked temperature function of molarity and soaked period (soaked temperature: 65 °C).

As shown in Figure 6, the constant soaked temperature shows the increments of the H$_2$S adsorption capacity with an increase in the ZnAc$_2$ molarity and the soaked period. The steepness of the increase in the H$_2$S adsorption capacity ranged from 0.2 M to 1.0 M for the soaked period of 30 min to 180 min. Figure 6 shows the effect of the interaction of the factors with the constant soaked temperature on the BET surface area. The decreases in the molarity with the lowest soaked period resulted in the highest BET surface area of 721.04 m^2/g, while at the lowest molarity (0.2 M) and a higher soaked period (>142.50 min), there was a reduction in the BET surface area.

Figure 7 presents the normal residual probability plot from the least squares fit, with both the predicted and the experimental data relatively similar to the straight line of 45° and the remaining points obeying the normal pattern of distribution. Hence, there was a high correlation and adequacy of the proposed model to predict the optimal conditions for preparing a highly efficient H$_2$S adsorbent.

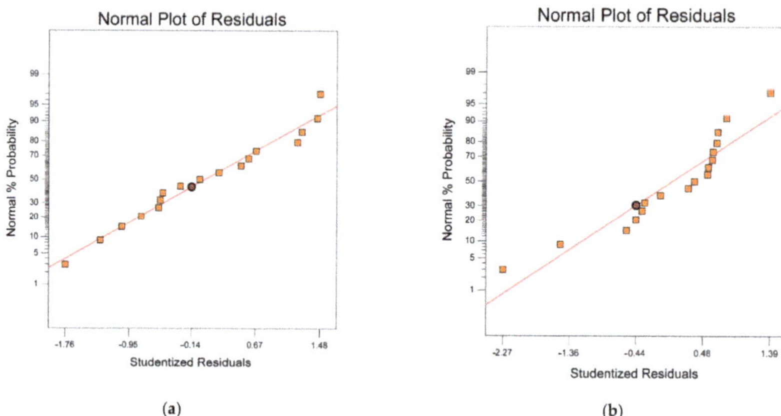

(a)   (b)

Figure 7. Normal probability plot: (**a**) adsorption capacity and (**b**) BET surface area response.

3.4. Optimization and Validation

The key goal of the optimization process was to identify the variable values at which the adsorption capacity of the H_2S and the BET surface area were optimal. Consequently, the Behnken configuration box was used to evaluate the best operating mode. Figure 8 displays the proposed model, showing that the highest adsorption capacity was 2.52 mg H_2S/g and the BET surface area was 620.55 m^2/g at the optimum molarity of 0.22 M, soaked time of 48.82 min, and soaked temperature of 95.08 °C. The desirability factor was 1.0, as shown in Figure 9, which reflected the most favorable or perfect response value [58].

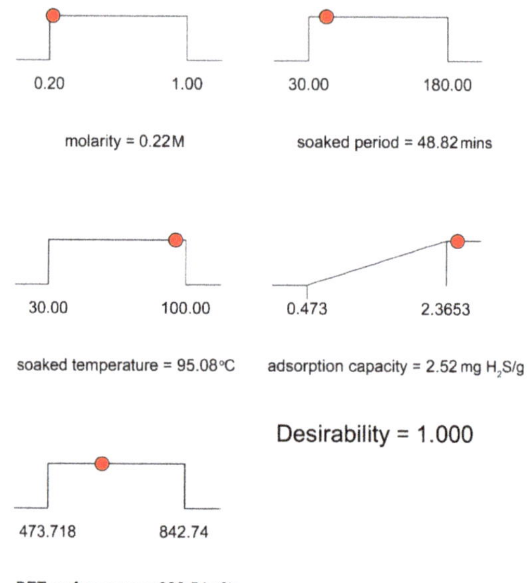

Figure 8. Optimum conditions according to the BBD statistical method.

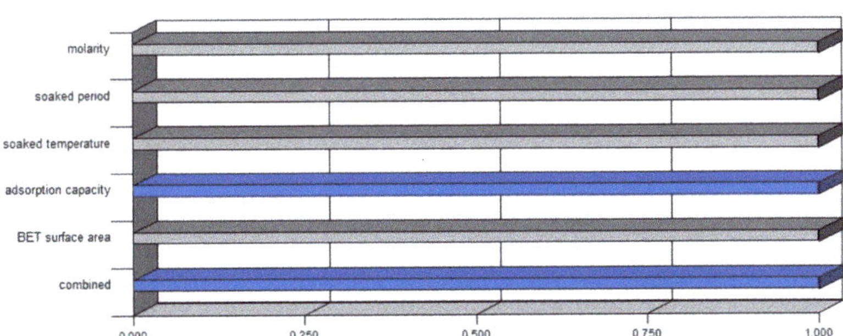

Figure 9. Individual and combined desirability functions.

For a comparison that quantified the acceptability of the model, an experimental study on H_2S adsorption was performed using the suggested optimum parameter conditions. The catalytic adsorbents were prepared using 350 g of CAC with a 0.22-M $ZnAc_2$ solution by soaking the CAC for up to 49 min at 95 °C. The experimental and theoretical verification was carried out using two responses, namely the H_2S adsorption capacity and the BET surface area, as shown in Figure 10.

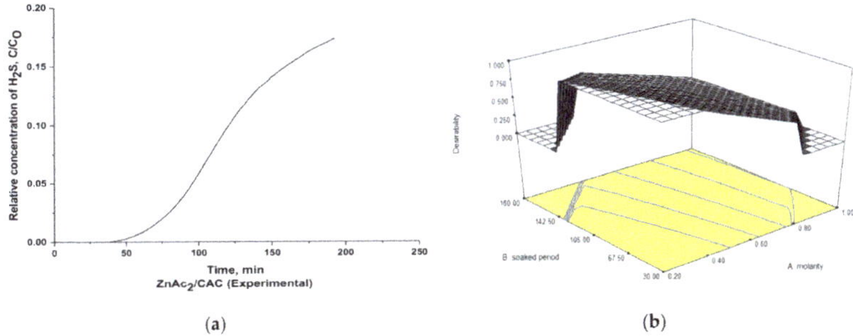

(a)   (b)

Figure 10. (a) Experimental plot for H_2S adsorption and (b) 3D contour plot for optimum theoretical condition for H_2S adsorption study.

The experimental data were collected through the synthesis of adsorbents based on the optimum parameter conditions as suggested at the end of BBD results. As a result, the adsorption capacity for H_2S was 2.12 mg H_2S/ g, whereas the theoretical data suggested a capacity of 2.52 mg H_2S/g. Moreover, the experimental BET surface area was 649.56 m^2/g, and the theoretical BET surface area was 620.55 m^2/g. Then, the relative error between the experimental and the theoretical values was approximately 16.2% for the adsorption capacity and 4.7% for the BET surface area. Therefore, the results obtained were in the range of acceptance, as the adsorption capacity and BET surface area were closer at the optimum condition of the variable for both the experimental and the theoretical data.

3.5. Adsorbent Characterization

Figure 11 presents the SEM images for the exhausted adsorbents for two types of adsorbents, namely the optimized ($ZnAc_2$/CAC_O (E)) and the non-optimized ($ZnAc_2$/CAC_N (E)) adsorbents. The optimized adsorbents were prepared under the optimum conditions from the Box–Behnken model suggestion. While the non-optimized adsorbents were prepared using a 0.2 M $ZnAc_2$ solution with a soaked period of 30 min at 65 °C. Both sample syntheses were tested with a commercial mixed gas up to the exhausted point and analyzed.

Next, the samples were assessed to visualize the details of the adsorbent properties in terms of the structural morphology images and the percentage of the elemental composition material presence on the adsorbent surface prepared.

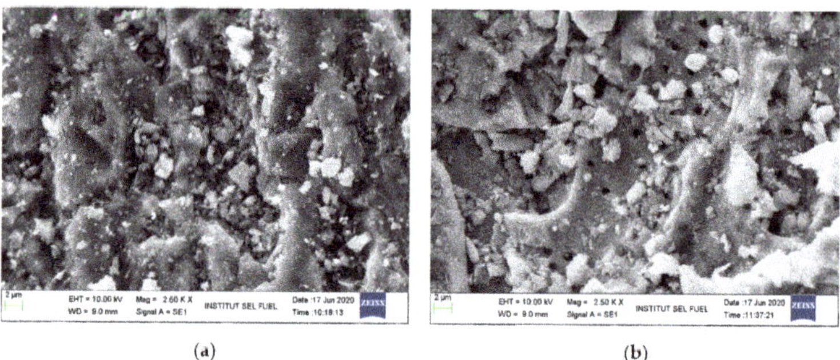

Figure 11. SEM analysis of exhausted $ZnAc_2/CAC$ adsorbent: (**a**) $ZnAc_2/CAC\_O$ (E) and (**b**) $ZnAc_2/CAC\_N$ (E) at 2.5 k × (10 μm).

Table 10 indicates the weight percentage (wt. percentage) of the element composition in a particular region of the optimized and non-optimized adsorbents for fresh (F), exhausted (E) and after desorption (D) compared to that fresh CAC (without impregnation). The EDX analysis was conducted on elements C, Ca, Na, K, Zn, O, and S (Table 10). The C content was different because of the composition of the volume of chemicals coated on a particular adsorbent surface. The presence of the Ca, Na, and K elements in the $ZnAc_2/CAC$ adsorbents normally observed on the activated carbon as similar data was obtained in a previous study by Zulkefli et al. [32] and Moradi et al. [48]. Meanwhile, the difference of concentration between $ZnAc_2$ for optimized and non-optimized adsorbents were 0.22 M and 0.2 M, respectively, which is about 10% difference. For soaked time and soaked temperature, the difference was about 18.8 min and 30 °C, respectively. Based on these optimized conditions for optimized adsorbent, the Zn element increased about 80%. As a result, the $ZnAc_2/CAC\_O$ (E) had a slightly higher S element by about 50% compared to that $ZnAc_2/CAC\_N$ (E) at exhausted adsorbent as shown in Table 10. This could be due to the presence of higher Zn which can help to improve the interaction between adsorbent and H_2S, and hence more H_2S can be adsorbed than the non-optimized adsorbents.

Table 10. Semi-quantitative chemical analysis of selected points in weight percent through EDX analysis.

Adsorbents	C	Ca	Na	K	Zn	O	S
CAC (F)	81.41	3.87	0.25	6.93	0.21	6.35	0.98
$ZnAc_2/CAC\_O$ (F)	85.75	0.75	0.02	0.43	7.72	5.14	0.19
$ZnAc_2/CAC\_O$ (E)	79.04	0.49	0.00	0.62	5.14	8.56	6.15
$ZnAc_2/CAC\_O$ (D)	87.21	0.63	0.00	0.39	6.01	5.53	0.23
$ZnAc_2/CAC\_N$ (F)	89.24	0.88	0.00	0.38	5.28	3.98	0.24
$ZnAc_2/CAC\_N$ (E)	86.04	0.37	0.00	0.52	2.84	4.81	5.42
$ZnAc_2/CAC\_N$ (D)	89.66	0.67	0.00	0.24	4.97	4.16	0.30

In the case of the exhausted adsorbents, the adsorbents were purged through the process of the desorption of air and N_2 gas, revealing the presence of sulfur (S) in the EDX analysis. The S element appeared on the surface for both optimized and non-optimized adsorbent as the H_2S adsorb during adsorption process. Similar findings were reported by Isik-Gulsac et al. [59] because of the inclusion of the S element on the surface of the adsorbents. Based on Table 10, it indicated the increment of S element from the fresh (F) to the exhausted (E) adsorbents. After the desorption process, the S element decreased to almost similar composition with the fresh adsorbent as indicated the physisorption occurred during the adsorption process.

Meanwhile, based on Table 10, the higher presence of the element composition of O on the adsorbent was obtained for optimized adsorbent which could be due to the increment of molarity of $ZnAc_2$. Based on Rodriguez et al., the composition of O normally had electrostatic interactions between the dipole of H_2S, and the ionic field generated by the charges in O might play a secondary role in accelerating and improving the adsorption process [60]. Hence, it would enhance the capability of the adsorbent to adsorb the H_2S gas as shown in the higher S element for optimized adsorbent in Table 10. Moreover, the presence of moisture and oxygen might affect the adsorption capacity of the activated carbons, and numerous studies have investigated their impact on the H_2S uptake. The presence of oxygen also increased the breakthrough time of H_2S adsorption for the latter adsorbents [61–63].

As shown in Table 11, the analysis of the BET surface area was conducted for the $ZnAc_2/CAC\_O$ and $ZnAc_2/CAC\_N$ adsorbents. In order to determine the specific surface area and the pore size distribution, an analysis was carried out of the N_2 adsorption/desorption for the fresh and the exhausted samples denoted as $ZnAc_2/CAC\_O$ (F), $ZnAc_2/CAC\_N$ (F), $ZnAc_2/CAC\_O$ (E), and $ZnAc_2/CAC\_N$ (E). The surface area was calculated using a BET isotherm calculation, while the pore volume and the average pore volume were calculated at P/Po of 0.98 through the N_2 adsorption isotherm. The pores included all the micropore, mesopore, and macropore volumes.

Table 11. Porous properties for regeneration of optimized and non-optimized adsorbents.

Adsorbents	BET Surface Area, m^2/g	Total Pore Volume, m^3/g ($\times 10^{-7}$)	V_{micro}/V_{total} (%)	Pore Size, Å
$ZnAc_2/CAC\_O$ (F)	713.81	3.49	0.78	19.33
$ZnAc_2/CAC\_O$ (E)	649.56	2.92	0.74	18.04
$ZnAc_2/CAC\_N$ (F)	717.41	3.48	0.77	19.26
$ZnAc_2/CAC\_N$ (E)	656.75	2.94	0.74	17.93

Upon the adsorption–desorption of H_2S, the BET surface area was influenced by the impregnation of $ZnAc_2$ as chemical compound in CAC and the presence of H_2S and its elements on the adsorbent which cause pores blocking by the H_2S components as previously observed [32]. The optimized adsorbents ($ZnAc_2/CAC\_O$) showed a slightly lower BET surface area than the non-optimized adsorbents ($ZnAc_2/CAC\_N$) because of the different parameter conditions for the prepared catalytic adsorbents. It is suggested the decrease in the BET surface area could be due to the increase of $ZnAc_2$ molarity used, hence blocking of some micropores on the adsorbent as mentioned previously. The exhausted adsorbents also showed a decrease in the BET surface area, total pore volume, volume ratio, and pore size, as a result of the interaction of H_2S with adsorbents which cause a blocking of the pores.

3.6. Performance of Adsorption–Desorption Cycle

The performance of the adsorbents was investigated through the adsorption degradation in the adsorption–desorption regeneration cycle. As $ZnAc_2$ composited as a catalyst on the adsorbents' surfaces was synthesis and observed the adsorption capacity performance through adsorption–desorption regeneration cycle. The adsorption capacity was calculated using the adsorption breakthrough time with the concentration change known as the mass transfer zone through downwards within the bed till further away from the inlet stream.

The H_2S adsorption capacity was compared between the optimized and the non-optimized adsorbents in order to observe the adsorbents' performance, as illustrated in Table 12 and Figure 12. Thus, the optimized adsorbents ($ZnAc_2/CAC\_O$) showed excellent performance based on the adsorption capacity, which was higher than that of the non-optimized adsorbents ($ZnAc_2/CAC\_N$) with an adsorption capacity difference of 49.3%. However, the performance of the optimized and the non-optimized adsorbents exhibited a degradation of up to 16% and 23% in the adsorption capacity throughout the regeneration cycle.

Table 12. Comparison of adsorption capacity in regeneration of adsorption–desorption H_2S.

Cycle	Adsorption Capacity, mg H_2S/g ZnAc$_2$/CAC_O	Adsorption Capacity, mg H_2S/g ZnAc$_2$/CAC_N
1	2.12	1.42
2	1.89	1.16
3	1.78	1.09

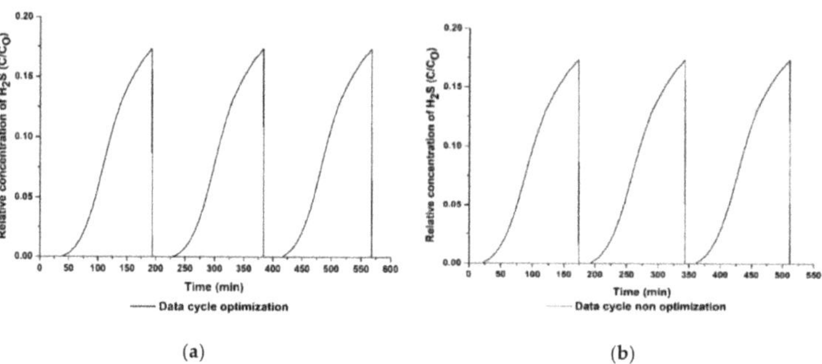

Figure 12. Regeneration adsorption–desorption curve for (**a**) optimized adsorbent and (**b**) non-optimized adsorbent.

In actual operation, the impregnation of activated carbon could involve physisorption and chemisorption which are both important for accelerating the adsorption process through physical forces or to catalyze the oxidation. Since the chemisorption probably can happen during the adsorption process, the degradation of the adsorbent could occur [64]. However, in this study, the adsorbent can still be regenerated in several adsorption–desorption cycles as shown in the capability of the adsorbent to adsorb the H_2S in the following cycles as shown in Figure 12. As discussed in a previous study, the presence of S elements throughout H_2S adsorption–desorption cycle can effectively remove the S elements on the adsorbent's surface up to 98% [30]. In this study, based on previous EDX analysis as shown in Table 10, after the desorption process, the S element decreased to almost similar composition with the fresh adsorbent as indicated the physisorption occurred during adsorption process as mentioned previously. However, there are slightly remaining S elements (as compared to the fresh adsorbent) which could be due to insufficient desorption process, i.e., non-optimized conditions for the desorption process, and probably due to complex mechanisms that happen during H_2S adsorption-desorption process. Hence, it probably could lead to a degradation of the H_2S adsorption capabilities for the following cycle of adsorption–desorption [62] as shown in Table 12 and Figure 12.

However, the degradation was low in each cycle, and the H_2S adsorption capacity could be enhanced by using a different desorption method in future works. Therefore, as proven by the previous characterization study, the performance of the optimized adsorbents improved the capabilities of the adsorbents as compared to the non-optimized adsorbents.

4. Conclusions

The performance of catalytic adsorbents for H_2S captured using the adsorption technique was examined through the impregnation of ZnAc$_2$ on the activated carbon surfaces. The optimization was carried out using RSM and the Box–Behnken experimental design to determine the optimum conditions for the adsorbent synthesis. Several factors and levels were evaluated, including the ZnAc$_2$ molarity, soaked period, and soaked temperature, along with the response of the H_2S adsorption capacity and the BET surface area. From the statistical analysis, the optimum points for $ZnAc_2$ molarity, soaked period, and soaked temperature were obtained as 0.22 M, 48.82 min, and 95.08 °C, respectively. Furthermore, the optimized adsorbents ($ZnAc_2$/CAC_O) improved the adsorbent efficiency by up to 49% of the adsorption capacity as compared to the non-optimized adsorbents ($ZnAc_2$/CAC_N). The optimized $ZnAc_2$ as the active catalyst dispersed onto the microporous materials of the activated carbon and improve the interaction of H_2S on adsorbent during the adsorption process. It was observed that the S element increase with the exhausted adsorbent from fresh adsorbent and the S element of optimized adsorbent was higher compared to that non-optimized. Based on the adsorption–desorption cycle, it was revealed the adsorbent slightly degraded by referring the calculated H_2S adsorption capacity up to 16% and 23% for optimized and non-optimized adsorbents, respectively throughout the cycles. It is suggested the degradation could be due to insufficient desorption process, i.e., non-optimized conditions, and probably due to complex mechanisms that happen during the adsorption–desorption process. Hence, comprehensive studies are required in the future to analyze the adsorbent degradation by optimizing the conditions of the desorption process and analyze the mechanism of adsorption-desorption of H_2S on the adsorbent.

Author Contributions: For research articles with several authors, the following statements should be used Conceptualization, N.N.Z. and M.S.M.; methodology, N.N.Z. and M.S.M.; software, N.N.Z. and S.N.H.A.B.; validation, M.S.M., H.A.H., and W.N.R.W.I.; formal analysis, N.N.Z., M.S.M. and N.M.S.; investigation N.N.Z.; resources, N.N.Z. and M.S.M.; data curation, N.N.Z. and M.S.M.; writing—original draft preparation, N.N.Z.; writing—review and editing, M.S.M., H.A.H., and W.N.R.W.I.; visualization, N.N.Z. and M.S.M.; supervision, M.S.M., H.A.H., and W.N.R.W.I.; project administration, M.S.M.; funding acquisition, M.S.M. All authors have read and agreed to the published version of the manuscript.

Funding: This research was funded by Ministry of Higher Education, Malaysia, grant number FRGS/1/2020/TK0/UKM/02/4 and Universiti Kebangsaan Malaysia, grant numbers GUP-2018-042 & PP-FKAB-2021.

Institutional Review Board Statement: Not applicable.

Informed Consent Statement: Not applicable.

Data Availability Statement: All relevant data are contained in the present manuscript. Other inherent data are available on request from the corresponding author.

Acknowledgments: This research was supported by the Ministry of Higher Education, Malaysia under research code FRGS/1/2020/TK0/UKM/02/4 and Universiti Kebangsaan Malaysia under research code GUP-2018-042 & PP-FKAB-2021.

Conflicts of Interest: The authors declare no conflict of interest.

Nomenclature List

χ_i	Encoded value of an independent variable
x_i	Actual value of an independent variable
x_0	Actual value of a center point's independent variable
Δx	Phase shift value of an independent variable
Y	Response variable
α_0	Model constant
α_i	Linear coefficient
α_{ii}	Quadratic coefficient
α_{ij}	Interaction coefficient
ε	Statistical error
N	Number of runs
k	Number of variables

C_N	Number of center points
MSS	Mean square
MSF	Mean square of factors (interactions)
MSE	Mean square of errors
A	Molarity
B	Soaked period
C	Soaked temperature
Q	Adsorption capacity, mg H_2S/g
R^2	Coefficient of determination
DF	Degree of freedom
Prob	Probability
PRESS	Predicted residual error sum of squares
C.V	Coefficient variation
F-value	Fisher's variance ratio
Prob > F	Probability value
C	Outlet concentration
C_O	Inlet concentration
BET	Brunauer–Emmett–Teller
SEM	Scanning electron microscopy
EDX	Energy dispersive X-ray analysis
CAC	Commercial coconut activated carbon
AC	Activated carbon

References

1. Sitthikhankaew, R.; Chadwick, D.; Assabumrungrat, S.; Laosiripojana, N. Effect of KI and KOH Impregnations over Activated Carbon on H_2S Adsorption Performance at Low and High Temperatures. *Sep. Sci. Technol.* **2014**, *49*, 354–366. [CrossRef]
2. Cornejo, C.; Wilkie, A.C. Greenhouse gas emissions and biogas potential from livestock in Ecuador. *Energy Sustain. Dev.* **2010**, *14*, 256–266. [CrossRef]
3. Wongsapai, W.; Thienburanathum, P.; Rerkkriengkrai, P. Biogas situation and development in Thai Swine Farm. *Renew. Energy Power Qual. J.* **2008**, *1*, 222–226. [CrossRef]
4. Tippayawong, N.; Promwungkwa, A.; Rerkkriangkrai, P. Long-term operation of a small biogas/diesel dual-fuel engine for on-farm electricity generation. *Biosyst. Eng.* **2007**, *98*, 26–32. [CrossRef]
5. Tippayawong, N.; Thanompongchart, P. Biogas quality upgrade by simultaneous removal of CO_2 and H_2S in a packed column reactor. *Energy* **2010**, *35*, 4531–4535. [CrossRef]
6. Arthur, R.; Baidoo, M.F.; Antwi, E. Biogas as a potential renewable energy source: A Ghanaian case study. *Renew. Energy* **2011**, *36*, 1510–1516. [CrossRef]
7. Angelidaki, I.; Ellegaard, L.; Ahring, B.K. Applications of The Anaerobic Digestion Process. *Adv. Biochem. Eng. BioTechnol.* **2003**, *82*, 1–33.
8. Balsamo, M.; Cimino, S.; de Falco, G.; Erto, A.; Lisi, L. ZnO-CuO Supported on Activated Carbon for H_2S Removal at Room Temperature. *Chem. Eng. J.* **2016**, *304*, 399–407. [CrossRef]
9. Siriwardane, I.W.; Udangawa, R.; de Silva, R.M.; Kumarasinghe, A.R.; Acres, R.G.; Hettiarachchi, A.; de Silva, K.M.N. Synthesis and Characterization of Nano Magnesium Oxide Impregnated Granular Activated Carbon Composite for H_2S Removal Applications. *Mater. Des.* **2017**, *136*, 127–136. [CrossRef]
10. Crisci, A.G.D.; Moniri, A.; Xu, Y. Hydrogen from hydrogen sulfide: Towards a more sustainable hydrogen economy. *Int. J. Hydrog. Energy* **2019**, *44*, 1299–1327. [CrossRef]
11. Zulkefli, N.N.; Masdar, M.S.; Jahim, J.; Harianto, E.H. Overview of H_2S Removal Technologies from Biogas Production. *Int. J. Appl. Eng. Res.* **2016**, *11*, 10060–10066.
12. Palma, V.; Vaiano, V.; Barba, D.; Colozzi, M.; Palo, E.; Barbato, L.; Cortese, S. H_2S Oxidative Decomposition for The Sim-ultaneous Production of Sulphur and Hydrogen. *Chem. Eng. Trans.* **2016**, *52*, 1201–1206.
13. Bandosz, T.J. On the Adsorption/Oxidation of Hydrogen Sulfide on Activated Carbons at Ambient Temperatures. *J. Colloid Interface Sci.* **2002**, *246*, 1–20. [CrossRef]
14. Sisani, E.; Cinti, G.; Discepoli, G.; Penchini, D.; Desideri, U.; Marmottini, F. Adsorptive Removal of H_2S In Biogas Conditions for High Temperature Fuel Cell Systems. *Int. J. Hydrog. Energy* **2014**, *39*, 21753–21766. [CrossRef]
15. Mescia, D.; Hernández, S.; Conoci, A.; Russo, N. MSW landfill biogas desulfurization. *Int. J. Hydrog. Energy* **2011**, *36*, 7884–7890. [CrossRef]
16. Bamdad, H.; Hawboldt, K.; MacQuarrie, S. A review on common adsorbents for acid gases removal: Focus on biochar. *Renew. Sustain. Energy Rev.* **2018**, *81*, 1705–1720. [CrossRef]
17. Khabazipour, M.; Mansoor, A. Removal of Hydrogen Sulfide from Gas Streams Using Porous Materials: A Review. *Ind. Eng. Chem. Res.* **2019**, *58*, 22133–22164. [CrossRef]

18. Sing, K.S.W.; Rouquerol, J.; Rouquerol, F. *Adsorption by Powders and Porous Solids*; Academic Press: San Diego, CA, USA, 1998.
19. Galuszka, J.; Iaquaniello, G.; Ciambelli, P.; Palma, V.; Brancaccio, E. Membrane-Assisted Catalytic Cracking of Hydrogen Sulphide (H_2S). In *Membrane Reactors for Hydrogen Production Processes*; Springer: London, UK, 2011; pp. 161–182.
20. Basile, A.; De Falco, M.; Centi, G.; Iaquaniello, G. *Membrane Reactor Engineering: Applications for a Greener Process Industry*; John Wiley & Sons: Hoboken, NJ, USA, 2016; p. 344.
21. Papurello, D.; Tomasi, L.; Silvestri, S.; Santarelli, M. Evaluation of the Wheeler-Jonas parameters for biogas trace compounds removal with activated carbons. *Fuel Process. Technol.* **2016**, *152*, 93–101. [CrossRef]
22. Pelaez-Samaniego, M.R.; Perez, J.F.; Ayiania, M.; Garcia-Perez, T. Chars from wood gasification for removing H_2S from biogas. *Biomass Bioenergy* **2020**, *142*, 105754. [CrossRef]
23. Surra, E.; Nogueira, M.C.; Bernardo, M.; Lapa, N.; Esteves, I.; Fonseca, I. New adsorbents from maize cob wastes and anaerobic digestate for H_2S removal from biogas. *Waste Manag.* **2019**, *94*, 136–145. [CrossRef]
24. Shah, M.S.; Tsapatsis, M.; Siepmann, J.I. Hydrogen Sulfide Capture: From Absorption in Polar Liquids to Oxide, Zeolite, and Metal–Organic Framework Adsorbents and Membranes. *Chem. Rev.* **2017**, *117*, 9755–9803. [CrossRef] [PubMed]
25. Ozekmekci, M.; Salkic, G.; Fellah, M.F. Use of zeolites for the removal of H_2S: A mini-review. *Fuel Process. Technol.* **2015**, *139*, 49–60. [CrossRef]
26. Kerr, G.T.; Johnson, G.C. Catalytic Oxidation of Hydrogen Sulfide to Sulfur Over A Crystalline Aluminosilicate. *J. Phys. Chem.* **1960**, *64*, 381–382. [CrossRef]
27. Steijns, M.; Derks, F.; Verloop, A.; Mars, P. ChemInform Abstract: The Mechanism of the Catalytic Oxidation of Hydrogen Sulfide. II. Kinetics and Mechanism of Hydrogen Sulfide Oxidation Catalyzed by Sulfur. *Chem. Inf.* **1976**, *7*, 87–95. [CrossRef]
28. Bashkova, S.; Baker, F.S.; Wu, X.; Armstrong, T.R.; Schwartz, V. Activated carbon catalyst for selective oxidation of hydrogen sulphide: On the influence of pore structure, surface characteristics, and catalytically-active nitrogen. *Carbon* **2007**, *45*, 1354–1363. [CrossRef]
29. Micoli, L.; Bagnasco, G.; Turco, M. H_2S removal from biogas for fuelling MCFCs: New adsorbing materials. *Int. J. Hydrog. Energy* **2014**, *39*, 1783–1787. [CrossRef]
30. Phooratsamee, W.; Hussaro, K.; Teekasap, S.; Hirunlabh, J. Increasing Adsorption of Activated Carbon from Palm Oil Shell for Adsorb H_2S From Biogas Production by Impregnation. *Am. J. Environ. Sci.* **2014**, *10*, 431–445. [CrossRef]
31. Yusuf, N.Y.M.; Masdar, M.S.; Isahak, W.N.R.W.; Nordin, D.; Husaini, T.; Majlan, E.H.; Wu, S.Y.; Rejab, S.A.M.; Lye, C.C. Impregnated carbon–ionic liquid as innovative adsorbent for H_2/CO_2 separation from biohydrogen. *Int. J. Hydrog. Energy* **2019**, *44*, 3414–3424. [CrossRef]
32. Zulkefli, N.N.; Masdar, M.S.; Isahak, W.N.R.W.; Jahim, J.M.; Rejab, S.A.M.; Lye, C.C. Removal of hydrogen sulfide from a biogas mimic by using impregnated activated carbon adsorbent. *PLoS ONE* **2019**, *14*, e0211713. [CrossRef]
33. Sidek, M.Z.; Cheah, Y.J.; Zulkefli, N.N.; Yusuf, N.M.; Isahak, W.N.R.W.; Sitanggang, R.; Masdar, M.S. Effect of impregnated activated carbon on carbon dioxide adsorption performance for biohydrogen purification. *Mater. Res. Express* **2018**, *6*, 015510. [CrossRef]
34. Feng, Y.; Dou, J.; Tahmasebi, A.; Xu, J.; Li, X.; Yu, J.; Yin, F. Regeneration of Fe–Zn–Cu Sorbents Supported on Activated Lignite Char for the Desulfurization of Coke Oven Gas. *Energy Fuels* **2015**, *29*, 7124–7134. [CrossRef]
35. Georgiadis, A.G.; Charisiou, N.D.; Goula, M.A. Removal of hydrogen sulfide from various industrial gases: A review of the most promising adsorbing materials. *Catalysts* **2020**, *10*, 521. [CrossRef]
36. Hunter, G.B.J.; Hunter, J.S. *Statistics for Experimenters*; John Wiley and Sons: New York, NY, USA, 2005.
37. Bruns, R.E.; Scarminio, I.S.; Neto, B.B. *Statistical Design—Chemometrics*; Elsevier: Amsterdam, The Netherlands, 2006.
38. Massart, D.L.; Vandeginste, B.G.M.; Buydens, L.M.C.; de Jong, S.; Lewi, P.J.; Smeyers-Verbeke, J. *Handbook of Chemometrics and Qualimetrics: Part A*; Elsevier: Amsterdam, The Netherlands, 1977.
39. Tong, L.-I.; Chang, Y.-C.; Lin, S.-H. Determining the optimal re-sampling strategy for a classification model with imbalanced data using design of experiments and response surface methodologies. *Expert Syst. Appl.* **2011**, *38*, 4222–4227. [CrossRef]
40. Myers, R.H.; Montgomery, D.C. *Response Surface Methodology: Process and Product Optimization Using Designed Experiments*, 2nd ed.; John Wiley & Sons, Inc: Hoboken, NJ, USA, 2002.
41. Papurello, D.; Iafrate, C.; Lanzini, A.; Santarelli, M. Trace compounds impact on SOFC performance: Experimental and modelling approach. *Appl. Energy* **2017**, *208*, 637–654. [CrossRef]
42. Kupecki, J.; Papurello, D.; Lanzini, A.; Naumovich, Y.; Motylinski, K.; Blesznowski, M.; Santarelli, M. Numerical model of planar anode supported solid oxide fuel cell fed with fuel containing H_2S operated in direct internal reforming mode (DIR SOFC). *Appl. Energy* **2018**, *230*, 1573–1584. [CrossRef]
43. Papurello, D.; Lanzini, A. SOFC single cells fed by biogas: Experimental tests with trace contaminants. *Waste Manag.* **2018**, *72*, 306–312. [CrossRef]
44. Zulkefli, N.N.; Khaia, T.Z.; Nadaraja, S.; Venugopal, N.R.; Yusri, N.A.M.; Sofian, N.M.; Masdar, M.S. Capabilities dual chemical mixture (DCM) adsorbents through metal anchoring in H_2S captured. *Solid State Technol.* **2020**, *63*, 181–191.
45. Baş, D.; Boyacı, I.H. Modeling and optimization I: Usability of response surface methodology. *J. Food Eng.* **2007**, *78*, 836–845. [CrossRef]
46. Sharma, P.; Singh, L.; Dilbaghi, N. Optimization of process variables for decolorization of Disperse Yellow 211 by Bacillus subtilis using Box–Behnken design. *J. Hazard. Mater.* **2009**, *164*, 1024–1029. [CrossRef]

47. Wu, L.; Yick, K.-L.; Ng, S.-P.; Yip, J. Application of the Box–Behnken design to the optimization of process parameters in foam cup molding. *Expert Syst. Appl.* **2012**, *39*, 8059–8065. [CrossRef]
48. Moradi, M.; Daryan, J.T.; Mohamadalizadeh, A. Response surface modeling of H_2S conversion by catalytic oxidation reaction over catalysts based on SiC nanoparticles using Box–Behnken experimental design. *Fuel Process. Technol.* **2013**, *109*, 163–171. [CrossRef]
49. Dong, C.H.; Xie, X.Q.; Wang, X.L.; Zhan, Y.; Yao, Y.J. Application of Box-Behnken design in optimisation for polysaccharides extraction from cultured mycelium of Cordyceps sinensis. *Food Bioprod. Process.* **2009**, *87*, 139–144. [CrossRef]
50. Sidek, M.Z.; Masdar, M.S.; Nik Dir, N.M.H.; Amran, N.F.A.; Ajit Sing, S.K.D.; Wong, W.L. Integrasi Sistem Penulenan Biohidrogen dan Aplikasi Sel Fuel. *J. Kejuruter.* **2018**, *1*, 41–48.
51. Arifina, R.A.; Hasana, H.A.; Kamarudina, N.H.N.; Ismailb, N.I. Synthesis of Mesoporous Silica for Ammonia Adsorption in Aqueous Solution. *J. Kejuruter.* **2018**, *1*, 59–64. [CrossRef]
52. Nakamura, T.; Kawasaki, N.; Hirata, M.; Oida, Y.; Tanada, S. Adsorption of Hydrogen Sulfide by Zinc-Containing Activated carbon. *Toxicol. Environ. Chem.* **2002**, *82*, 93–98. [CrossRef]
53. Bazrafshan, E.; Al-Musawi, T.J.; Silva, M.F.; Panahi, A.H.; Havangi, M.; Mostafapur, F.K. Photocatalytic degradation of catechol using ZnO nanoparticles as catalyst: Optimizing the experimental parameters using the Box-Behnken statistical methodology and kinetic studies. *Microchem. J.* **2019**, *147*, 643–653. [CrossRef]
54. Pimenta, C.D.; Silva, M.B.; de Morais Campos, R.L.; de Campos, W.R., Jr. Desirability and design of experiments applied to the optimization of the reduction of decarburization of the process heat treatment for steel wire SAE 51B35. *Am. J. Theor. Appl. Stat.* **2018**, *7*, 35–44. [CrossRef]
55. Feng, W.; Kwon, S.; Borguet, E.; Vidic, R. Adsorption of Hydrogen Sulfide onto Activated Carbon Fibers: Effect of Pore Structure and Surface Chemistry. *Environ. Sci. Technol.* **2005**, *39*, 9744–9749. [CrossRef]
56. Keyvanloo, K.; Towfighi, J.; Sadrameli, S.; Mohamadalizadeh, A. Investigating the effect of key factors, their interactions and optimization of naphtha steam cracking by statistical design of experiments. *J. Anal. Appl. Pyrolysis* **2010**, *87*, 224–230. [CrossRef]
57. Sen, R.; Swaminathan, T. Response surface modeling and optimization to elucidate and analyze the effects of inoculum age and size on surfactin production. *Biochem. Eng. J.* **2004**, *21*, 141–148. [CrossRef]
58. Mohamadalizadeh, A.; Towfighi, J.; Rashidi, A.; Manteghian, M.; Mohajeri, A.; Arasteh, R. Nanoclays as nano adsorbent for oxidation of H_2S into elemental sulfur. *Korean J. Chem. Eng.* **2011**, *28*, 1221–1226. [CrossRef]
59. Isik-Gulsac, I. Investigation of Impregnated Activated Carbon Properties Used in Hydrogen Sulfide Fine Removal. *Braz. J. Chem. Eng.* **2016**, *33*, 1021–1030. [CrossRef]
60. Rodriguez, J.A.; Jirsak, T.; Chaturvedi, S. Reaction of H_2S with MgO(100) and Cu/MgO(100) surfaces: Band-gap size and chemical reactivity. *J. Chem. Phys.* **1999**, *111*, 8077–8087. [CrossRef]
61. Preece, D.A.; Montgomery, D.C. Design and Analysis of Experiments. *Int. Stat. Rev.* **1978**, *46*, 120. [CrossRef]
62. Primavera, A.; Trovarelli, A.; Andreussi, P.; Dolcetti, G. The effect of water in the low-temperature catalytic oxidation of hydrogen sulfide to sulfur over activated carbon. *Appl. Catal. A Gen.* **1998**, *173*, 185–192. [CrossRef]
63. Bagreev, A.; Bandosz, T.J. H_2S adsorption/oxidation on unmodified activated carbons: Importance of prehumidification. *Carbon* **2001**, *39*, 2303–2311. [CrossRef]
64. Vinodhini, V.; Das, N. Packed bed column studies on Cr (VI) removal from tannery wastewater by neem sawdust. *Desalination* **2010**, *264*, 9–14. [CrossRef]

 *catalysts*

Article

Polyoxometalate Dicationic Ionic Liquids as Catalyst for Extractive Coupled Catalytic Oxidative Desulfurization

Jingwen Li, Yanwen Guo, Junjun Tan and Bing Hu *

School of Materials and Chemical Engineering, Hubei University of Technology, Wuhan 430068, China; lijingwen@hbut.edu.cn (J.L.); guoyanwen@hbut.edu.cn (Y.G.); tanjunjun2011@hbut.edu.cn (J.T.)
* Correspondence: hubing@hbut.edu.cn; Tel.: +86-13667257353

Abstract: Wettability is an important factor affecting the performance of catalytic oxidative desulfurization. In order to develop an efficient catalyst for the extractive coupled catalytic oxidative desulfurization (ECODS) of fuel oil by H_2O_2 and acetonitrile, a novel family of imidazole-based polyoxometalate dicationic ionic liquids (POM-DILs) $[C_n(MIM)_2]PW_{12}O_{40}$ (n = 2, 4, 6) was synthesized by modifying phosphotungstic acid ($H_3PW_{12}O_{40}$) with double imidazole ionic liquid. These kinds of catalysts have good dispersity in oil phase and H_2O_2, which is conducive to the deep desulfurization of fuel oil. The catalytic performance of the catalysts was studied under different conditions by removing aromatic sulfur compound dibenzothiophene (DBT) from model oil. Results showed that $[C_2(MIM)_2]PW_{12}O_{40}$ had excellent desulfurization efficiency, and more than 98% of DBT was removed under optimum conditions. In addition, it also exhibited good recyclability, and activity with no significant decline after seven reaction cycles. Meanwhile, dibenzothiophene sulfone ($DBTO_2$), the only oxidation product of DBT, was confirmed by Gas Chromatography-Mass Spectrometry (GC-MS), and a possible mechanism of the ECODS process was proposed.

Keywords: polyoxometalate; dicationic ionic liquids; extraction; oxidative desulfurization; dibenzothiophene

Citation: Li, J.; Guo, Y.; Tan, J.; Hu, B. Polyoxometalate Dicationic Ionic Liquids as Catalyst for Extractive Coupled Catalytic Oxidative Desulfurization. *Catalysts* **2021**, *11*, 356. https://doi.org/10.3390/catal11030356

Academic Editor: Oxana Kholdeeva

Received: 1 February 2021
Accepted: 6 March 2021
Published: 9 March 2021

Publisher's Note: MDPI stays neutral with regard to jurisdictional claims in published maps and institutional affiliations.

1. Introduction

With the development of the economy, traditional fossil fuels still occupy a large proportion of supply and demand in the market [1]. Some sulfur compounds contained in fuel oil, such as mercaptan, thioether, thiophene and their derivatives, will produce sulfur oxides during combustion, which can lead to a series of environmental problems such as acid rain and haze [2–4]. Therefore, many countries are constantly strengthening the control standards of sulfur content in fuel oil. Improving the technology to produce high quality fuel oil in accordance with the standards has become a top priority for refineries [5,6]. Hydrodesulfurization (HDS) is the most mature technique and has been applied in industry [7–10]. It can efficiently eliminate aliphatic sulfur compounds, such as mercaptan and thioether. However, in addition to the harsh operation conditions, HDS is not effective for removing aromatic sulfur compounds and their derivatives with steric hindrance [11,12]. In this context, as a nonhydrodesulfurization technology that can achieve deep desulfurization of fuel oil under mild conditions, the ECODS process has become a main focus due to its simplicity and effectiveness [13–16]. Although various types of solvents and oxidants have been used in the desulfurization process, and also play an important role, the biggest challenge for a successful ECODS process is to use catalysts with high activity.

Polyoxometalates (POMs), represented by $H_3PW_{12}O_{40}$, have been widely used as the catalysts for oxidative desulfurization under mild conditions of the model oil system due to their strong Bronsteic acidities and redox properties [17–20]. On the other hand, ionic liquids (ILs), as a prominent catalyst/extractant with low vapor pressure, good thermal stability, recyclability and environmental friendliness, are also usually used in desulfurization reactions [21]. However, the shortcomings of POMs and ILs are the main obstacles to their industrial application. For example, the small specific surface area of POMs (<10 m^2/g) makes their catalytic activity low [22], and the liquid properties of ILs make them difficult to separate and recover. In order to solve those problem, according to the characteristics that specific ILs with different properties can be designed by the combination of different cations and anions [23], and the catalytic performance of POMs can be regulated by the electrostatic interaction and hydrogen bonding between the cations of specific ionic liquids and the anions of POMs [24], a new type of POM-IL with different physical and chemical properties, which is formed by the combination of heteropolyanions and organic cations, has attracted widespread attention. Huang et al., synthesized a heteropolyanionic-based ionic liquid catalyst [(3-sulfonic acid) propylpyridine]$_3$PW$_{12}$O$_{40}$·2H$_2$O ([PSPy]$_3$PW$_{12}$O$_{40}$·2H$_2$O) by the reaction of N-Propanesulfone pyridinium with an aqueous solution of H$_3$PW$_{12}$O$_{40}$, which showed high catalytic activity and excellent recyclability in the oxidative desulfurization of fuel oil [25]. Our groups successively synthesized a kind of POM-IL, [Hmim]$_5$PMo$_{10}$V$_2$O$_{40}$ [26], [C$_3$H$_3$N$_2$(CH$_3$)(C$_n$H$_{2n}$)]$_5$PMo$_{10}$V$_2$O$_{40}$ ([C$_n$mim]PMoV n = 2, 4 and 6) [27], by the reaction of molybdovanadophosphoric acid (H$_5$PMo$_{10}$O$_{40}$) with N-methylimidazole and imidazole bromides, respectively, and applied it as a catalyst to the desulfurization process with H$_2$O$_2$ as the oxidant. The results showed that 99.1% and 100% of dibenzothiophene (DBT) in the model oil are removed, respectively, and the catalytic activity of POM-ILs decreased slightly after six cycles. However, these systems still need more catalysts and a relatively long reaction time to achieve ideal desulfurization efficiency. Therefore, it is necessary to find other methods to improve the economic applicability and effectiveness of the catalysts. In recent years, many other types of POM-ILs have been used as catalysts for oxidative desulfurization, such as [C$_{11}$H$_9$N(CH$_2$)$_4$SO$_3$H]$_3$PW$_{12}$O$_{40}$ (PhPyBs-PW) [28], [3-(pyridine-1-ium-1-yl)propane-1-sulfonate]$_3$(NH$_4$)$_3$Mo$_7$O$_{24}$·4H$_2$O ([PyPS]$_3$(NH$_4$)$_3$Mo$_7$O$_{24}$) [C$_6$H$_5$NO$_2$CH$_2$(CH$_2$)$_2$CH$_3$]$_7$PMo$_{12}$O$_{40}$ ([29] and (NKBu)$_7$PMo$_{12}$O$_{42}$) [30], which can effectively improve the desulfurization efficiency in a short period of time with a small amount of catalyst. However, it is rarely reported that POM-based dicationic ionic liquids with higher thermal stability, good wettability and high activity are used as catalysts for oxidative desulfurization. DILs are a new type of ionic liquid compound with higher stability and lower toxicity, which consists of two monomers linked by alkyl or aryl groups [31–34]. Compared with the traditional monocationic ILs, DILs have a larger cationic volume, which makes the π-π interaction between cations and aromatic sulfides stronger, and can effectively remove aromatic sulfides in fuel oil. In addition, through electrostatic interaction and hydrogen bonding, the double cation can be well connected with the anion, so as to improve the overall catalytic activity of the catalyst. Due to their unique properties, DILs have been successfully used and achieved ideal effects in esterification, supercapacitor, biodiesel catalysis and extractive desulfurization as eutectic solvents/catalysts, electrolyte additives, catalysts and extractant, respectively [35–38]. In our group's recent research results, a series of novel binuclear magnetic ionic liquids (MILs) [C$_n$(MIM)$_2$]Cl$_2$/mFeCl$_3$ (n = 2, 4 or 6 and m = 1, 2 or 3) were synthesized and used as catalysts for the desulfurization with oxidant H$_2$O$_2$ and extractant acetonitrile [39]. The results showed that 97.07% desulfurization efficiency can be achieved in 10 min, showing ultrahigh catalytic activity of MILs.

Inspired by the above research, we are deeply interested in the preparation of novel POM-DILs and their application in the field of catalytic oxidative desulfurization. In this work, in order to give full play to the advantages of POMs and DILs, we prepared a new kind of POM-DIL catalyst [C$_n$(MIM)$_2$]PW$_{12}$O$_{40}$ (n = 2, 4, 6) with H$_3$PW$_{12}$O$_{40}$ modified by imidazole-based DILs, and the catalyst was further applied in ECODS system, which was constructed with H$_2$O$_2$ as the oxidant and acetonitrile as the extractant. The optimum reaction conditions and parameters were determined. In addition to their high thermal stability, the catalysts also showed high catalytic activity for the removal of DBT from model oil due to their excellent wettability. At the same time, the recycling performance of the catalyst was also explored. Finally, based on the corresponding characterization results, the possible mechanism of the desulfurization process was proposed.

2. Results and Discussion

2.1. Characterization of Catalyst

Fourier transform infrared spectroscopy (FT-IR) is a suitable technique to prove the success of catalyst synthesis. The FT-IR spectra (wavelength range from 4000 to 500 cm^{-1}) of catalysts $[C_2(MIM)_2]PW_{12}O_{40}$, $[C_4(MIM)_2]PW_{12}O_{40}$, $[C_6(MIM)_2]PW_{12}O_{40}$ are shown in Figure 1a. The absorption peak near 2950 cm^{-1} was attributed to the C-H stretching vibration of imidazole ring. The stretching vibration absorption peaks around 1630 and 1564 cm^{-1} were attributed to C=C and C=N skeleton vibrations on the aromatic ring, respectively. The peaks at 1463, 1408 cm^{-1}, 1341 and 1253 cm^{-1} were related to C–N heterocycles. In the range of 870-1100 cm^{-1}, four characteristic absorption peaks of Keggin structure were observed in all three catalysts, which were 1084 (P-O), 983 (W=O), 898 (W-O$_c$-W corner-sharing) and 800 cm^{-1} (W-O$_e$-W edge-sharing), respectively. This was consistent with the characteristic peaks of $H_3PW_{12}O_{40}$ in Figure 1b, which showed that the catalysts still had the Keggin structure of $PW_{12}O_{40}{}^{3-}$.

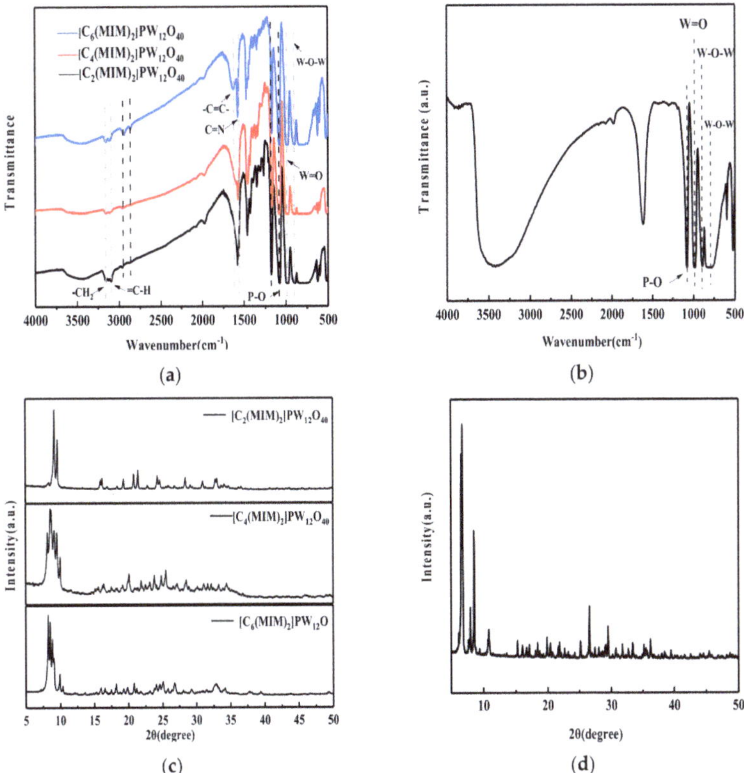

Figure 1. *Cont.*

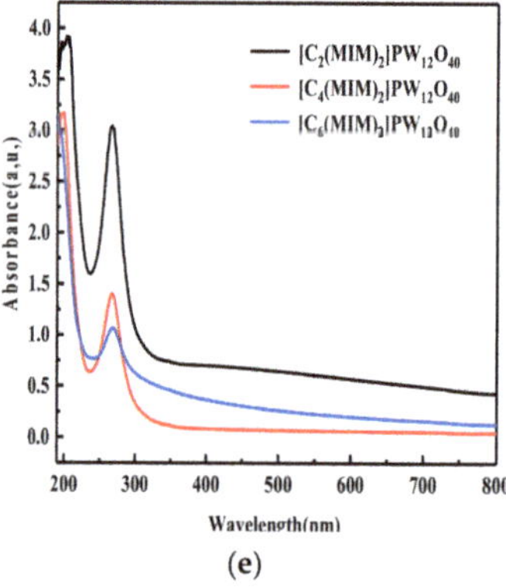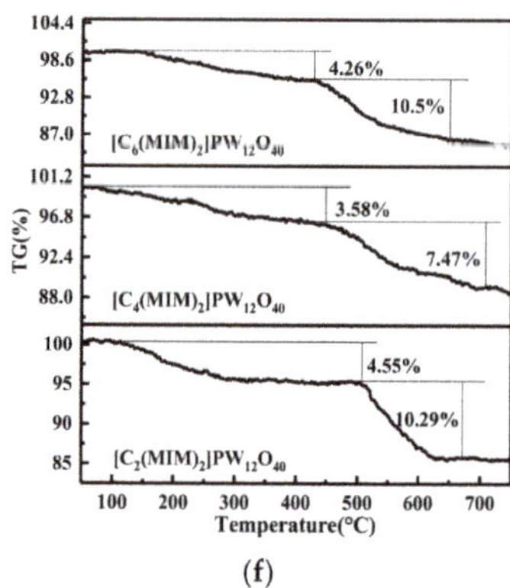

Figure 1. Characterizations of the samples. (**a**) FT-IR spectra of catalysts; (**b**) FT-IR spectra of $H_3PW_{12}O_{40}$; (**c**) XRD patterns of catalysts; (**d**) XRD patterns of $H_3PW_{12}O_{40}$; (**e**) Ultraviolet visible (UV-vis) spectra of catalysts; (**f**) Thermogravimetric (TG) curve of catalysts.

X-ray diffraction (XRD) also provided strong evidence to support the successful synthesis of catalysts. Compared with the XRD pattern (2θ from $5°$ to $50°$) of catalysts (Figure 1c) and $H_3PW_{12}O_{40}$ (Figure 1d), some diffractive peaks of the three catalysts were obtained at near $10°$, $21°$ and $32.8°$. This was due to the disappearance of coordination water $H_5O_2^+$ and H_3O^+ which interact with the anions of the Keggin structure by replacing the secondary structure proton of $H_3PW_{12}O_{40}$ with $[C_n(MIM)_2]^{2+}$ (n = 2, 4, 6).

UV-vis spectroscopy is a rapid and accurate method to determine the molecular structure of organic compounds and the charge transfer behavior of catalysts. The UV-vis spectra of the catalysts are shown in Figure 1e. There were two characteristic peaks in the range of 190–400 nm near the ultraviolet region, which were related to the electronic properties of the center metal atoms in the anions of the catalyst structure. This structure was similar to $[PW_{12}O_{40}]^{3-}$ [40]. The absorption peaks at 203, 196 and 191 nm of the three catalysts were caused by O→P transition, and the strong absorption peaks at 266, 266 and 267 nm of the three catalysts were considered to charge the transfer of metal atoms ($O^{2-} \rightarrow W^{6+}$), where W atoms were located in W-O_e-W intrabridges between edge-sharing WO_6 octahedra in the Keggin units.

By recording the TG curve of the synthesized catalysts, the thermal stability of the catalysts can be clearly displayed in Figure 1f. The first mass loss occurred at 100 °C, which was caused by the disappearance of physical water and crystallization water in the catalyst. With the increase in temperature, the weight loss within the range of 300 to 800 °C was related to the decomposition of catalysts. The $[C_n(MIM)_2]^{2+}$ cation was decomposed first; the initial decomposition temperature of the three POM-DIL catalysts was 495 °C for $[C_2(MIM)_2]PW_{12}O_{40}$, 420 °C for $[C_4(MIM)_2]PW_{12}O_{40}$ and 427 °C for $[C_6(MIM)_2]PW_{12}O_{40}$. Then, the $PW_{12}O_{40}^{3-}$ anion was decomposed at about 600 °C [25]. From the results, it can be concluded that the reasons for the high thermal stability of POM-DIL catalysts were not only the cation symmetric structure of the double imidazole ring but also the Keggin structure of the anion. At the same time, the carbon chain length of cations had no obvious effect on the thermal stability of the catalysts.

In order to further accurately determine the moisture in the samples, the content of free water in the samples was determined by the Karl Fischer Titrator (KFT) Ipol method with the Karl Fischer Moisture Titrator (870 KF Titrino plus), and the results are listed in Table 1. From the test results, the moisture content in the sample was extremely small, and the purity of each sample could reach 99%.

Table 1. The moisture in the samples.

Samples	First Determination (%)	Second Determination (%)	Average Value (%)
$[C_2(MIM)_2]PW_{12}O_{40}$	0.43	0.47	0.45
$[C_4(MIM)_2]PW_{12}O_{40}$	0.83	1.32	1.08
$[C_6(MIM)_2]PW_{12}O_{40}$	0.19	0.19	0.19

X-ray photoelectron spectroscopy (XPS) characterization is a powerful technique to determine the composition, content and molecular structure of catalysts [41]. The composition of the catalyst $[C_2(MIM)_2]PW_{12}O_{40}$ was analyzed by XPS, as can be seen from the survey spectrum of the sample (Figure 2a), the catalyst was mainly composed of C, N, O, P and W elements. The contents of each element are listed in Table 2. The results showed that $[C_2(MIM)_2]PW_{12}O_{40}$ was mainly composed of C and O elements. At the same time, the higher O content indicated that there were abundant oxygen-containing groups in the catalyst, which was consistent with the characterization results of FT-IR. The XPS spectrum of C1s (Figure 2b) can be fitted into three peaks with binding energies of 285.3, 284.5 and 283.3 eV, respectively, which were attributed to C=N–C, C–C/C=C and C=N. In the XPS spectra of N 1s, O 1s and P 2s (Figure 2c–e), the corresponding binding energies were 400, 529.5 and 132.9 eV, which correspond to C-N, O^{2-} and P-O, respectively. The two peaks at 36.7 and 34.6 eV (Figure 3f) were attributed to W $4f_{5/2}$ and W $4f_{7/2}$. Compared with the existing literature [42,43], due to the electrostatic interaction between the cation and the anion, significant electron shift was observed in the W4f spectra. The results showed that the surface electron density of the catalysts was effectively enhanced, which was conducive to the formation of hydroxyl radicals, thus improving the overall ECODS performance.

Table 2. Element composition and content of $[C_2(MIM)_2]PW_{12}O_{40}$.

Elements	Atomic (%)
C	27.29
N	10.03
O	47.93
P	3.12
W	11.63

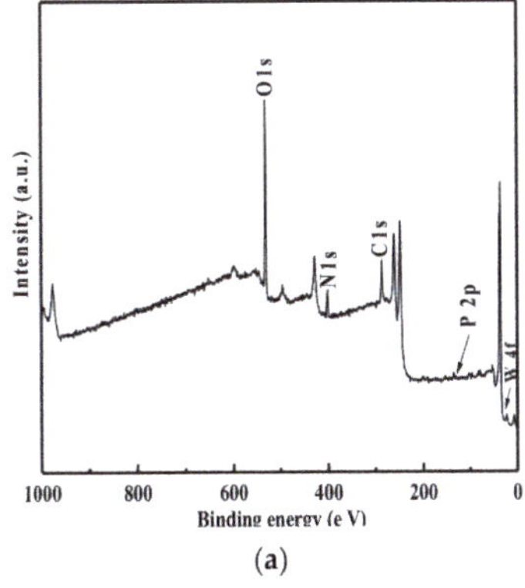

(a)

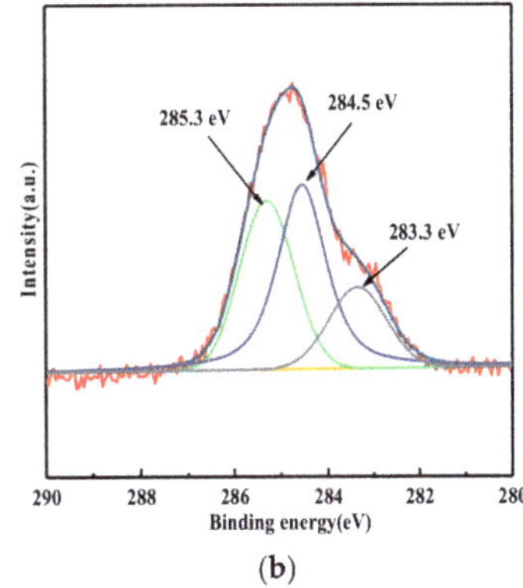

(b)

Figure 2. *Cont.*

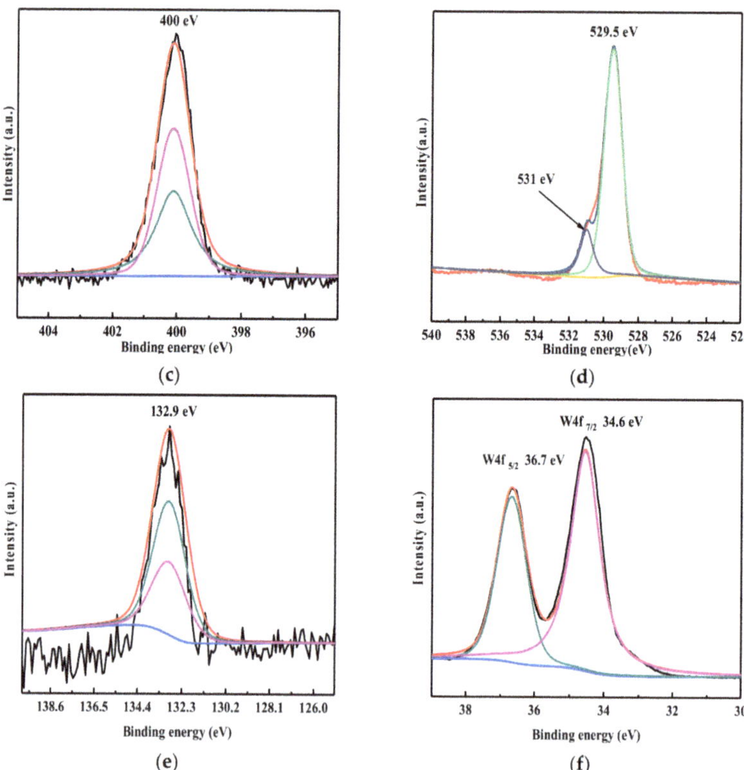

Figure 2. XPS spectra of [C$_2$(MIM)$_2$]PW$_{12}$O$_{40}$; (**a**) Survey of the catalyst; (**b**) C 1s; (**c**) N 1s; (**d**) O 1s; (**e**) P 2p; (**f**) W 4f.

Figure 3 shows the results of hydrophilicity and hydrophobicity tests of the catalysts. The instantaneous contact angles of a water droplet on the three POM-DIL catalysts were all less than 90°. The contact angle of [C$_2$(MIM)$_2$]PW$_{12}$O$_{40}$ was also measured with n-octane as the testing droplet. The results showed that the instantaneous contact angle between the oil droplet and the catalyst surface was almost 0. These results indicated that the catalysts have good wettability for both H$_2$O$_2$ and n-octane, which can effectively improve the utilization rate of the oxidant and the overall desulfurization efficiency.

2.2. Catalytic Activity of Catalyst

Table 3 summarizes and compares the effects of different kinds of POM-IL catalysts on the DBT removal effect in fuel oil under major reaction parameters, such as the H$_2$O$_2$/DBT molar ratio (n(H$_2$O$_2$)/n(S)), reaction temperature and reaction time. The results showed that increasing the length of the carbon chain in catalyst cation has no obvious effect on the desulfurization effect under the same reaction conditions. However, when the volume of cation increased, the desulfurization effect was obviously improved. On one hand, the catalyst had good wettability in the oil phase and H$_2$O$_2$, which can rapidly interaction with H$_2$O$_2$ and oil. When the catalyst was fully contacted with the oxidant, it could decompose more active substances [44], which was more conducive to the removal of sulfide. On the other hand, the large cations had higher aromatic π electron density, which could effectively enhance the π-π interaction between the double imidazole ring and the thiophene ring, thus making POM-DILs have excellent desulfurization performance. Considering the economic factors and desulfurization effect, [C$_2$(MIM)$_2$]PW$_{12}$O$_{40}$ was selected for the subsequent desulfurization research.

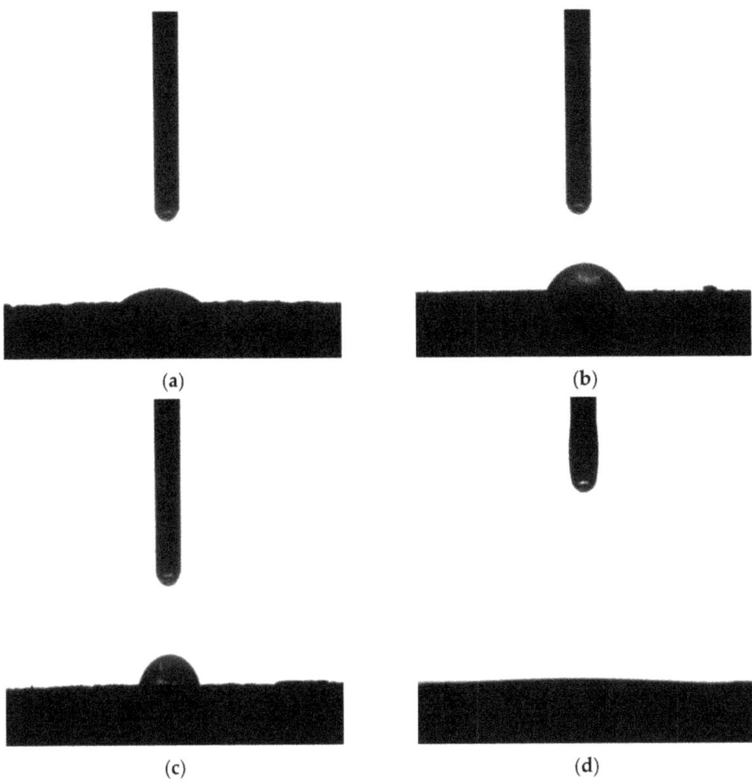

Figure 3. Contact angles of water droplets on the surface of the catalyst: (**a**) $[C_2(MIM)_2]PW_{12}O_{40}$; (**b**) $[C_4(MIM)_2]PW_{12}O_{40}$ and (**c**) $[C_6(MIM)_2]PW_{12}O_{40}$. Contact angles of n-octane droplets on the surface of the catalyst: (**d**) $[C_2(MIM)_2]PW_{12}O_{40}$.

Table 3. Comparison of different catalysts for removal of dibenzothiophene (DBT) in model fuel.

Catalyst	Reaction Conditions	S-Removal (%)	Ref
$[C_2(MIM)_2]PW_{12}O_{40}$	n(catalyst)/n(S) = 0.025; n(H$_2$O$_2$)/n(S) = 6; 50 °C; 60 min; acetonitrile = 0.5 mL; V(model oil) = 5 mL	98.4	This work
$[C_4(MIM)_2]PW_{12}O_{40}$	n(catalyst)/n(S) = 0.025; n(H$_2$O$_2$)/n(S) = 6; 50 °C; 60 min; acetonitrile = 0.5 mL; V(model oil) = 5 mL	97.0	This work
$[C_6(MIM)_2]PW_{12}O_{40}$	n(catalyst)/n(S) = 0.025; n(H$_2$O$_2$)/n(S) = 6; 50 °C; 60 min; acetonitrile = 0.5 mL; V(model oil) = 5 mL	95.5	This work
$[C_4mim]_3PW_{12}O_{40}$	m(catalyst) = 0.03 g, 60 °C, 30 min, n(H$_2$O$_2$)/n(DBT) = 3	11.4	[45]
$[C_8mim]_3PW_{12}O_{40}$	m(catalyst) = 0.03 g, 60 °C, 30 min, n(H$_2$O$_2$)/n(DBT) = 3	10.3	[45]
$[C_{16}mim]_3PW_{12}O_{40}$	m(catalyst) = 0.03 g, 60 °C, 30 min, n(H$_2$O$_2$)/n(DBT) = 3	12.5	[45]
$Cs_{2.5}H_{0.5}PW_{12}O_{40}$	60 min; 60 °C; O/S = 15; acetonitrile = 60 mL	70.5	[46]
$[C_{16}mim]_3PW_{12}O_{40}$	m(catalyst) = 0.01 g; 60 °C; 1 h; O/S = 3; [Bmim]BF$_4$ = 1 mL	21.4	[47]

The solubility of catalysts in different solvents is an important factor to be considered in their application. Therefore, the solubility of the catalyst in model oil and acetonitrile was tested. According to the results in Table 4, the catalysts were slightly dissolved in n-octane and acetonitrile, and tended to be dissolved in acetonitrile with relatively strong polarity. Combined with the desulfurization effect in Table 3, the partial dissolution of the catalyst in solvent had little effect on desulfurization efficiency and oil quality, which can be almost ignored.

Table 4. Solubility of catalyst in n-octane and acetonitrile.

Catalyst	Solvent	Solubility (g/100 g)	Solvent	Solubility (g/100 g)
$[C_2(MIM)_2]PW_{12}O_{40}$	n-octane	0.0276	Acetonitrile	0.0325
$[C_4(MIM)_2]PW_{12}O_{40}$	n-octane	0.0193	Acetonitrile	0.0275
$[C_6(MIM)_2]PW_{12}O_{40}$	n-octane	0.0166	Acetonitrile	0.0250

Condition: m(catalyst) = 0.005 g; V(solvent) = 5 mL; T=50 °C; t = 60 min.

After the ECODS system was established, the initial reaction conditions were optimized to ensure the best desulfurization efficiency. The effect of reaction temperature on the removal of sulfide DBT from model oil was investigated, and the results are displayed in Figure 4. When the temperature increased from 20 to 80 °C, the desulfurization efficiency showed a trend of increasing first and then decreasing. This was because the oxidation reaction was limited by kinetics and POM-DILs could not effectively catalyze desulfurization at low temperature [48]. With the increase in reaction temperature, the activity of the catalyst and oxidant was gradually enhanced, and the oxidation rate of DBT into $DBTO_2$ was accelerated. However, with the further increase in reaction temperature, the desulfurization efficiency would decrease slightly due to the gradual decomposition of H_2O_2 and the deactivation of active components [16,49]. Therefore, the best reaction temperature was determined to be 50 °C.

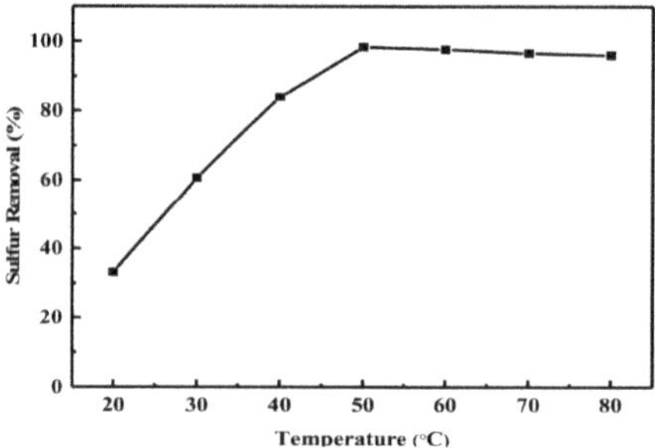

Figure 4. Effect of reaction temperature on desulfurization. Reaction conditions: V(oil) = 5 mL; n(catalyst)/n(S) = 0.025; $n(H_2O_2)/n(S)$ = 6; V(acetonitrile) = 0.5 mL; t = 60 min.

It can be seen from Figure 5, the desulfurization efficiency increased greatly in the initial stage of the reaction. After the reaction for 60 min, the desulfurization efficiency reached the maximum. This may be because the sulfide content in the model oil was highest in the initial reaction stage, so both the extraction rate and oxidation rate were much higher than the other reaction stage. Then, with the increase in reaction time, the overall desulfurization reaction tended to equilibrium, and the desulfurization efficiency did not increase significantly, or even slightly decreased, which was due to the partial volatilization of n-octane in a longer reaction time. Therefore, taking into account the effect of reaction time, the t = 60 min was selected to be used in the rest of the experiments.

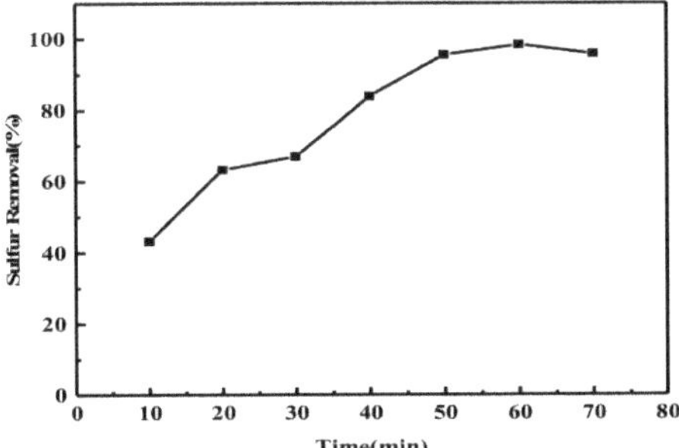

Figure 5. Effect of reaction time on desulfurization. Reaction conditions: V(oil) = 5 mL; n(catalyst)/n(S) = 0.025; n(H$_2$O$_2$)/n(S) = 6; V(acetonitrile) = 0.5 mL; T = 50 °C.

The influence of the catalyst amount on the desulfurization effect was also considered. As shown in Figure 6, without catalyst, the desulfurization effect was 67.11%. When the molar ratio of catalyst to sulfide was increased to 0.025, the desulfurization effect was increased to 98.35%. According to the reported literature [50], the high catalytic efficiency of H$_3$PW$_{12}$O$_{40}$ composites in oxidative desulfurization (ODS) was mainly due to the existence of catalytic active center W=O. Therefore, we speculated that with the increase in the amount of catalyst, the active sites provided for oxidative desulfurization increased, which promoted the effective removal of DBT. When the molar ratio of catalyst to sulfide was further increased, the desulfurization effect did not change obviously. Therefore, 0.025 was a suitable molar ratio of catalyst to sulfide.

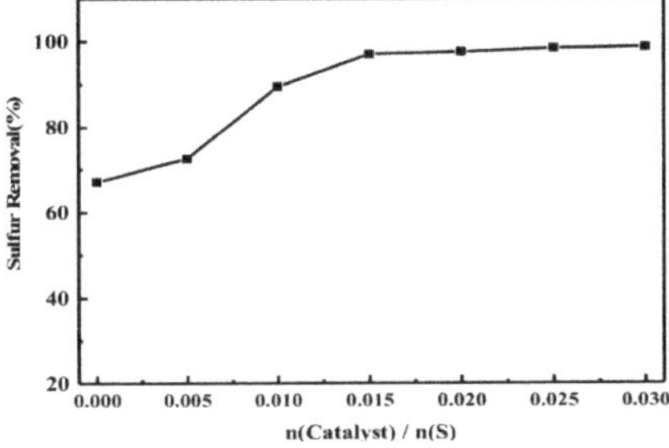

Figure 6. Effect of the mole ratio of catalyst to sulfide on desulfurization. Reaction conditions: V(oil) = 5 mL; n(H$_2$O$_2$)/n(S) = 6; V(acetonitrile) = 0.5 mL; t = 60 min; T=50 °C.

The effect of oxidant dosage on the removal of sulfide DBT from model oil was presented in Figure 7. When the desulfurization system was carried out in the absence of H$_2$O$_2$, the desulfurization efficiency was only about 67.48%. However, when the n(H$_2$O$_2$)/n(S) was increased from 0 to 4, the conversion of DBT was improved significantly. After further increasing the n(H$_2$O$_2$)/n(S) to 6, the removal of DBT could reach 98.35%. According to the stoichiometric reaction, the oxidation of 1 mol DBT to the corresponding sulfones requires 2 mol of H$_2$O$_2$ [51]. In theory, the excessive amount of H$_2$O$_2$ was beneficial to fully oxidize DBT to sulfones. In practice, considering the desulfurization efficiency and excessive oxidant may cause oil pollution, n(H$_2$O$_2$)/n(S) = 6 was appropriate.

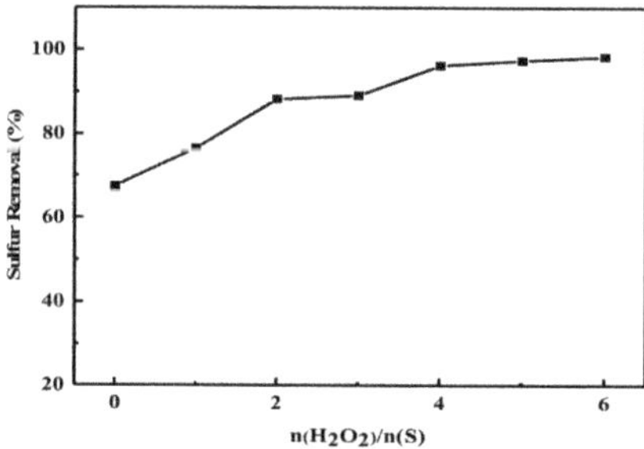

Figure 7. Effect of the mole ratio of oxidant to sulfide on desulfurization. Reaction conditions: V(oil) = 5 mL; n(catalyst)/n(S) = 0.025; V(acetonitrile) = 0.5 mL; t = 60 min; T = 50 °C.

The effect of the extractant dosage on DBT removal from model oil is shown in Figure 8. When the desulfurization test was conducted without acetonitrile as the extractant, the removal rate of DBT in the model oil reached 69.49%, which was the result of the combination of oxidant and catalyst. With the increase in the dosage of extractant, the desulfurization effect was significantly improved. This was because when the catalyst, oxidant and extractant were added to the reactor, an environment similar to the emulsion was formed. This environment could effectively increase the contact among the catalyst, oxidant and sulfide, so as to improve the desulfurization efficiency. When the amount of extractant became greater than 0.5 mL, the desulfurization efficiency was slightly improved. Therefore, 0.5 mL extractant was suitable for sulfide extraction.

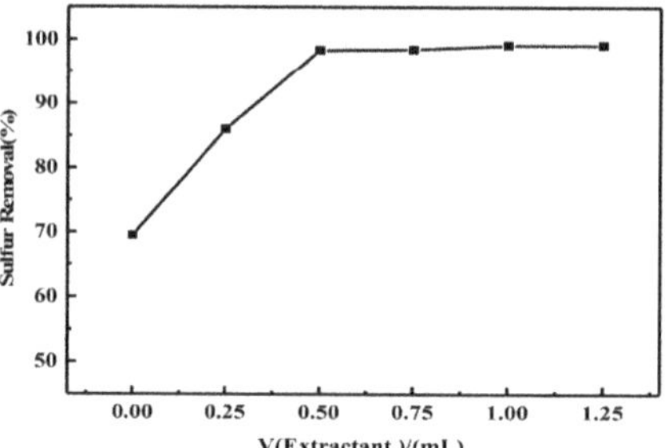

Figure 8. Effect of extractant dosage on desulfurization efficiency. Reaction conditions: V(oil) = 5 mL; n(catalyst)/n(S) = 0.025; n(H$_2$O$_2$)/n(S) = 6; t = 60 min; T = 50 °C.

Many ILs have been developed which are used as both the catalyst and extractant. Usually, the cation side of ILs influences the extraction ability of this material for DBT removal [38]. Therefore, the extraction ability of DILs was investigated. The results in Table 5 show that the DILs also exhibited a certain extraction ability. Due to the higher aromatic π-electron density of DILs, a stronger π-π interaction could be formed between the cations of DILs and aromatic sulfides, so DILs have a higher extraction efficiency than monocationic ionic liquids and ordinary organic solvents.

Table 5. Effect of different extractants on the removal of DBT.

Entry	Extractants	Sulfur Removal (%)
1	$[C_2(MIM)_2]Cl_2$	68.67
2	$[C_4(MIM)_2]Cl_2$	60.66
3	$[C_6(MIM)_2]Cl_2$	57.21
4	Acetonitrile	48.37
5	Methanol	31.68
6	$BMIMBF_4$	55.17
7	$BMIMPF_6$	56.86

Reaction conditions: V(oil) = 5 mL; V(extractant) = 0.5 mL; t = 60 min; T = 50 °C.

2.3. Recycling of Catalysts

The recyclability of the POM-DIL catalyst in the reaction system was further investigated from the perspective of economic cost. After the reaction, the upper oil phase was taken for analysis. The solvent in the system was separated by the decanting method, and the catalyst was dried at 100 °C to remove the residual H_2O_2, acetonitrile and model oil. Then, the catalyst was reused for the next cycle by the addition of fresh model oil, oxidant and extraction agent. Each group of data was repeated at least three times, and the standard deviation was less than 1.2. As results show in Figure 9, the desulfurization rate was still up to 89.42% after seven cycles of the catalyst. Compared with the catalytic performance of the fresh catalyst, the desulfurization effect was only slightly reduced (<10%), indicating that the catalyst $[C_2(MIM)_2]PW_{12}O_{40}$ had good recycling performance. In addition, the compounds in the oil phase before and after desulfurization were tested by gas chromatography-flame ionization detection (GC-FID). It can be seen from the results in Figure 10 that the oxidation product $DBTO_2$ of DBT was detected in the oil phase after the reaction, which indicated that some oxidation products were still in the oil phase. It could be inferred from the results that with the increase in the number of cycles, the decrease in desulfurization efficiency might be partly due to the presence of oxidation product in the oil phase, which resulted in the blocking of mass transfer in the reaction process. Figure 11 shows the infrared spectrum analysis of the recycled catalyst and fresh catalyst. After recycled tests, the structure of the catalyst was not destroyed, indicating that the catalyst has excellent stability.

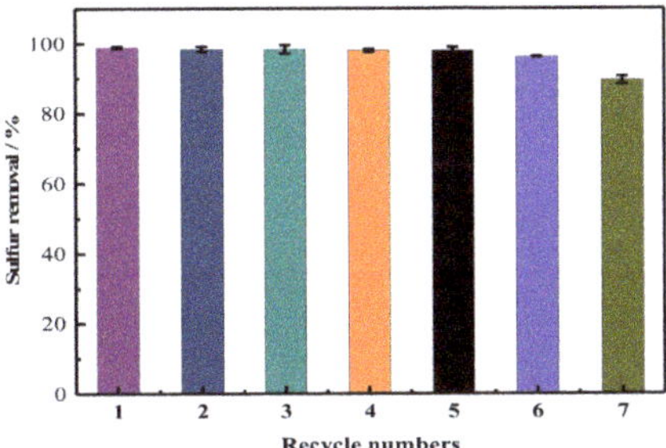

Figure 9. Desulfurization efficiency of recycling system. Reaction conditions: V(oil) = 5 mL; n(catalyst)/n(S) = 0.025; n(H_2O_2)/n(S) = 6; t = 60 min; T = 50 °C; V(acetonitrile) = 0.5 mL.

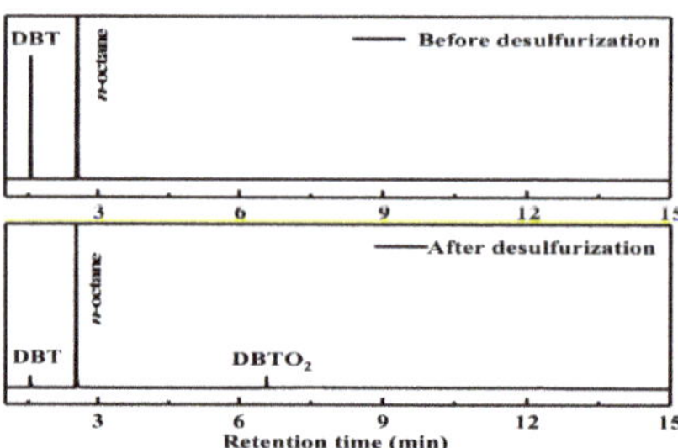

Figure 10. GC-FID of the model oil containing DBT before and after desulfurization in the proposed system.

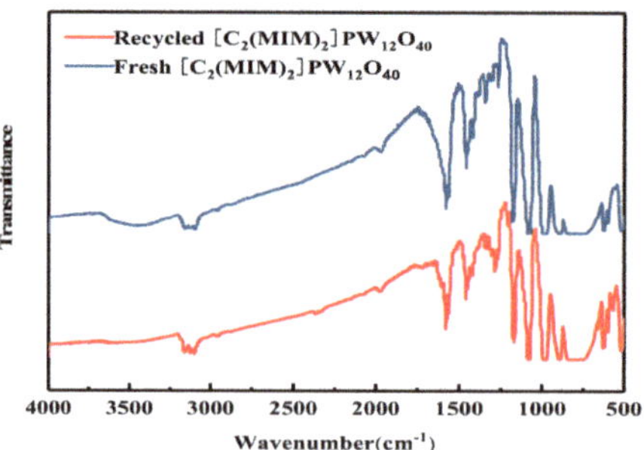

Figure 11. FT-IR spectra of fresh and recycled $[C_2(MIM)_2]PW_{12}O_{40}$.

2.4. Oxidation Product and Reaction Mechanism

The oxidation product of DBT was further verified by Gas Chromatography-Mass Spectrometry (GC-MS), and the results are shown in Figure 12. From the analysis results, it can be concluded that the oxidation product of DBT in model oil was $DBTO_2$ ($m/z = 216.0$). Based on our research and the related literature reports [52], we hypothesized the reaction mechanism of ECODS, as shown in Scheme 1. It was assumed that DBT was first extracted into the extraction phase by acetonitrile and POM-DILs with the extraction function. In the process of catalytic oxidation, $[PW_{12}O_{40}]^{3-}$ in the catalyst was oxidized by H_2O_2 to the intermediate product $[PO_4\{W(O)(O_2)_2\}_4]^{3-}$ with strong oxidation, and H_2O_2 was activated to form $\cdot OH$ (the catalytically active O species). Then, by a redox reaction between DBT and the intermediate product and hydroxyl radical, DBT was selectively oxidized to $DBTO_2$, and $[PO_4\{W(O)(O_2)_2\}_4]^{3-}$ was reduced to the original $[PW_{12}O_{40}]^{3-}$. As the reaction proceeded, DBT was continuously oxidized and extracted, and the content of DBT in the oil phase was continuously reduced, while the oxidation product $DBTO_2$ was continuously accumulated in the extraction phase until the end of the reaction.

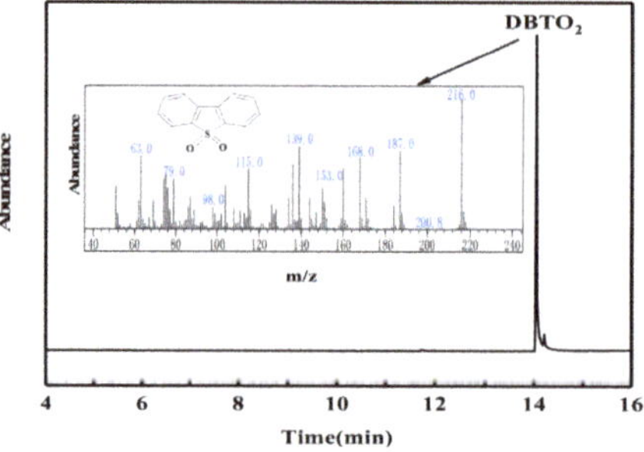

Figure 12. Gas Chromatography-Mass Spectrometry (GC-MS) analysis of catalyst after reaction.

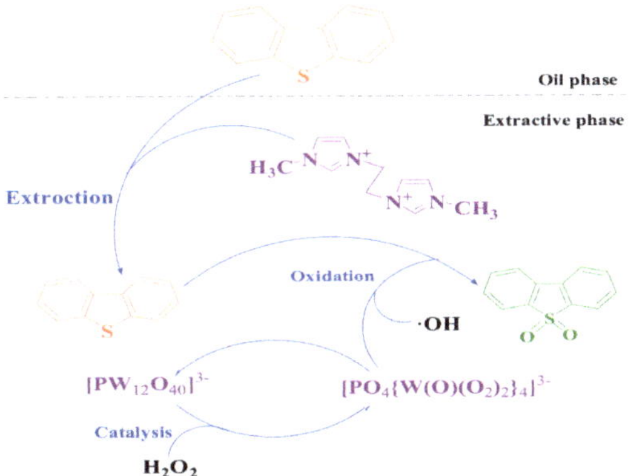

Scheme 1. Proposed mechanism for the oxidation of DBT in the ECODS.

3. Materials and Methods

3.1. Materials

1-Methylimidazole, 1,2-dichloroethane, 1,4-dichlorobutane, 1,6-dichlorohexane, DBT and $H_3PW_{12}O_{40}$ were purchased from MACKLIN (Shanghai, China). Acetone, H_2O_2 (30%), acetonitrile, ethanol and n-octane were purchased from Sinopharm Chemical Reagent Co., Ltd. (Shanghai, China). All the reagents and chemicals were directly used in experiments without any purification.

3.2. Synthesis of Catalyst

POM-DIL catalysts were prepared by a two-step method. Take $[C_2(MIM)_2]PW_{12}O_{40}$ as an example, and the synthesis steps are shown in Scheme 2. 1-methylimidazole (3.284 g, 0.04 mol) and 1,2-dichloroethane (1.9792 g, 0.02 mol) were charged into a 100 mL round-bottomed flask with a condensation reflux device. Under solvent-free conditions, the reaction mixture was stirred at 90 °C for 2–4 h until white solid appeared. The white solid intermediate product $[C_2(MIM)_2]Cl_2$ was washed repeatedly with acetone to remove nonionic residues and dried at 80 °C in a vacuum for 6 h. Then, $[C_2(MIM)_2]Cl_2$ (0.3946 g, 1.5 mmol) and $H_3PW_{12}O_{40}$ (2.8800 g, 1.0 mmol) were dissolved in 30 mL distilled water and stirred for 2 h at room temperature. After centrifugation and drying, the final catalyst $[C_2(MIM)_2]PW_{12}O_{40}$ was obtained. Preparation of $[C_4(MIM)_2]PW_{12}O_{40}$ and $[C_6(MIM)_2]PW_{12}O_{40}$ was similar to that of $[C_2(MIM)_2]PW_{12}O_{40}$. The results of ^1H nuclear magnetic resonance (^1H NMR) spectroscopy (400 MHz, DMSO) characterization were as follows: $[C_2(MIM)_2]PW_{12}O_{40}$: δ 8.98 (s, 2H), 7.72 (s, 2H), 7.57 (s, 2H), 4.65 (s, 4H), 3.86 (s, 6H), 3.37 (s, 6H), 2.52 (s, 2H). $[C_4(MIM)_2]PW_{12}O_{40}$: δ 9.07 (s, 2H), 7.72 (s, 2H), 4.20 (s, 4H), 3.86 (s, 6H), 3.37 (s, 6H), 2.50 (s, 2H). 1.79 (s, 2H). $[C_6(MIM)_2]PW_{12}O_{40}$: δ 9.07 (s, 2H), 7.72 (s, 2H), 4.20 (s, 4H), 3.86 (s, 6H), 3.48 (s, 6H), 2.52 (s, 4H), 1.81 (s, 2H), 1.30 (s, 2H).

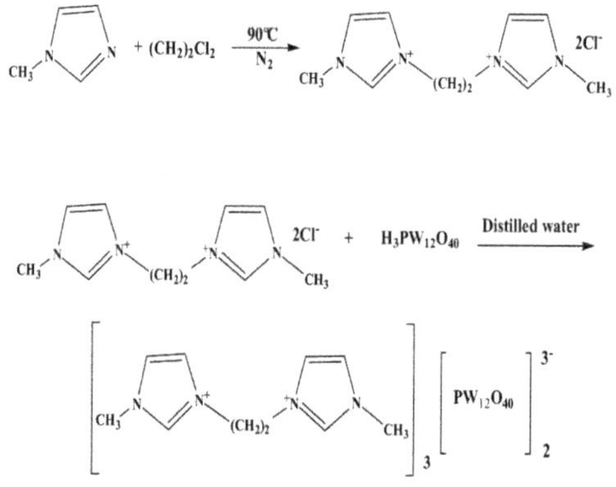

Scheme 2. Synthetic route of $[C_2(MIM)_2]PW_{12}O_{40}$.

3.3. Characterization

^{1}H NMR spectra were obtained on BRUKER AVANCE 400 (Karlsruhe, Germany) using dimethyl sulfoxide (DMSO) as the solvent. Fourier transform infrared spectroscopy (FT-IR) analyses were performed on a Nicolet 6700 FT-IR spectrometer (Thermo Fisher, Waltham, MA, USA) using KBr pellets at room temperature. XRD was performed on an Empyrean X-ray diffractometer (PANalytical B.V., Almelo, The Netherlands) equipped with Cu-Kα source. The scan speed and step size were 5°/min and 0.02°, respectively. UV-vis spectra were obtained with a UV-vis spectrometer (UVmini-1280, Shimadzu, Suzhou, China) in acetonitrile. TG analyses were carried out on Microcomputer differential thermal balance HCT-3 instrument (Beijing Hengjiu Scientific Instrument Factory, Beijing, China) from 35 to 800 °C, with a heating rate of 10 °C /min in N_2 atmosphere. The moisture content was determined on a Karl Fischer Moisture Titrator (870 KF Titrino plus, Heirishau, Switzerland). The main components of KF reagent were I_2, SO_2, pyridine (buffer) and methanol (solvent). XPS was carried out on ESCALAB250xi (Thermo Scientifle, Waltham, MA, USA) with a monochromatic Mg-Kα source with 1487 eV of energy to explore the surface composition. The contact angle tests were conducted on a contact angle instrument (JC2000D, Shanghai Zhongchen Digital Technic Apparatus Co. Ltd., Shanghai, China). The oxidation product of DBT in the model oil was measured by GC-MS (Agilent 5975C, Santa Clara, CA, USA) and the signals were collected from 4 to 16 min.

3.4. Oxidative Desulfurization Process

Model oil containing specific sulfide was selected for the test. In this experiment, DBT (500 mg/L) was used as model oil substrate and n-octane as solvent to form model oil. Firstly, a certain amount of catalyst, H_2O_2 (30%) and acetonitrile were added to a 100 mL round bottomed flask containing 5 mL of model oil. Then, the mixture was magnetically stirred for a period of time in a thermostatic water bath. After the reaction, the mixture precipitated for a certain time, the upper oil phase was taken and the sulfide content was determined by GC-FID (VF-1column type; 30 m × 0.25 mm × 0.25 µm; column temperature: 230 °C; the temperature was raised from 100 to 230 °C at the rate of 20 °C /min and kept for 2 min; injection temperature: 300 °C; detector temperature: 320 °C). Each group of data was repeated at least three times. According to the initial and final sulfur content in the model oil, the desulfurization efficiency can be calculated as follows: S-removal efficiency (%) = $(1 - S_1/S_0) \times 100\%$, where S_0 is the initial sulfur content in model oil (mg/L); S_1 is the final sulfur content in the model oil (mg/L).

4. Conclusions

In this study, three kinds of imidazole-based POM-DIL catalysts were synthesized by a two-step method and applied to the removal of DBT from model oil with acetonitrile as the extractant and H_2O_2 as the oxidant. Compared with the previous POM-IL desulfurization system, the experimental conditions of the present work were greatly optimized. Under the optimum reaction conditions—n([C_2(MIM)_2]PW_{12}O_{40})/n(S) = 0.025; n(H_2O_2)/n(S) = 6; V(acetonitrile) = 0.5 mL; T = 50 °C; t = 60 min—the removal rate of DBT reached above 98%. Meanwhile, $[C_2(MIM)_2]PW_{12}O_{40}$ could be reused seven times without obvious catalytic activity loss, and the original Keggin structure of the POM-DIL catalyst was undamaged. The results showed that the catalyst displayed good catalytic activity, stability and recycling performance. This is because the double cation can effectively regulate the interaction between the catalyst and sulfide, oxidant, which has a great impact on enhancing its desulfurization performance. In addition, increasing the carbon chain lengths of the catalysts with short carbon chain displayed no significant effect on the desulfurization efficiency due to the good wettability for both H_2O_2 and model oil. Finally, the oxidation product of DBT was proved to be $DBTO_2$ by GC-MS. Detailed analysis of the ECODS mechanism found that the W=O in the catalyst was the active center to activate H_2O_2 to generate ·OH, which was beneficial to desulfurization process. This study provided a new reference for the development of an efficient catalyst for catalytic oxidative desulfurization.

Author Contributions: Conceptualization, B.H. and J.T.; methodology, B.H.; software, J.L.; validation, B.H.; formal analysis, J.L.; investigation, J.L. and Y.G.; resources, J.L.; data curation, J.L.; writing—original draft preparation, J.L.; writing—review and editing, B.H. and J.T.; visualization, B.H.; supervision, B.H.; project administration, B.H.; funding acquisition, B.H. All authors have read and agreed to the published version of the manuscript.

Funding: This research was funded by Hubei Natural Science Foundation Project (2013CKB032) and The Doctoral Foundation Project of Hubei University of Technology (0701).

Data Availability Statement: The data presented in this study are available in article.

Acknowledgments: We would like to thank the teachers of the School of Materials and Chemical Engineering for their support in the inspection equipment used for experiments.

Conflicts of Interest: The authors declare no conflict of interest.

References

1. Zhou, M.; Wang, D. Generational differences in attitudes towards car, car ownership and car use in Beijing. *Transp. Res. D Transp. Environ.* **2019**, *72*, 261–278. [CrossRef]
2. Summers, J.C.; Baron, K. The effects of SO_2 on the performance of noble metal catalysts in automobile exhaust. *J. Catal.* **1979**, *57*, 380–389. [CrossRef]
3. Srivastava, V.C. An evaluation of desulfurization technologies for sulfur removal from liquid fuels. *RSC Adv.* **2012**, *2*, 759–783. [CrossRef]
4. Yun, J.; Zhu, C.; Wang, Q.; Hu, Q.L.; Yang, G. Strong affinity of mineral dusts for sulfur dioxide and catalytic mechanisms towards acid rain formation. *Catal. Commun.* **2018**, *114*, 79–83. [CrossRef]
5. Song, C.; Ma, X. New design approaches to ultra-clean diesel fuels by deep desulfurization and deep dearomatization. *Appl. Catal. B Environ.* **2003**, *41*, 207–238. [CrossRef]
6. Song, C. An overview of new approaches to deep desulfurization for ultra-clean gasoline, diesel fuel and jet fuel. *Catal. Today* **2003**, *86*, 211–263. [CrossRef]
7. Stanislaus, A.; Marafi, A.; Rana, M.S. Recent advances in the science and technology of ultra low sulfur diesel (ULSD) production. *Catal. Today* **2010**, *153*, 1–68. [CrossRef]
8. Tanimu, A.; Alhooshani, K. Advanced hydrodesulfurization catalysts: A review of design and synthesis. *Energy Fuels* **2019**, *33*, 2810–2838. [CrossRef]
9. Yankov, D.S.V.; Shishkova, I.; Chavdarov, I.; Petkov, P.; Palichev, T. Opportunity to Produce Near Zero Sulphur Gasoline and Improve Refining Profitability by Combining FCC Feed Hydrotreatment and Gasoline Post Treatment. *Oil Gas Eur. Mag.* **2013**, *4*, 1–7.
10. Xi, Y.B.; Zhang, D.Q.; Chu, Y.; Gao, X.D. Development of RSDS-III Technology for Ultra-Low-Sulfur Gasoline Production. *China Pet. Process Petrochem. Technol.* **2015**, *17*, 46–49.
11. Gutiérrez, O.Y.; Singh, S.; Schachtl, E.; Kim, J.; Kondratieva, E.; Hein, J.; Lercher, J.A. Effects of the support on the performance and promotion of (Ni) MoS_2 catalysts for simultaneous hydrodenitrogenation and hydrodesulfurization. *ACS Catal.* **2014**, *4*, 1487–1499. [CrossRef]
12. Jin, M.; Guo, Z.; Lv, Z. Synthesis of Convertible {PO_4[WO_3⇌W(O)_2(O_2)]_4}-DMA16 in SBA-15 Nanochannels and Its Catalytic Oxidation Activity. *Catal. Lett.* **2019**, *149*, 2794–2806. [CrossRef]

13. Dizaji, A.K.; Mortaheb, H.R.; Mokhtarani, B. Extractive-catalytic oxidative desulfurization with graphene oxide-based heteropoly-acid catalysts: Investigation of affective parameters and kinetic modeling. *Catal. Lett.* **2019**, *149*, 259–271. [CrossRef]

14. Bertleff, B.; Goebel, R.; Claußnitzer, J.; Korth, W.; Skiborowski, M.; Wasserscheid, P.; Jess, A.; Albert, J. Investigations on catalyst stability and product isolation in the extractive oxidative desulfurization of fuels using polyoxometalates and molecular oxygen. *ChemCatChem* **2018**, *10*, 4602–4609. [CrossRef]

15. Liu, H.; Xu, H.H.; Hua, M.Q.; Chen, L.L.; Wei, Y.C.; Wang, C.; Wu, P.W.; Zhu, F.X.; Chu, X.Z.; Li, H.M.; et al. Extraction combined catalytic oxidation desulfurization of petcoke in ionic liquid under mild conditions. *Fuel* **2020**, *260*, 116200. [CrossRef]

16. Xun, S.H.; Yu, Z.D.; He, M.Q.; Wei, Y.C.; Li, X.W.; Zhang, M.; Zhu, W.S.; Li, H.M. Supported phosphotungstic-based ionic liquid as an heterogeneous catalyst used in the extractive coupled catalytic oxidative desulfurization in diesel. *Res. Chem. Intermediat.* **2019**, *45*, 1–20. [CrossRef]

17. Dizaji, A.K.; Mokhtarani, B.; Mortaheb, H.R. Deep and fast oxidative desulfurization of fuels using graphene oxide-based phosphotungstic acid catalysts. *Fuel* **2019**, *236*, 717–729. [CrossRef]

18. Abdalla, Z.E.A.; Li, B.S.; Asma, T. Direct synthesis of mesoporous $(C_{19}H_{42}N)_4H_3(PW_{11}O_{39})/SiO_2$ and its catalytic performance in oxidative desulfurization. *Colloid Surf. A.* **2009**, *341*, 86–92. [CrossRef]

19. Li, L.Y.; Lu, Y.Z.; Meng, H.; Li, C.X. Lipophilicity of amphiphilic phosphotungstates matters in catalytic oxidative desulfurization of oil by H_2O_2. *Fuel* **2019**, *253*, 802–810. [CrossRef]

20. Pham, X.N.; Van Doan, H. Activity and stability of amino-functionalized SBA-15 immobilized 12-tungstophosphoric acid in the oxidative desulfurization of a diesel fuel model with H_2O_2. *Chem. Eng. Commun.* **2019**, *206*, 1139–1151. [CrossRef]

21. Andevary, H.H.; Akbari, A.; Omidkhah, M. High efficient and selective oxidative desulfurization of diesel fuel using dual-function [Omim]FeCl$_4$ as catalyst/extractant. *Fuel Process Technol.* **2019**, *185*, 8–17. [CrossRef]

22. Leng, K.Y.; Sun, Y.Y.; Zhang, X.; Yu, M.; Xu, W. Ti-modified hierarchical mordenite as highly active catalyst for oxidative desulfurization of dibenzothiophene. *Fuel* **2016**, *174*, 9–16. [CrossRef]

23. Made, M.; Liu, J.F.; Pang, L. Environmental application, fate, effects, and concerns of ionic liquids: A review. *Environ. Sci. Technol.* **2015**, *49*, 12611–12627.

24. Li, S.W.; Gao, R.M.; Zhao, J.S. Deep oxidative desulfurization of fuel catalyzed by modified heteropolyacid: The comparison performance of three kinds of ionic liquids. *ACS Sustain. Chem. Eng.* **2018**, *6*, 15858–15866. [CrossRef]

25. Huang, W.L.; Zhu, W.S.; Li, H.M.; Shi, H.; Zhu, G.P.; Liu, H.; Chen, G.Y. Heteropolyanion-based ionic liquid for deep desulfurization of fuels in ionic liquids. *Ind. Eng. Chem. Res.* **2010**, *49*, 8998–9003. [CrossRef]

26. Li, J.L.; Hu, B.; Hu, C.Q. Deep desulfurization of fuels by heteropolyanion-based ionic liquid. *Bull. Korean Chem. Soc.* **2013**, *34*, 225–230. [CrossRef]

27. Zhuang, J.Z.; Hu, B.; Tan, J.J.; Jin, X.Y. Deep oxidative desulfurization of dibenzothiophene with molybdovanadophosphoric heteropolyacid-based catalysts. *Transit. Metal. Chem.* **2014**, *39*, 213–220. [CrossRef]

28. Rafiee, E.; Mirnezami, F. Keggin-structured polyoxometalate-based ionic liquid salts: Thermoregulated catalysts for rapid oxidation of sulfur-based compounds using H_2O_2 and extractive oxidation desulfurization of sulfur-containing model oil. *J. Mol. Liq.* **2014**, *199*, 156–161. [CrossRef]

29. Hao, L.W.; Sun, L.L.; Su, T.; Hao, D.M.; Liao, W.P.; Deng, C.L.; Ren, W.Z.; Zhang, Y.M.; Lü, H.Y. Polyoxometalate-based ionic liquid catalyst with unprecedented activity and selectivity for oxidative desulfurization of diesel in [Omim]BF$_4$. *Chem. Eng. J.* **2019**, *358*, 419–426. [CrossRef]

30. Akopyan, A.; Eseva, E.; Polikarpova, P.; Kedalo, A.; Vutolkina, A.; Glotov, A. Deep oxidative desulfurization of fuels in the presence of brönsted acidic polyoxometalate-based ionic liquids. *Molecules* **2020**, *25*, 536. [CrossRef] [PubMed]

31. Montalbán, M.G.; Villora, G.; Licence, P. Ecotoxicity assessment of dicationic versus monocationic ionic liquids as a more environmentally friendly alternative. *Ecotox. Environ. Saf.* **2018**, *150*, 129–135. [CrossRef] [PubMed]

32. Gindri, I.M.; Siddiqui, D.A.; Bhardwaj, P.; Rodriguez, L.C.; Palmer, K.L.; Frizzo, C.P.; Martins, M.A.P.; Rodrigues, D.C. Dicationic imidazolium-based ionic liquids: A new strategy for non-toxic and antimicrobial materials. *RSC Adv.* **2014**, *4*, 62594–62602. [CrossRef]

33. Li, D.; Kang, Y.; Li, J.; Wang, Z.; Yan, Z.; Sheng, K. Chemically tunable DILs: Physical properties and highly efficient capture of low-concentration SO_2. *Sep. Purif. Technol.* **2020**, *240*, 116572. [CrossRef]

34. Zhao, D.; Liu, M.; Zhang, J.; Li, J.; Ren, P. Synthesis, characterization, and properties of imidazole dicationic ionic liquids and their application in esterification. *Chem. Eng. J.* **2013**, *221*, 99–104. [CrossRef]

35. Ji, Y.A.; Hou, Y.C.; Ren, S.H.; Yao, C.F.; Wu, W.Z. Highly efficient separation of phenolic compounds from oil mixtures by imidazolium-based dicationic ionic liquids via forming deep eutectic solvents. *Energ. Fuel.* **2017**, *31*, 10274–10282. [CrossRef]

36. Zakharov, M.A.; Ivanov, A.S.; Arkhipova, E.A.; Desyatov, A.V.; Savilov, S.V.; Lunin, V.V. Structure and properties of new dicationic ionic liquid DBTMEDA(BF$_4$)$_2$. *Struct. Chem.* **2019**, *30*, 451–456. [CrossRef]

37. Mei, X.Y.; Yue, Z.; Ma, Q.; Dunya, H.; Mandal, B.K.; Dunya, H.; Mandal, B.K. Synthesis and electrochemical properties of new dicationic ionic liquids. *J. Mol. Liq.* **2018**, *272*, 1001–1018. [CrossRef]

38. Li, J.J.; Lei, X.J.; Tang, X.D.; Zhang, X.P.; Wang, Z.Y. Acid dicationic ionic liquids as extractants for extractive desulfurization. *Energy Fuels* **2019**, *33*, 4079–4088. [CrossRef]

39. Wang, T.; Yu, W.H.; Li, T.X.N.; Wang, Y.T.; Tan, J.J.; Hu, B.; Nie, L.H. Synthesis of novel magnetic ionic liquids as high efficiency catalysts for extraction-catalytic oxidative desulfurization in fuel oil. *New J. Chem.* **2019**, *43*, 19232–19241. [CrossRef]

40. Jalil, P.A.; Faiz, M.; Tabet, N.; Hamdan, N.M.; Hussain, Z. A study of the stability of tungstophosphoric acid, $H_3PW_{12}O_{40}$, using synchrotron XPS, XANES, hexane cracking, XRD, and IR spectroscopy. *J. Catal.* **2003**, *217*, 292–297. [CrossRef]

41. Kuhn, B.L.; Osmari, B.F.; Heinen, T.M.; Bonacorso, H.G.; Zanatta, N.; Nielsen, S.O.; Ranathunga, D.T.S.; Villetti, M.A.; Frizzo, C.P. Dicationic imidazolium-based dicarboxylate ionic liquids: Thermophysical properties and solubility. *J. Mol. Liq.* **2020**, *308*, 112983. [CrossRef]

42. Xun, S.H.; Zheng, D.; Yin, S.; Qin, Y.J.; Zhang, M.; Jiang, W.; Zhu, W.S.; Li, H.M. TiO_2 microspheres supported polyoxometalate-based ionic liquids induced catalytic oxidative deep-desulfurization. *RSC Adv.* **2016**, *6*, 42402–42412. [CrossRef]

43. Lu, G.; Li, X.Y.; Qu, Z.P.; Zhao, Q.D.; Li, H.; Shen, Y.; Chen, G.H. Correlations of WO_3 species and structure with the catalytic performance of the selective oxidation of cyclopentene to glutaraldehyde on WO_3/TiO_2 catalysts. *Chem. Eng. J.* **2010**, *159*, 242–246. [CrossRef]

44. Zhu, W.S.; Zhu, G.P.; Li, H.M.; Chao, Y.H.; Zhang, M.; Du, D.L.; Wang, Q.; Zhao, Z. Catalytic kinetics of oxidative desulfurization with surfactant-type polyoxometalate-based ionic liquids. *Fuel Process Technol.* **2013**, *106*, 70–76. [CrossRef]

45. Yu, Z.D.; Huang, X.Y.; Xun, S.H.; He, M.Q.; Zhu, L.H.; Wu, L.L.; Yuan, M.M.; Zhu, W.S.; Li, H.M. Synthesis of carbon nitride supported amphiphilic phosphotungstic acid based ionic liquid for deep oxidative desulfurization of fuels. *J. Mol. Liq.* **2020**, *308*, 113059. [CrossRef]

46. Li, M.; Zhang, M.; Wei, A.M.; Zhu, W.S.; Xun, S.H.; Li, Y.N.; Li, H.P.; Li, H.M. Facile synthesis of amphiphilic polyoxometalate-based ionic liquid supported silica induced efficient performance in oxidative desulfurization. *J. Mol. Catal. A Chem.* **2015**, *406*, 23–30. [CrossRef]

47. Wang, R.; Zhang, G.F.; Zhao, H.X. Polyoxometalate as effective catalyst for the deep desulfurization of diesel oil. *Catal. Today* **2010**, *149*, 117–121. [CrossRef]

48. Yang, H.W.; Jiang, B.; Sun, Y.L.; Zhang, L.H.; Sun, Z.N.; Wang, J.Y.; Tantai, X.W. Polymeric cation and isopolyanion ionic self-assembly: Novel thin-layer mesoporous catalyst for oxidative desulfurization. *Chem. Eng. J.* **2017**, *317*, 32–41. [CrossRef]

49. Zheng, H.Q.; Zeng, Y.N.; Chen, J.; Lin, R.G.; Zhuang, W.E.; Cao, R.; Lin, Z.J. Zr-Based Metal-Organic Frameworks with Intrinsic Peroxidase-Like Activity for Ultradeep Oxidative Desulfurization: Mechanism of H_2O_2 Decomposition. *Inorg. Chem.* **2019**, *58*, 6983–6992. [CrossRef]

50. Li, S.W.; Wang, W.; Zhao, J.S. Highly effective oxidative desulfurization with magnetic MOF supported W-MoO3 catalyst under oxygen as oxidant. *Appl. Catal. B Environ.* **2020**, *277*, 119224. [CrossRef]

51. Qi, Z.Y.; Huang, Z.X.; Wang, H.X.; Li, L.; Ye, C.S.; Qiu, T. In situ bridging encapsulation of a carboxyl-functionalized phospho-tungstic acid ionic liquid in UiO-66: A remarkable catalyst for oxidative desulfurization. *Chem. Eng. Sci.* **2020**, *225*, 115818. [CrossRef]

52. Akbari, A.; Chamack, M.; Omidkhah, M. Reverse microemulsion synthesis of polyoxometalate-based heterogeneous hybrid catalysts for oxidative desulfurization. *J. Mater. Sci.* **2020**, *55*, 6513–6524. [CrossRef]

catalysts

MDPI

Article

Valorization of Raw and Calcined Chicken Eggshell for Sulfur Dioxide and Hydrogen Sulfide Removal at Low Temperature

Waseem Ahmad [1], Sumathi Sethupathi [1,*], Yamuna Munusamy [1] and Ramesh Kanthasamy [2]

1   Faculty of Engineering and Green Technology, Universiti Tunku Abdul Rahman, Jalan Universiti,
    Bandar Barat, Kampar 31900, Perak, Malaysia; waseemssb@gmail.com (W.A.); yamunam@utar.edu.my (Y.M.)
2   Chemical and Materials Engineering Department, Faculty of Engineering Rabigh, King Abdulaziz University,
    P.O. Box 344, Rabigh 21911, Saudi Arabia; rsampo@kau.edu.sa
*   Correspondence: sumathi@utar.edu.my; Tel.: +60-54-688-888; Fax: +60-54-667-449

Abstract: Chicken eggshell (ES) is a waste from the food industry with a high calcium content produced in substantial quantity with very limited recycling. In this study, eco-friendly sorbents from raw ES and calcined ES were tested for sulfur dioxide (SO_2) and hydrogen sulfide (H_2S) removal. The raw ES was tested for SO_2 and H_2S adsorption at different particle size, with and without the ES membrane layer. Raw ES was then subjected to calcination at different temperatures (800 °C to 1100 °C) to produce calcium oxide. The effect of relative humidity and reaction temperature of the gases was also tested for raw and calcined ES. Characterization of the raw, calcinated and spent sorbents confirmed that calcined eggshell CES (900 °C) showed the best adsorption capacity for both SO_2 (3.53 mg/g) and H_2S (2.62 mg/g) gas. Moreover, in the presence of 40% of relative humidity in the inlet gas, the adsorption capacity of SO_2 and H_2S gases improved greatly to about 11.68 mg/g and 7.96 mg/g respectively. Characterization of the raw and spent sorbents confirmed that chemisorption plays an important role in the adsorption process for both pollutants. The results indicated that CES can be used as an alternative sorbent for SO_2 and H_2S removal.

Keywords: chicken eggshell; waste valorization; adsorption; biogas; flue gas

Citation: Ahmad, W.; Sethupathi, S.; Munusamy, Y.; Kanthasamy, R. Valorization of Raw and Calcined Chicken Eggshell for Sulfur Dioxide and Hydrogen Sulfide Removal at Low Temperature. *Catalysts* **2021**, *11*, 295. https://doi.org/10.3390/catal 11020295

Academic Editor: Daniela Barba

Received: 18 January 2021
Accepted: 19 February 2021
Published: 23 February 2021

Publisher's Note: MDPI stays neutral with regard to jurisdictional claims in published maps and institutional affiliations.

1. Introduction

Sulfur dioxide (SO_2) and hydrogen sulfide (H_2S) are both considered toxic gases. SO_2 is mainly part of the flue gases while H_2S is naturally present in many fossil fuels and quickly oxidizes to SO_2 upon burning. Direct release of these acidic gases to the open air can cause serious environmental repercussions [1,2]. SO_2 can be removed from flue gas in many ways and this process is named as flue gas desulfurization (FGD). The adsorption processes are already used and are well known for SO_2 removal as well as for H_2S removal. Common sorbents for SO_2 removal are mostly calcium-based oxide/hydroxide ($CaO/Ca(OH)_2$), zinc oxide-based (ZnO), sodium hydroxide based (NaOH) and ammonia-based [3]. Limestone, slaked lime or a mixture of slaked lime with fly ash is commercially used in FGD systems [4]. There is a lot of room for improvement in the traditional FGD technologies as they consume a large quantity of water and at times CO_2 leakage to the environment [5,6]. Therefore, recently, many alternative sorbents such as red mud and various modified carbonaceous catalysts, have been developed with the aim to reduce the cost, promote principles of circular economy, and improve energy efficiency [7,8].

The removal of H_2S, on the other hand, is considered a crucial step in the biogas industry because of its toxic and corrosive nature [9]. Effective utilization of biogas as biomethane is a challenge because of its costly purification steps [10]. Many technologies such as adsorption, alkaline washing (absorption), membrane separation, and cryogenic distillation have been tested to efficiently removes H_2S [11]. The most common type of sorbent used for adsorption process is impregnated activated carbon [10]. Activated carbon has been reported to have H_2S removal capacities in the range of 150–650 mg/g [12].

However, adsorption of SO_2 or H_2S using activated carbon generates secondary waste, which is acidic and difficult for landfilling. Apart from activated carbons, various types of waste-based sorbents derived from municipal waste sludge, fly ash, forestry, slaughter-house, etc. have been also tested for H_2S removal [10]. Nevertheless, these sorbents are either not re-generable or has a very low removal efficiency. Thus, recently researchers are focusing on developing new cost-effective and regenerative sorbents for H_2S removal.

Chicken eggshell (ES) is a waste product from the food industry and is mostly disposed in landfills in Malaysia. It contains about 90–95% of calcium carbonate in the form of calcite, 1% magnesium carbonate, 1% calcium phosphate and some organic compounds [13,14]. According to Food and Agriculture Organization (FAO) of the United Nations, approximately 70.4 million tons of chicken eggs were produced worldwide in 2015 and the production is estimated to increase to 90 million tons by 2030 [15]. In Malaysia, about 642,600 tonnes of chicken eggs are produced annually which produces approximately 70,686 tonnes of ES waste [16]. Considering the volume of ES waste produced, its reutilization is still very limited and a greater part of it is disposed in landfills. ES has been reported to be used occasionally as a soil conditioner, fertilizer, and additive for animal feed [17]. Recently in the literature, ES waste has been valorized in many innovative applications such as special materials for bone tissue restoration, as a sorbent for metal ions in wastewater treatment and also as a catalyst in different applications [18]. Recently, ES based sorbent was used for CO_2 adsorption and the removal capacity was reported as 10.47 mg/g at 1 bar and 30 °C [19]. Sethupathi et al. (2017) had also carried out a preliminary study on the SO_2 removal from the gaseous stream using CES (950 °C) and achieved maximum adsorption capacity of 2.15 mg/g [20]. One of the latest literatures reported a carbonized hybrid sorbent (ES and lignin) to remove SO_2 from the air [21]. In this study, the possibility of replacing conventional calcite-based sorbents with raw and calcined chicken eggshell for acidic gases removal at low temperature was appraised. Adsorption experiments were conducted in a lab-scale adsorption rig and SO_2 and H_2S were mixed with nitrogen gas with fixed concentration separately. Various characterization techniques like FTIR, XRD, EDX, and FESEM were used to further investigate the adsorption mechanism.

2. Results and Discussions

2.1. Characterization of Raw and Calcined Eggshell before and after Adsorption Tests

The morphology of ES sorbents was examined with Field Emission Scanning Electron Microscope (FESEM) images. Figure 1a shows the outer shell of ES and its FESEM images at different magnification. It can be seen that the outer shell of ES has a smooth surface with cracks. Figure 1b is the inner ES with the membrane. FESEM images clearly show fibrous network morphology of protein which is very porous in nature. Figure 1c shows the actual image of powdered (<90 μm) raw eggshell (RES) and FESEM images at different magnification. The porous nature of RES can be clearly seen with the pore hole like structures on the particles. Pore structures of ES sorbents are categorized as Type II as per Brunauer, Deming and Teller classification, stating their characteristics belong to macroporous material, nonporous materials, or materials with open voids.

Figure 1. Digital camera and FESEM images of raw eggshell. (**a**) Outer at 300× and 5000× magnification, (**b**) inner at 300× and 5000× magnification, and (**c**) particle size of <90 µm at 10,000× and 30,000× magnification.

Figure 2a,b show FESEM images of (CES 900 °C) and (CES 1100 °C). The images for 900 °C show a stable and structured particle compared to the one calcined at 1100 °C. (CES 900 °C) shows well-arranged particles with smooth surfaces on each particle. (CES 1100 °C) was totally the opposite, particles lose their shapes, and each particle shows intensive surface cracks due to the sintering process.

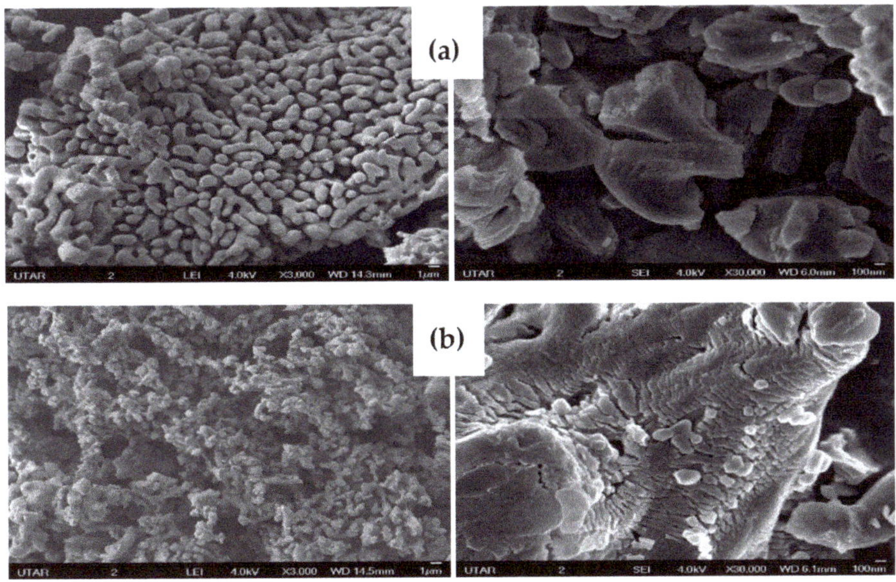

Figure 2. FESEM images of calcined eggshell at (**a**) 900 °C and (**b**) 1100 °C with 3000× and 30,000× magnification.

BET surface area values of the CES sorbents in comparison to RES are shown in Table 1. BET surface area of RES was reported as 0.56 m^2/g and the readings were increasing as RES was calcined. However, the values decrease when the temperature was further increased up to 1000 °C and 1100 °C. The highest BET surface area of 6.74 m^2/g was recorded at 900 °C. The BET surface area of CES was low compared to commercial-grade CaO whose BET surface area is in the range of 11–25 m^2/g [22].

Table 1. BET surface area of calcined eggshell at different temperature.

Calcination Temperature (°C)	BET Surface Area (m^2/g)
800	2.98
900	6.74
950	6.54
1000	6.30
1100	2.68

Figure 3 shows the nitrogen-adsorption isotherm for CES. The isotherm is of Type IV. As per IUPAC standard, this kind of isotherm is obtained for a combination of microporous and mesoporous structure which is formed at a higher relative pressure. The low BET surface area of CES could be due to eggshell structure and impurities.

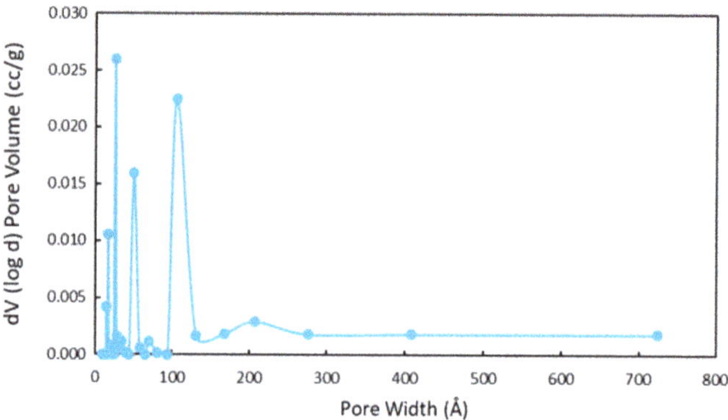

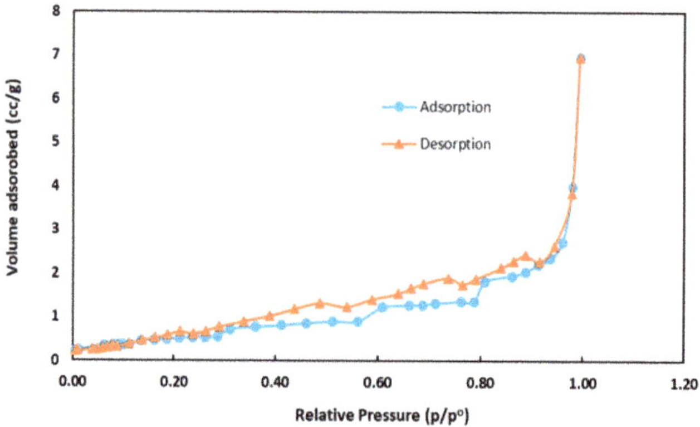

Figure 3. Pore size distribution and N2 adsorption-desorption isotherm of (CES 900 °C).

The elemental content of RES, CES, and their respective spent sorbents are shown in Table 2. The presence of the sulfur element in the spent RES and CES affirms the adsorption of SO_2 and H_2S and the occurrence of chemisorption.

Table 2. The differences in elemental content of raw eggshell and (CES 900 °C) before and after adsorption.

Sample	EDX Figure	Elements	Percentage
RES		Ca	16.85
		C	31.80
		O	49.13
		Others	2.22
RES Spent SO_2		Ca	12.08
		C	27.68
		O	57.84
		S	0.55
		Others	1.85
RES Spent H_2S		Ca	11.35
		C	56.25
		O	30.95
		S	0.11
		Others	1.34
(CES 900 °C)		Ca	24.93
		C	35.87
		O	38.23
		Others	0.47
(CES 900 °C) Spent SO_2		Ca	14.65
		C	41.33
		O	42.4
		S	0.65
		Others	1.06
(CES 900 °C) Spent H_2S		Ca	16.05
		C	36.8
		O	46.25
		S	0.21
		Others	0.69

The Ca content in the CES is more compared to the one in raw eggshell (RES) because, during the calcination, CO_2 and other volatile matters are released. It is also evident that the contents of other impurities in ES such as Zn, Mg, Al, and Cu remain almost the same in RES, CES and spent sorbents. This shows that these impurities were not involved in the sorption process. In the spent adsorbent, the content of Ca was reduced. This indicated the conversion of Ca into sulfite complex.

FTIR spectra of RES, (CES 900 °C) and the spent sorbents are presented in Figure 4. The wideband approximately at 3430 cm^{-1} in the RES is attributed to the stretching of the OH bond [23]. The two well-defined bands at 1413 cm^{-1} and 874 cm^{-1} are distinctive to the bending of C–O bond of $CaCO_3$ while the band at 712 cm^{-1} is related to Ca–O bond [24]. These indicate that RES comprises of calcite [25]. For (CES 900 °C), the well-defined band at approximately 3630 cm^{-1} corresponds to the vibration of OH bonds probably attached to the surface of CaO [24]. The peak at 1413 cm^{-1} is sharper in (CES 900 °C) showing a higher percentage of CaO and more prevailing than RES. New peaks were not detected in the spent RES sorbents. Nevertheless, there were changes in the intensity of the peaks. This could be due to the contact of the acidic gases. However, a new peak at 1080 cm^{-1} was visible for (CES 900 °C) spent sorbents. This affirms the presence of sulfite [26].

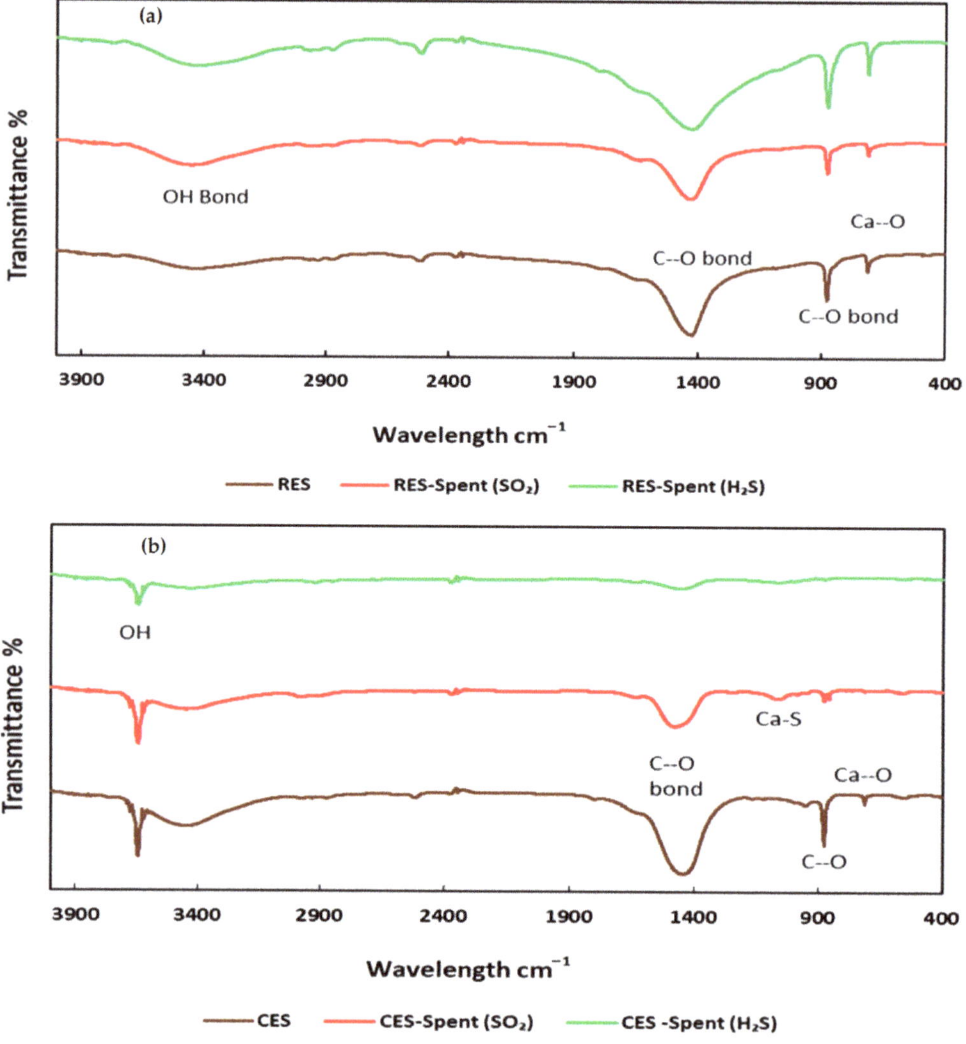

Figure 4. FTIR spectra of (**a**) raw eggshell and (**b**) calcined eggshell (900 °C) before and after adsorption.

The proximate analyses of RES and CES are listed in Table 3. The moisture content and volatile matter of CES were much lesser than RES due to the high-temperature calcination process. The residue for CES was far greater than RES which shows the stability of CES at high temperature and amount of CaO produced.

Table 3. Proximate analysis of raw and calcined eggshell (900 °C).

Temperature (°C)	Proximate Analysis	RES (%)	CES (%)
25–120	Moisture	1.03	0.28
120–450	Volatile content	4.18	8.01
450–800	CO_2	43.17	1.63
800–900	Residue (CaO)	51.62	90.08

Figure 5 show X-ray diffraction (XRD) of RES, (CES 900 °C) and the spent sorbents. RES showed a major peak at $2\theta = 29.5°$ which indicates that $CaCO_3$ is a major constituent of the waste ES. In the (CES 900 °C), regular peaks were obtained at $2\theta = 32°, 34°, 37.5°$, and $54°$, showing the conversion of $CaCO_3$ to CaO [23]. It is noted that peak at $2\theta = 29.5°$ is no longer visible in the (CES 900 °C), which implies a complete conversion of $CaCO_3$ to CaO.

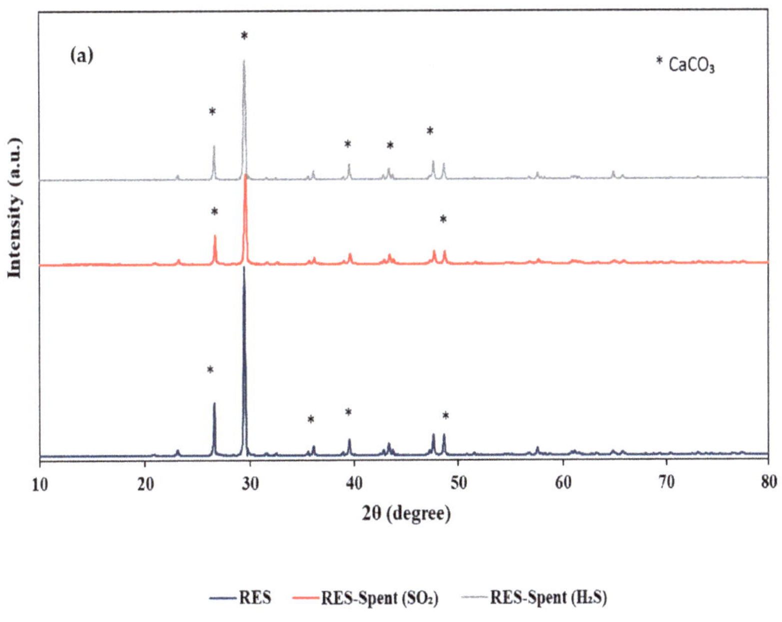

Figure 5. *Cont.*

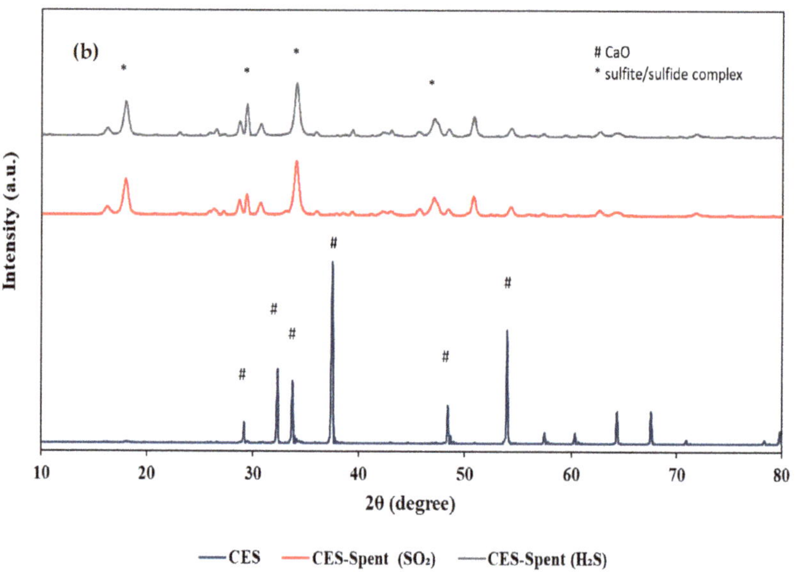

Figure 5. XRD pattern of (**a**) raw eggshell and (**b**) calcined eggshell (900 °C) before and after adsorption.

For the spent RES, it can be seen that there was no significant difference in the crystalline structure after the adsorption of SO_2 and H_2S. It shows no chemical interaction between the sorbent and the gases. Thus, it can be concluded that for RES the adsorption was merely physical. However, for spent (CES 900 °C), the initial peaks at $2\theta = 32°$, $34°$, $37.5°$, and $54°$ have disappeared and new peaks were formed at $2\theta = 17°$ and $29°$. The peak at $34°$ in the (CES 900 °C) corresponds to CaO. This peak reduced in the spent sorbent, indicating the presence of unreacted CaO in the spent sorbent. The small peaks at $2\theta = 17°$, $28°$, $28°$, and $2\theta = 34°$, $47°$, $52°$, and $54°$ may correspond to $CaSO_3$ and $Ca(OH)_2$ respectively for the spent (CES 900 °C). These spread-out peaks show that the crystallinity of (CES 900 °C) after adsorption has dropped. Similar peaks were reported by others as well [27].

pH values of the RES and CES before and after adsorption are tabulated in Tables 4 and 5 to show the reactivity of the acidic gases on RES and CES. There was a clear increase in pH with the increase in calcination temperature because of the formation of CaO which is basic in nature. Whereas for RES, the sorbents with membrane have slightly higher pH compared to the one without membrane. This is due to the different types of protein in the membrane. It was noticed that for all cases, spent sorbent's pH values dropped one level, indicating successful adsorption of acidic gases.

Table 4. Adsorption capacity and saturation time of raw eggshell with and without membrane at different particle size.

Sample	Particle Size (μm)	Saturation Time (min)		Adsorption Capacity (mg/g)		Original pH	pH (after Adsorption)	
		SO₂	H₂S	SO₂	H₂S		SO₂	H₂S
Raw eggshell with membrane	<90	32.2	44.0	1.09	0.65			
	90–125	29.0	20.5	0.89	0.19	9.2 ± 0.2	8.7 ± 0.2	8.9 ± 0.2
	125–180	15.3	8.0	0.61	0.08			
Raw eggshell without membrane	<90	29.8	33.0	0.98	0.26			
	90–125	25.5	17.5	0.66	0.14	8.9 ± 0.2	8.4 ± 0.2	8.6 ± 0.2
	125 180	13.3	7.5	0.54	0.05			

Table 5. Adsorption capacity and pH of calcined eggshell at different temperature.

Calcination Temperature (°C)	BET Surface Area (m²/g)	Adsorption Capacity (mg/g)		Original pH	pH (After Adsorption)	
		SO_2	H_2S		SO_2	H_2S
800	2.98	2.12	1.30	10.9 ± 0.2	10.1 ± 0.2	10.3 ± 0.2
900	6.74	3.53	2.62	11.8 ± 0.2	10.9 ± 0.2	11.1 ± 0.2
950	6.54	3.22	2.35	12.0 ± 0.2	11.2 ± 0.2	11.4 ± 0.2
1000	6.30	2.69	1.85	12.1 ± 0.2	11.3 ± 0.2	11.5 ± 0.2
1100	2.68	2.55	1.66	12.3 ± 0.2	11.4 ± 0.2	11.5 ± 0.2

2.2. Effect of the Eggshell Membrane and Particle Size

Three different particle sizes (<90 μm, 90–125 μm, and 125–180 μm) of RES with and without membrane were tested for the SO_2 and H_2S removal. During this analysis, other parameters such as sorbent dosage (1 g), flow rate (300 mL/min), humidity (0%), gas inlet concentration (300 ppm), pressure (1 bar), and reaction temperature (ambient temperature, 29 °C) were all kept constant. Table 4 shows the adsorption capacity and saturation time of SO_2 and H_2S adsorption by RES with and without membrane. It was noticed that for both H_2S and SO_2 the adsorption capacities were less than 1.1 mg/g on dry basis. The breakthrough point was not detected which indirectly shows that there was no immediate chemical interaction between the gas and sorbent. It is known that $CaCO_3$ is a highly stable material at ambient conditions. Thus, RES with or without membrane could have similar behavior. RES with membrane recorded higher adsorption capacity and longer saturation time for both H_2S and SO_2 gas compared to the one without membrane. According to Tsai et al. (2006), ES membrane comprises of a grid of fibrous proteins which contributes to its large surface areas and these fibrous proteins have higher BET surface area compared to the shell itself [28]. In the literature, the role of ES membrane in the removal of reactive dyes, heavy metals, phenols, and various other substances was reported and in most of the cases, it was reported that the adsorption capacity was better with membrane compared to one without membrane [29,30]. Thus, in this study, it was found that RES with membrane enhanced the sorption of H_2S and SO_2.

As for the effect of particle size, as anticipated, the smallest particle size i.e., <90 μm shows the best results for both RES with and without membrane. The calculated adsorption capacity values followed the following sequence for both gases: 90 μm > 125 μm > 180 μm. Witoon (2011) had stated that the smaller particle size of calcined ES had higher CO_2 capture capacity because it provides a greater exposed surface for the adsorption [23]. Similarly, in this study, smaller particles can offer a greater surface area for gas–solid interactions. The macro-pores and pits are irregularly dispersed over the surface of the RES, which could be one of the factors for low adsorption capacity as evident from the FESEM image. At high temperature, $CaCO_3$ breaks down to $CaO_{(s)}$ and release CO_2. The Ca^{+2} of CaO is unstable and reacts with $SO_{2(g)}$ replacing oxygen at high temperature. The reaction of SO_2 gas on $CaCO_3$ can be shown by the following chemical reaction [31];

$$CaCO_{3(s)} + SO_{2(g)} \rightarrow CaSO_{3(s)} + CO_{2(g)}. \tag{1}$$

However, calcium sulfate ($CaSO_4$) is only formed from the reaction between calcium carbonate ($CaCO_3$) and SO_2 gas when the temperature is more than 750 K in presence of oxygen (O_2) as illustrated below [31];

$$CaCO_{3(s)} + SO_{2(g)} + \frac{1}{2} O_{2(g)} \rightarrow CaSO_{4(s)} + CO_{2(g)}. \tag{2}$$

As the experiments were carried out in room temperature and O_2 was not induced in this study, the only possible reaction would be the formation of calcium sulfite ($CaSO_3$)

rather than $CaSO_4$. For the reaction between $CaCO_3$ of RES and H_2S, the following direct sulfidation reaction is expected [32]:

$$CaCO_{3(s)} + H_2S_{(g)} \rightarrow CaS_{(s)} + CO_{2(g)} + H_2O_{(v)}. \tag{3}$$

However, at room temperature, $CaCO_3$ is very stable and the chances of the above reactions are very slim. However, for RES only physical adsorption has happened for both SO_2 and H_2S which is the main reason for its low adsorption capacities. As physical adsorption is happening, so the particle size and surface porosity played an important role.

2.3. Effect of Calcination Temperature

The influence of various calcination temperature on RES and their effect on SO_2 and H_2S adsorption were tested using dried and powdered (<90 μm) RES with membrane. Other adsorption parameters were kept the same as in section "Effect of the eggshell membrane and particle size". Figure 6a,b shows the breakthrough curves of SO_2 and H_2S versus calcination temperature of ES. Among the calcination temperature, 900 °C shows a prominent outstanding curve. Breakthrough points were very short however, it has a longer saturation time (84 min and 77 min for SO_2 and H_2S, respectively) for both gases.

Table 5 tabulates the adsorption capacity of the CES at different temperatures. (CES 900 °C) shows the highest adsorption capacity i.e., 2.63 mg/g and 3.53 mg/g for H_2S and SO_2 respectively. High adsorption capacity at 900 °C calcination could be due to the complete conversion of calcium carbonate ($CaCO_3$) to calcium oxide (CaO). It is noted that complete conversion takes places around 930 °C [32,33]. Thus, at temperature 800 °C, lower adsorption capacity was noticed for both gases. Moreover, the 800 °C calcined ES was visibly grayish in color compared to the ones calcined at a higher temperature which were whitish confirming the incomplete calcination. At higher temperature i.e., >900 °C, the adsorption capacities for both SO_2 and H_2S decreased. Moreover, the saturation time was shorter and there was no breakthrough point. The decrease in adsorption capacity is due to the sintering effect. Similar work on SO_2 adsorption by calcined limestone reported that pore size distribution is greatly affected by the calcination temperature.

It was reported that 950 °C was the optimized calcination temperature for limestone and at this temperature, the pores of CaO have the least diffusion resistance and highest activity for SO_2 removal [34]. Another similar work mentioned that sintering of CaO derived from pure $CaCO_3$ starts at 800–900 °C and it becomes more severe after 950 °C. ES is less pure than the commercial limestone, where for each 100 g of air-dried of ES waste only 88 g are of $CaCO_3$ [35], which corresponds to an increase in the rate of sintering [36,37]. Also, it was proven that a breakdown of the pores occurs for 1100 °C. These statements can be confirmed using BET surface area readings, and FESEM images.

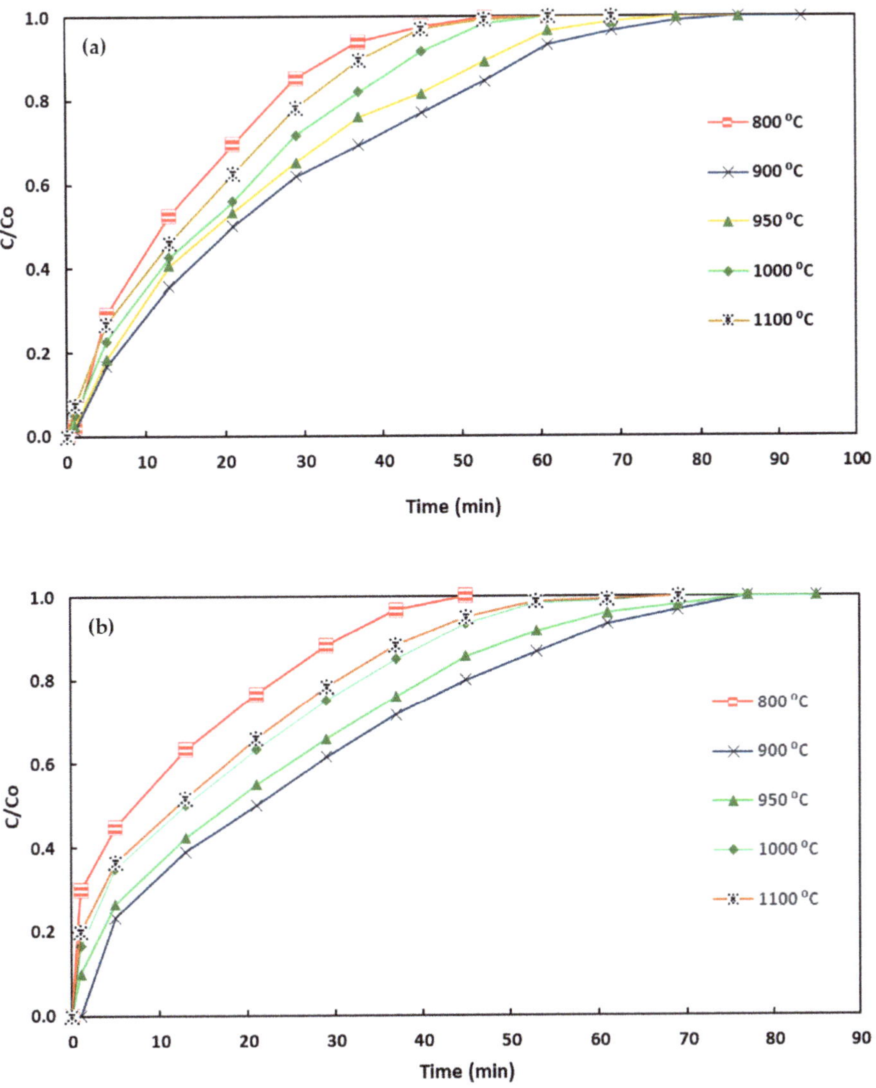

Figure 6. Adsorption breakthrough curves of (**a**) SO_2 and (**b**) H_2S by eggshell at different calcination.

2.4. Effect of Reaction Temperature and Humidity

The influence of reaction temperature on SO_2 and H_2S adsorption by RES and (CES 900 °C) was evaluated using two different reactor temperatures (100 °C and 200 °C). Other parameters such as sorbent dosage (1 g), flow rate (300 mL/min), gas inlet concentration (300 ppm), and pressure (1 bar) were all kept constant. Table 6 shows the adsorption capacity for the effect of reaction temperature and humidity. Figure 7 shows the breakthrough curve for the effect of reaction temperature on H_2S and SO_2 removal. The trend was comparatively better than the ones done at the room temperature (29 °C) earlier. However, the impact was very minimal for (CES 900 °C). Meanwhile, for RES, the adsorption capacity was doubled when the reactor temperature was increased from 30 °C to 200 °C. It was claimed that an increase in reaction temperature, increases the chemical interaction of SO_2 and H_2S with limestone-based $CaCO_3$ or CaO or $Ca(OH)_2$ sorbents [38–40]. The effects of reactor temperature in the range of room temperature to 200 °C is considered low in magnitude. At a temperature below 200 °C, only calcium sulfite will be formed during the reaction of SO_2 with $CaCO_3$ and CaO [41]. Calcium sulfite, however, is stable at around 200 °C and only becomes unstable at a temperature above 727 °C decomposing to form calcium sulfate and calcium carbide [42]. Though, at a temperature above 100 °C the rate of calcium sulfite formation tends to increase [41]. There are limited data on SO_2 and H_2S by $CaCO_3$ and CaO at temperature < 250 °C. Most of the studies have been done at elevated temperatures, i.e., >400 °C.

Table 6. Adsorption capacity of raw and calcined (900 °C) eggshell at different reaction temperature and 40% relative humidity.

Reaction Condition	Raw Eggshell (mg/g)		Calcined Eggshell (900 °C) (mg/g)	
	SO_2	H_2S	SO_2	H_2S
Reaction Temperature (100 °C)	1.43	0.77	3.93	3.37
Reaction Temperature (200 °C)	2.03	1.20	4.30	4.22
Humidity (40%)	8.89	1.42	11.68	7.96

These analyses illustrated that there are some interactions between SO_2 and H_2S with RES and CES even at low temperature. A lower temperature will be favorable for SO_2 and H_2S removal because the temperature of flue gas going to the stack is around 150 °C and the working temperature of is about 55 °C [43,44]. However, at a higher temperature, the reaction could be further enhanced. A study of SO_2 adsorption by lime (80% CaO) reported that the conversion rate gets double when the temperature increases from 400 °C to 800 °C [45]. Moreover, during the reaction of $CaCO_3$ with H_2S, complete conversion of $CaCO_3$ to CaS is only feasible if sulfidation is carried out at a temperature above its calcination temperature [46]. Thus, it can be concluded that a more conducive environment is created for the sulfidation of both RES and (CES 900 °C) by increasing the reaction temperature up to 200 °C.

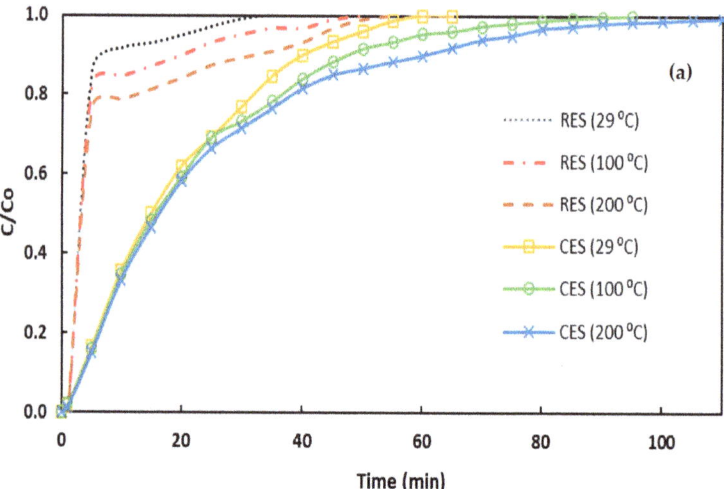

Figure 7. *Cont.*

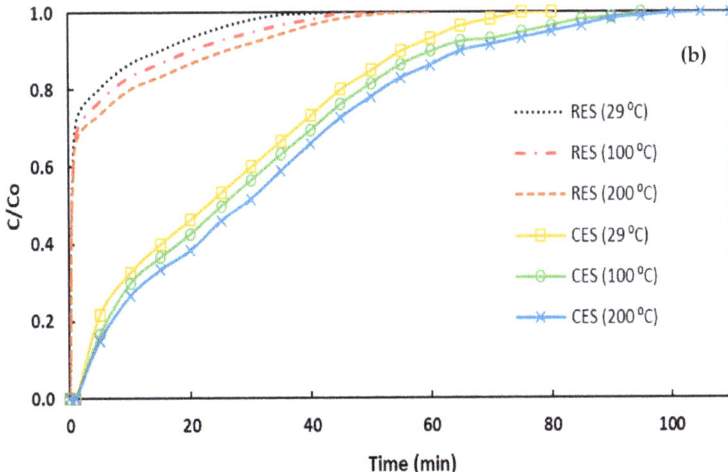

Figure 7. Breakthrough curves of (**a**) SO_2 and (**b**) H_2S at different reaction temperature.

Figure 8 shows the breakthrough curves of RES and (CES 900 °C) with a response to the effect of humidity. All other parameters were kept constant. It was noticed that with the addition of 40% relative humidity (RH), the performance of both RES and (CES 900 °C) improved significantly. The adsorption capacity of SO_2 and H_2S by (CES 900 °C) increased almost triple with humidity. There were significant improvements in the breakthrough time as well as saturation time for both SO_2 and H_2S. The breakthrough point was clearly noticeable for (CES 900 °C). For RES, a great increase was noticed for SO_2, however, for H_2S, only small improvement was noticed. Results can be compared from Tables 4–6. The presence of humidity in the inlet gas improved the adsorption capacity of SO_2 more than the H_2S for RES. This could be due to the solubility of SO_2. The solubility of SO_2 in water is about 16 times more than the solubility of H_2 as per the data published by [47]. The chemical reaction of SO_2 and $CaCO_3$ in the presence of RH is as shown below;

$$CaCO_3 \text{ (s)} + H_2O \text{ (g)} \rightarrow Ca(OH)(CO_3H) \tag{4}$$

$$Ca(OH)(CO_3H) + SO_2 \text{ (g)} \rightarrow CaSO_3 + H_2CO_3. \tag{5}$$

Carbonic acid (H_2CO_3) is a product of this surface reactions, not alike the one without RH which produces carbon dioxide [31]. Between 30% and 85% RH, SO_2 and $CaCO_3$ reaction are improved significantly, approximately by 5 to 10 fold for single crystal $CaCO_3$ (calcite) in the presence of moisture [48]. A slight improvement is noticed because of the humidity which made the contact possible between $CaCO_3$ and the acidic gases [49]. It is known that CaO reaction in the presence of water vapor will form $Ca(OH)_2$. At low temperature, it is expected that only sulfite hemihydrate will be formed when SO_2 is present [41]. The following reaction could occur;

$$Ca(OH)_2 + SO_2 \rightarrow CaSO_3 \cdot 0.5H_2O + 0.5H_2O. \tag{6}$$

The chemisorption process of SO_2 onto the sorbent surface as described Equation (6) chemical reaction increases with increasing RH [41]. It has been reported that moisture can enhance the adsorption capacity of carbonates and oxides for atmospheric gases [31]. For example, in a humid air condition, the deposition velocity of SO_2 gas onto calcite and dolomite increases [31]. Moreover, SO_2 can oxidize to SO_3, to form sulfuric acid [50]. Similarly, for H_2S, there was a small increase in the adsorption capacity although the gas is not readily soluble in water. This increase is attributed to the contact time between CaO sorbent and water vapor. H_2S partly dissolves in water to form a weak acid and CaO would readily attract a water molecule to convert to a more stable form of calcium hydroxide $(CaOH)_2$. Yet, the chances of additional reaction between CaOH and H_2S to form CaS are low as this reaction could only happen at high temperature, i.e., above 900 °C [51].

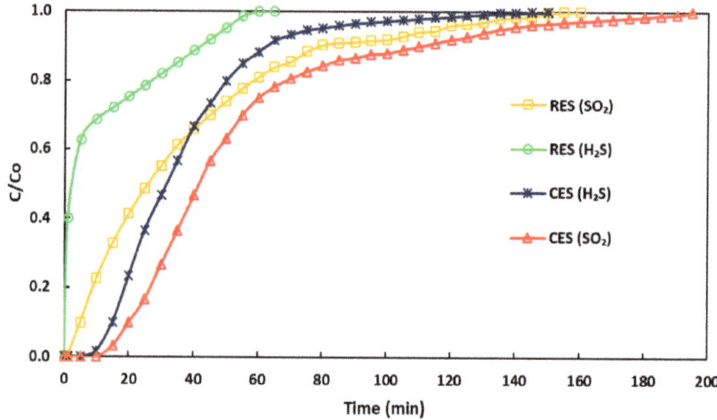

Figure 8. Breakthrough curves of SO_2 and H_2S with 40% relative humidity.

2.5. Comparison Study

Table 7 shows a comparison of various sources of Ca-based sorbents and its potential to adsorb SO_2 and H_2S respectively. It can be seen that the adsorption capacity of the CES is comparable to that of the Ca-based sorbents. Lower adsorption capacity could be due to the low BET surface area, impurities in ES, and the unstable nature of ES. Some of the Ca-sorbents reported in the literature were modified with chemicals which further enhanced the adsorption capacity. No one has reported ES in the form of $CaCO_3$. If it has been tested in the raw form maybe the reported adsorption capacity readings would be much lower compared to ES. Thus, it is expected that if ES is further modified to calcined ES, it could definitely perform better than the ones reported in the literature. The only one who has done similar work on ES is Witoon (2011) who tried for CO_2 capture at various temperatures by TGA method [23]. The carbonation rate was around 35% at 750 °C of calcination temperature. Thus, there is a big potential for calcined ES to be used for pollutant gases removal.

Table 7. Comparison of raw and calcined (900 °C) eggshell with other Ca-based sorbents for SO_2 and H_2S.

Gas	Sorbent	Conditions	Adsorption Capacity/ Removal Efficiency	Ref.
	Ethanol treated calcined limestone (CaO) (BET = 35.5 m²/g)	Reaction temperature = 1050 °C SO_2—2000 ppm by volume Residence time = 1 s Ca/S ratio = 3	Around 55%	[52]
SO_2	Commercial $Ca(OH)_2$ mixed with rice husk ash (BET = 106.10 m²/g)	Reaction temperature—100 °C SO_2 = 1000 ppm NO =500 ppm CO_2 = 12% N_2 = balance	10.72 mg/g	[53]
	Commercial $Ca(OH)_2$ mixed oil palm ash (BET = 88.3 m²/g)		5.36 mg/g	
	Commercial $Ca(OH)_2$ mixed with coal fly ash (BET = 62.2 m²/g)		4.29 mg/g	

Table 7. *Cont.*

Gas	Sorbent	Conditions	Adsorption Capacity/ Removal Efficiency	Ref.
	$Ca(OH)_2$ obtained from oyster shell (BET = 12.9 m^2/g)—CaO content = 53.8%	Reaction temperature—150 °C) SO_2 = 1800 ppm NO_x = 250 ppm O_2 = 6% Water vapors = 10%	0.78 mmol/g	[54]
	Hydrated Lime (BET = 8.76 m^2/g)	Reaction temperature = 70 °C NO_2 = 249 ppm SO_2 = 906 ppm Relative humidity = 60%	25% (SO_2)	[55]
	$Ca(OH)_2$ mixed with fly ash in ratio of 1:3 with additional treatment with $KMnO_4$ (BET = 19.04 m^2/g)	Reaction temperature = 80 °C SO_2 = 500 ppm NO = 200 ppm O_2 = 5% Relative Humidity = 80%	60–80%	[56]
	Oxidant enriched $Ca(OH)_2$ (BET = 11 to 14 m^2/g)	Reaction temperature—80 °C SO_2—600 ppm NO—300 ppm O_2—8% H_2O—10.5%	80–100 mg/g	[57]
	Calcined eggshell (BET = 6.74 m^2/g)	Reaction temperature—30 °C SO_2—300 ppm N_2-balance Relative humidity—40%	11.68 mg/g	This work
	CaO from waste $CaCO_3$ (BET < 3 m^2/g)	Reaction temperature—20 °C H_2S—50 ppm in air TGA Analysis Reaction time 55 min	55%	
	$CaCO_3$ from waste (BET < 3.98 m^2/g)	Reaction temperature-Ambient Biogas with 200 ppmv H_2S Reaction time 400 min	85%	
	Dried fly ash (BET = 17.71 m^2/g)	Reaction temperature-Not Given H_2S—300 ppm in air	10.4 mg/g	[10]
H_2S	Australian red soil (BET = not given)	Reaction temperature-Not Given H_2S = 2000 ppm N_2-balance	18.8 mg/g	
	Calcined eggshell (BET 6.74 m^2/g)	Reaction temperature—30 °C H_2S—300 ppm N_2-balance Relative humidity—40%	7.96 mg/g	This work

3. Materials and Methods

3.1. Sorbent Preparation

Chicken eggshell waste (brown in color classified as Grade B and C) was collected from the student's food court in the university campus. It was thoroughly soaked and washed with tap water until a clean eggshell were obtained. Two types of raw eggshell were prepared i.e., with the membrane (RES) and without the membrane. For the samples without the membrane, the membrane was carefully removed after soaking in water. The ES samples were then dried in an UF 110 oven (Memmert, Schwabach, Germany) at 105 °C for 24 h to remove the excessive moisture. Finally, the samples were ground to different particle sizes using a MX-GM1011H dry blender (Panasonic, Selangor, Malaysia). The powdered ES samples were then sieved using a WS TYLER RX29 vibrating sieve

(Fisher Scientific, Pittsburgh, PA, USA). Three different particle sizes were isolated (<90 μm, 90–125 μm and 125–180 μm). Calcined ES (CES) samples were prepared by heating the powdered RES at different temperatures (800 °C, 900 °C, 950 °C, 1000 °C, 1100 °C) for 2 h using a LEF-103S model muffle furnace (LabTech, Debrecen, Hungary). The retention time for calcination was chosen based on preliminary studies.

3.2. Adsorption Tests

The adsorption tests were performed in a lab-scale adsorption reactor as shown in Figure S1 (Supplementary Material). SO_2 and H_2S gas flow from the gas cylinders (2000 ppm, 99% purity) were controlled automatically using SDPROC mass flow controllers and flow meters (Aalborg, New York, NY, USA). The gas flow rates of both gases into the reactor were kept at 300 mL/min throughout the experiments. H_2S and SO_2 concentrations were kept constant at 300 ppm by introducing nitrogen gas as balance. Low flowrate and concentration were chosen for the overall safety of the lab. A down-flow fixed bed reactor made of stainless steel with an internal diameter of 9 mm and a height of 180 mm was prepared to fill the sorbents. The reactor was fixed inside an oven (Memmert, Schwabach, Germany) for temperature study. In each run, 1 g of sorbent was placed inside the reactor. The initial and outlet concentrations of H_2S and SO_2 gas were measured using Biogas 5000 Portable Gas Analyzer (Geotech, Chelmsford Essex, UK) and Vario-Plus Industrial Gas-Analyzer (MRU Instruments Inc., Humble, TX, USA) respectively. SO_2 and H_2S analyzers recorded the gas concentrations every second. The adsorption capacity was calculated from the following equation [58] using the breakthrough curve generated during the experiment:

$$Q = \frac{CoMw\,q}{1000w\,Vm} \int_0^t \left(1 - \frac{C}{Co}\right) dt \tag{7}$$

where Q is adsorption capacity (mg/g), c_o is the initial inlet concentration (ppm), M_w is the gas molecular weight (g/mol), q is the total flow rate (L/min), w is the weight of sorbent (g), V_m is the molar volume (L/mol), c is the outlet concentration of the gas (ppm) at time t (min). The adsorption capacity calculation was based on an average of 3 repetitions of the adsorption breakthrough curve. The differences between the 3 readings were less than 3%.

3.3. Process Parameters Study

In the process study, the effects of the relative humidity in the inlet gas and reaction temperature were evaluated for both RES and CES sorbents. Only optimized sorbents (both RES and CES) were selected for the process study. The reactor was fixed inside the oven and the temperature of the oven was varied from room temperature (approximately 30 °C) to 200 °C to study the effect of temperature on adsorption tests. Whereas the humidity in the inlet gas was created by passing the inlet gas through an airtight conical flask which was submerged in a temperature-controlled water bath before it could enter the reactor at room temperature. Required humidity in the inlet gas was created as the gas passing through the water bath was saturated with water vapor at the set temperature. The temperature of the water bath was set based on the steam tables formulation to create 40% relative humidity in the inlet gas. 40% of relative humidity was selected as it can be generated at room temperature and also to protect the gas sensors in the analyzer.

3.4. Characterization of the Sorbents

Morphology of the RES and CES were analysed by field emission scanning electron microscope (FESEM), model JEOL JSM-6701F (JEOL, Akishima City, Japan). The energy-dispersive X-ray spectroscopy (EDS) (JEOL, Akishima City, Japan) was employed to detect the specific elements on the surface of the materials. Fourier Transform Infrared (FTIR) Lambda 35 (Perkin Elmer, Waltham, MA, USA) was used to determine the surface functional groups of the sorbents. The spectra were recorded in the spectral range of 400–4000 cm^{-1} with a resolution of 4 cm^{-1} by mixing a small quantity of the sorbent with potassium bromide. pH was measured by preparing a solution in the ratio of 0.1 g of sorbent in 20 mL deionized water and stirred for 1.5 hr. A digital pH meter was used (Hanna Instruments, Woonsocket, RI, USA). X-ray diffraction model Lab X XRD-6000 (Shimadzu, Tokyo, Japan) was used to identify the XRD patterns of the sorbents at room temperature at 2θ with a step size of 0.02. The Brunauer–Emmett–Teller (BET) surface area and pore size distribution were calculated using nitrogen adsorption and desorption isotherms conducted at 77 K with micromeritics, ASAP 2020 V4.02 (Micromeritis, Norcross, GA, USA) volumetric gas adsorption instrument. Thermogravimetric analysis (TGA) of RES and CES was done with TGA/DSC 3+ (Mettler Toledo, Ohio, OH, USA) for proximate analysis. Nitrogen gas was used at 20 mL/min at a heating rate of 10 °C /min until 900 °C. Original images of ES were taken with a smartphone.

4. Conclusions

RES and CES sorbents were tested for SO_2 and H_2S adsorption. It was found that RES with membrane and having the smallest particle size i.e., <90 μm showed the best adsorption capacity for both SO_2 and H_2S. (CES 900 °C) showed the best adsorption capacity among the other calcination temperature and RES. It can be concluded that physical adsorption was dominant over the chemical adsorption for both RES and CES sorbents. The characterization study shows the existences of sulfur element in the spent adsorbents which further verifies the adsorption of SO_2 and H_2S by CES. The presence of the relative humidity in the inlet gas and increasing reaction temperature improved the performance of both RES and CES sorbents. (CES 900 °C) showed a greater adsorption capacity compared to RES with the addition of humidity. The best adsorption capacity of SO_2 and H_2S was recorded as 11.68 mg/g and 7.96 mg/g respectively using (CES 900 °C) with 40% RH. These results indicate that chicken eggshell have great potential to be used as sorbents upon modification for the removal of pollutant gases such as SO_2 and H_2S from contaminated air.

Supplementary Materials: The following are available online at https://www.mdpi.com/2073-434 4/11/2/295/s1, Figure S1: Schematic diagram of adsorption experimental setup.

Author Contributions: W.A. and S.S. contributed equally. Conceptualization, S.S. and Y.M.; methodology, R.K.; validation, R.K., Y.M., and S.S.; formal analysis, W.A. and S.S.; investigation, W.A. and S.S.; resources, S.S. and Y.M.; data curation, W.A. and R.K.; writing—original draft preparation, W.A.; writing—review and editing, W.A., S.S., R.K., and Y.M.; visualization, W.A. and Y.M.; supervision, S.S.; project administration, S.S. and Y.M.; funding acquisition, S.S. All authors have read and agreed to the published version of the manuscript.

Funding: This research was funded by Universiti Tunku Abdul Rahman Research Fund number UTARRF/2017-C1/S07 and the APC was funded by Universiti Tunku Abdul Rahman under Financial Support for Journal Paper Publication Scheme and the authors.

Data Availability Statement: Data is contained within the article or Supplementary Material.

Acknowledgments: The authors gratefully acknowledge the financial support received from Universiti Tunku Abdul Rahman.

Conflicts of Interest: The authors declare no conflict of interest.

References

1. Awe, O.W.; Zhao, Y.; Nzihou, A.; Minh, D.P.; Lyczko, N. A Review of Biogas Utilisation, Purification and Upgrading Technologies. *Waste Biomass Valorization* **2017**, *8*, 267–283. [CrossRef]
2. Du, E.; Dong, D.; Zeng, X.; Sun, Z.; Jiang, X.; de Vries, W. Direct Effect of Acid Rain on Leaf Chlorophyll Content of Terrestrial Plants in China. *Sci. Total Environ.* **2017**, *605*, 764–769. [CrossRef]
3. Zhang, Y.; Wang, T.; Yang, H.; Zhang, H.; Zhang, X. Experimental Study on SO_2 Recovery Using a Sodium—Zinc Sorbent Based Flue Gas Desulfurization Technology. *Chin. J. Chem. Eng.* **2015**, *23*, 241–246. [CrossRef]
4. Córdoba, P. Status of Flue Gas Desulphurisation (FGD) Systems from Coal-Fired Power Plants: Overview of the Physic-Chemical Control Processes of Wet Limestone FGDs. *Fuel* **2015**, *144*, 274–286. [CrossRef]
5. Qu, Z.; Sun, F.; Liu, X.; Gao, J.; Qie, Z.; Zhao, G. The Effect of Nitrogen-Containing Functional Groups on SO_2 Adsorption on Carbon Surface: Enhanced Physical Adsorption Interactions. *Surf. Sci.* **2018**, *677*, 78–82. [CrossRef]

6. Li, P.; Wang, X.; Allinson, G.; Li, X.; Stagnitti, F.; Murray, F.; Xiong, X. Effects of Sulfur Dioxide Pollution on the Translocation and Accumulation of Heavy Metals in Soybean Grain. *Environ. Sci. Pollut. Res.* **2011**, *18*, 1090–1097. [CrossRef] [PubMed]

7. Li, C.; Sheng, Y.; Sun, X. Simultaneous Removal of SO_2 and NOX by a Combination of Red Mud and Coal Mine Drainage. *Environ. Eng. Sci.* **2019**, *36*, 444–452. [CrossRef]

8. Silas, K.; Ghani, W.A.; Choong, T.S.; Rashid, U. Carbonaceous Materials Modified Catalysts for Simultaneous SO_2/NOx Removal from Flue Gas: A Review. *Catal. Rev.* **2019**, *61*, 134–161. [CrossRef]

9. El Asri, O.; Hafidi, I.; elamin Afilal, M. Comparison of Biogas Purification by Different Substrates and Construction of a Biogas Purification System. *Waste Biomass Valorization* **2015**, *6*, 459–464. [CrossRef]

10. Ahmad, W.; Sethupathi, S.; Kanadasan, G.; Lau, L.C.; Kanthasamy, R. A Review on the Removal of Hydrogen Sulfide from Biogas by Adsorption Using Sorbents Derived from Waste. *Rev. Chem. Eng.* **2019**. [CrossRef]

11. Shah, M.S.; Tsapatsis, M.; Siepmann, J.I. Hydrogen Sulfide Capture: From Absorption in Polar Liquids to Oxide, Zeolite, and Metal-Organic Framework Adsorbents and Membranes. *Chem. Rev.* **2017**, *117*, 14. [CrossRef]

12. Abatzoglou, N.; Boivin, S. A Review of Biogas Purification Processes. *Biofuels Bioprod. Biorefining* **2009**, *3*, 42–71. [CrossRef]

13. Baláž, M. Ball Milling of Eggshell Waste as a Green and Sustainable Approach: A Review. *Adv. Colloid Interface Sci.* **2018**, *256*, 256–275. [CrossRef]

14. Oliveira, D.A.; Benelli, P.; Amante, E.R. A Literature Review on Adding Value to Solid Residues: Egg Shells. *J. Clean. Prod.* **2013**, *46*, 42–47. [CrossRef]

15. Food and Agriculture Organization (FAO); U.N. Livestock Production. 2015. Available online: http://www.fao.org/docrep/005/y4252e/y4252e07.htm (accessed on 30 August 2019).

16. Rohim, R.; Ahmad, R.; Ibrahim, N.; Hamidin, N.; Abidin, C.Z.A. Characterization of Calcium Oxide Catalyst from Eggshell Waste. *Adv. Environ. Biol.* **2014**, *8*, 35–38.

17. De Angelis, G.; Medeghini, L.; Conte, A.M.; Mignardi, S. Recycling of Eggshell Waste into Low-Cost Adsorbent for Ni Removal from Wastewater. *J. Clean. Prod.* **2017**, *164*, 1497–1506. [CrossRef]

18. Laca, A.; Laca, A.; Díaz, M. Eggshell Waste as Catalyst: A Review. *J. Environ. Manag.* **2017**, *197*, 351–359. [CrossRef] [PubMed]

19. Hosseini, S.; Eghbali Babadi, F.; Masoudi Soltani, S.; Aroua, M.K.; Babamohammadi, S.; Mousavi Moghadam, A. Carbon Dioxide Adsorption on Nitrogen-Enriched Gel Beads from Calcined Eggshell/Sodium Alginate Natural Composite. *Process Saf. Environ. Prot.* **2017**, *109*, 387–399. [CrossRef]

20. Sethupathi, S.; Kai, Y.C.; Kong, L.L.; Munusamy, Y.; Bashir, M.J.K.; Iberahim, N. Preliminary Study of Sulfur Dioxide Removal Using Calcined Egg Shell. *Malays. J. Anal. Sci.* **2017**, *21*, 719–725. [CrossRef]

21. Sun, Y.; Yang, G.; Zhang, L. Hybrid Adsorbent Prepared from Renewable Lignin and Waste Egg Shell for SO_2 Removal: Characterization and Process Optimization. *Ecol. Eng.* **2018**, *115*, 139–148. [CrossRef]

22. Cho, Y.B.; Seo, G.; Chang, D.R. Transesterification of Tributyrin with Methanol over Calcium Oxide Catalysts Prepared from Various Precursors. *Fuel Process. Technol.* **2009**, *90*, 1252–1258. [CrossRef]

23. Witoon, T. Characterization of Calcium Oxide Derived from Waste Eggshell and Its Application as CO_2 Sorbent. *Ceram. Int.* **2011**, *37*, 3291–3298. [CrossRef]

24. Gergely, G.; Weber, F.; Lukacs, I.; Toth, A.L.; Horvath, Z.E.; Mihaly, J.; Balazsi, C. Preparation and Characterization of Hydroxyapatite from Eggshell. *Ceram. Int.* **2010**, *36*, 803–806. [CrossRef]

25. Engin, B.; Demirtaş, H.; Eken, M. Temperature Effects on Egg Shells Investigated by XRD, IR and ESR Techniques. *Radiat. Phys. Chem.* **2006**, *75*, 268–277. [CrossRef]

26. Lee, K.T.; Mohamed, A.R.; Bhatia, S.; Chu, K.H. Removal of Sulfur Dioxide by Fly Ash/CaO/$CaSO_4$ Sorbents. *Chem. Eng. J.* **2005**, *114*, 171–177. [CrossRef]

27. Zhao, Y.; Han, Y.; Chen, C. Simultaneous Removal of SO_2 and NO from Flue Gas Using Multicomposite Active Absorbent. *Ind. Eng. Chem. Res.* **2011**, *51*, 480–486. [CrossRef]

28. Tsai, W.T.; Yang, J.M.; Lai, C.W.; Cheng, Y.H.; Lin, C.C.; Yeh, C.W. Characterization and Adsorption Properties of Eggshells and Eggshell Membrane. *Bioresour. Technol.* **2006**, *97*, 488–493. [CrossRef] [PubMed]

29. Al-ghouti, M.A.; Khan, M. Eggshell Membrane as a Novel Bio Sorbent for Remediation of Boron from Desalinated Water. *J. Environ. Manag.* **2017**, *207*, 405–416. [CrossRef]

30. Pramanpol, N.; Nitayapat, N. Adsorption of Reactive Dye by Eggshell and Its Membrane. *Kasetsart J.* **2006**, *40*, 192–197.

31. Al-Hosney, H.A.; Grassian, V.H. Water, Sulfur Dioxide and Nitric Acid Adsorption on Calcium Carbonate: A Transmission and ATR-FTIR Study. *Phys. Chem. Chem. Phys.* **2005**, *7*, 1266–1276. [CrossRef]

32. De-Diego, L.; Abad, A.; Garcia-Lebiano, F.; Adanez, J.; Gayan, P. Simultaneous Calcination and Sulfidation of Calcium-Based Sorbents. *Ind. Eng. Chem. Res.* **2004**, *43*, 3261–3269. [CrossRef]

33. Valverde, J.M.; Medina, S. Reduction of Calcination Temperature in the Calcium Looping Process for CO_2 Capture by Using Helium: In Situ XRD Analysis. *Sustain. Chem. Eng.* **2016**, *4*, 7090–7097. [CrossRef]

34. Doğu, T.I. The Importance of Pore Structure and Diffusion in the Kinetics of Gas-Solid Non-Catalytic Reactions: Reaction of Calcined Limestone with SO_2. *Chem. Eng. J.* **1981**, *21*, 213–222. [CrossRef]

35. Quina, M.J.; Soares, M.A.R.; Quinta-ferreira, R. Applications of Industrial Eggshell as a Valuable Anthropogenic Resource. *Resour. Conserv. Recycl.* **2017**, *123*, 176–186. [CrossRef]

36. Hartman, M.; Svoboda, K.; Trnka, O.; Čermák, J. Reaction between Hydrogen Sulfide and Limestone Calcines. *Ind. Eng. Chem. Res.* **2002**, *41*, 2392–2398. [CrossRef]
37. Borgwardt, R.H. Sintering of Nascent Calcium Oxide. *Chem. Eng. Sci.* **1989**, *44*, 53–60. [CrossRef]
38. Slack, A.V.; Falkenberry, H.L.; Harrington, R.E. Sulfur Oxide Removal from Waste Gases: Lime-Limestone Scrubbing Technology. *J. Air Pollut. Control Assoc.* **1972**, *22*, 159–166. [CrossRef]
39. Liu, C.-F.; Shih, S.-M.; Lin, R.-B. Kinetics of the Reaction of $Ca(OH)_2$/Fly Ash Sorbent with SO_2 at Low Temperatures. *Chem. Eng. Sci.* **2002**, *57*, 93–104. [CrossRef]
40. Agnihotri, R.; Chauk, S.S.; Mahuli, S.K.; Fan, L.S. Mechanism of CaO Reaction with H_2S: Diffusion through CaS Product Layer. *Chem. Eng. Sci.* **1999**, *54*, 3443–3453. [CrossRef]
41. Krammer, G.; Brunner, C.; Khinast, J.; Staudinger, G. Reaction of $Ca(OH)_2$ with SO_2 at Low Temperature. *Ind. Eng. Chem. Res.* **1997**, *36*, 1410–1418. [CrossRef]
42. Galloway, B.D.; MacDonald, R.A.; Padak, B. Characterization of Sulfur Products on CaO at High Temperatures for Air and Oxy-Combustion. *Int. J. Coal Geol.* **2016**, *167*, 1–9. [CrossRef]
43. Song, C.; Pan, W.; Srimat, S.T.; Zheng, J.; Li, Y.; Wang, Y.H.; Xu, B.Q.; Zhu, Q.M. Tri-Reforming of Methane over Ni Catalysts for CO_2 Conversion to Syngas with Desired H_2CO Ratios Using Flue Gas of Power Plants without CO_2 Separation. *Stud. Surf. Sci. Catal.* **2004**, *153*, 315–322.
44. Weiland, P. Biogas Production: Current State and Perspectives. *Appl. Microbiol. Biotechnol.* **2010**, *85*, 849–860. [CrossRef] [PubMed]
45. Shi, L.; Xu, X. Study of the Effect of Fly Ash on Desulfurization by Lime. *Fuel* **2001**, *80*, 1969–1973. [CrossRef]
46. Fenouil, L.A.; Towler, G.P.; Lynn, S. Removal of H_2S from Coal Gas Using Limestone: Kinetic Considerations. *Ind. Eng. Chem. Res.* **1994**, *33*, 265–272. [CrossRef]
47. Wilhelm, E.; Battino, R.; Wilcock, R.J. Low-Pressure Solubility of Gases in Liquid Water. *Chem. Rev.* **1977**, *77*, 219–262. [CrossRef]
48. Baltrusaitis, J.; Usher, C.R.; Grassian, V.H. Reactions of Sulfur Dioxide on Calcium Carbonate Single Crystal and Particle Surfaces at the Adsorbed Water Carbonate Interface. *Phys. Chem. Chem. Phys.* **2007**, *9*, 3011. [CrossRef] [PubMed]
49. Martínez, M.G.; Minh, D.P.; Nzihou, A.; Sharrock, P. Valorization of Calcium Carbonate-Based Solid Wastes for the Treatment of Hydrogen Sulfide in a Semi-Continuous Reactor. *Chem. Eng. J.* **2019**, *360*, 1167–1176. [CrossRef]
50. Ma, S.; Yao, J.; Gao, L.; Ma, X.; Zhao, Y. Experimental Study on Removals of SO_2 and NO_x Using Adsorption of Activated Carbon/Microwave Desorption. *J. Air Waste Manag. Assoc.* **2012**, *62*, 1012–1021. [CrossRef]
51. Garcia-Labiano, F.; De Diego, L.F.; Adánez, J. Effectiveness of Natural, Commercial, and Modified Calcium-Based Sorbents as H_2S Removal Agents at High Temperatures. *Environ. Sci. Technol.* **1999**, *33*, 288–293. [CrossRef]
52. Adanez, J.; Fierro, V.; Garcia-Labiano, F.; Palacios, J. Study of Modified Calcium Hydroxides for Enhancing SO_2 Removal during Sorbent Injection in Pulverized Coal Boilers. *Fuel* **1997**, *76*, 257–265. [CrossRef]
53. Dahlan, I.; Mohamed, A.R.; Kamaruddin, A.H.; Lee, K.T. Dry SO_2 Removal Process Using Calcium/Siliceous-Based Sorbents: Deactivation Kinetics Based on Breakthrough Curves. *Chem. Eng. Technol. Ind. Chem. Equip. Process Eng.* **2007**, *30*, 663–666. [CrossRef]
54. Jung, J.; Yoo, K.; Kim, H.; Lee, H.; Shon, B.-H. Reuse of Waste Oyster Shells as a SO_2/NO_x Removal Absorbent. *J. Ind. Eng. Chem.* **2007**, *13*, 512–517.
55. Nelli, C.H.; Rochelle, G.T. Simultaneous Sulfur Dioxide and Nitrogen Dioxide Removal by Calcium Hydroxide and Calcium Silicate Solids. *J. Air Waste Manag. Assoc.* **1998**, *48*, 819–828. [CrossRef]
56. Zhang, H.; Tong, H.; Wang, S.; Zhuo, Y.; Chen, C.; Xu, X. Simultaneous Removal of SO_2 and NO from Flue Gas with Calcium-Based Sorbent at Low Temperature. *Ind. Eng. Chem. Res.* **2006**, *45*, 6099–6103. [CrossRef]
57. Ghorishi, S.B.; Singer, C.F.; Jozewicz, W.S.; Sedman, C.B.; Srivastava, R.K. Simultaneous Control of Hg0, SO_2, and NO_x by Novel Oxidized Calcium-Based Sorbents. *J. Air Waste Manag. Assoc.* **2002**, *52*, 273–278. [CrossRef] [PubMed]
58. Iberahim, N.; Sethupathi, S.; Bashir, M.J.K. Optimization of Palm Oil Mill Sludge Biochar Preparation for Sulfur Dioxide Removal. *Environ. Sci. Pollut. Res.* **2018**. [CrossRef]

Article

Tailoring Properties of Metal-Free Catalysts for the Highly Efficient Desulfurization of Sour Gases under Harsh Conditions

Cuong Duong-Viet [1,2], Jean-Mario Nhut [1], Tri Truong-Huu [3], Giulia Tuci [4], Lam Nguyen-Dinh [3], Charlotte Pham [5], Giuliano Giambastiani [1,4,*] and Cuong Pham-Huu [1,*]

[1] Institute of Chemistry and Processes for Energy, Environment and Health (ICPEES), UMR 7515 CNRS-University of Strasbourg (UdS), 25, rue Becquerel, CEDEX 02, 67087 Strasbourg, France; duongviet@unistra.fr (C.D.-V.); nhut@unistra.fr (J.-M.N.)

[2] Ha-Noi University of Mining and Geology, 18 Pho Vien, Duc Thang, Bac Tu Liem, Ha-Noi 100000, Vietnam

[3] Faculty of Chemical Engineering, The University of Da-Nang, University of Science and Technology, 54 Nguyen Luong Bang, Da-Nang 550000, Vietnam; thtri@dut.udn.vn (T.T.-H.); ndlam@dut.udn.vn (L.N.-D.)

[4] Institute of Chemistry of OrganoMetallic Compounds, ICCOM-CNR Via Madonna del Piano, 10, Sesto F.no, 50019 Florence, Italy; giulia.tuci@iccom.cnr.it

[5] SICAT SARL, 20 Place des Halles, 67000 Strasbourg, France; pham@sicatcatalyst.com

* Correspondence: giambastiani@unistra.fr (G.G.); cuong.pham-huu@unistra.fr (C.P.-H.)

Abstract: Carbon-based nanomaterials, particularly in the form of N-doped networks, are receiving the attention of the catalysis community as effective metal-free systems for a relatively wide range of industrially relevant transformations. Among them, they have drawn attention as highly valuable and durable catalysts for the selective hydrogen sulfide oxidation to elemental sulfur in the treatment of natural gas. In this contribution, we report the outstanding performance of N-C/SiC based composites obtained by the surface coating of a non-oxide ceramic with a mesoporous N-doped carbon phase, starting from commercially available and cheap food-grade components. Our study points out on the importance of controlling the chemical and morphological properties of the N-C phase to get more effective and robust catalysts suitable to operate H_2S removal from sour (acid) gases under severe desulfurization conditions (high GHSVs and concentrations of aromatics as sour gas stream contaminants). We firstly discuss the optimization of the SiC impregnation/thermal treatment sequences for the N-C phase growth as well as on the role of aromatic contaminants in concentrations as high as 4 vol.% on the catalyst performance and its stability on run. A long-term desulfurization process (up to 720 h), in the presence of intermittent toluene rates (as aromatic contaminant) and variable operative temperatures, has been used to validate the excellent performance of our optimized N-C^2/SiC catalyst as well as to rationalize its unique stability and coke-resistance on run.

Keywords: mesoporous N-doped carbon coating; silicon carbide composites; gas-tail desulfurization treatment; BTX contaminants; elemental sulfur

check for **updates**

Citation: Duong-Viet, C.; Nhut, J.-M.; Truong-Huu, T.; Tuci, G.; Nguyen-Dinh, L.; Pham, C.; Giambastiani, G.; Pham-Huu, C. Tailoring Properties of Metal-Free Catalysts for the Highly Efficient Desulfurization of Sour Gases under Harsh Conditions. *Catalysts* **2021**, *11*, 226. https://doi.org/10.3390/catal11020226

Academic Editor: Daniela Barba

Received: 19 January 2021
Accepted: 4 February 2021
Published: 9 February 2021

1. Introduction

Natural gas (NG) is certainly the cleanest fossil fuel employed for energy purposes. At odds with other natural sources (i.e., petroleum and charcoal) [1], NG holds important environmental merits because it can produce more heat and light energy by mass while keeping its environmental impact significantly lower in terms of carbon footprint and other pollutants that contribute to smog and unhealthy air. The primary constituent of NG is methane (CH_4) but, depending on its origin, it may also contain higher hydrocarbons, including aromatics and sulfur-containing compounds, N_2, CO_2, He, H_2S and noble gases to various extents [2]. In this regard, a complex but effective sequence of thermal and catalytic transformations have been implemented for NG processing as to remove undesired compounds and preventing the diffusion of corrosive and potentially hazardous substances for human health and environment. In particular, organic and inorganic S-compounds removal from natural gas is a mandatory step before any gas manipulation/processing [3–5]. Indeed, their presence besides being highly risky for the toxicity of selected species (e.g., H_2S) can led to undesired phenomena such as the fouling or even the permanent deactivation of catalysts employed in the gas processing. Although current catalytic technologies allow a selective and almost complete (up to 99.9%) hydrogen sulfide conversion into elemental sulfur [6], they still suffer from technical limitations linked to a progressive catalyst deactivation when desulfurization process is operated under harsh reaction conditions (gas hourly space velocity (GHSV) close to those employed in industrial plants), or in the presence of contaminants such as heavy hydrocarbons and aromatics (i.e., benzene, toluene and xylene (BTX)) [2] that are commonly present in untreated natural gas streams. Such impurities deeply impact the performance, stability, and lifetime of catalysts and detrimentally burden on the overall process economy balance. BTX ($C \geq 7$) in particular can lead, throughout the desulfurization process, to the formation of carbonaceous or heavy carbon-sulfur deposits [7] whose incomplete removal translates into catalyst fouling phenomena with the subsequent alteration of its performance or—in worst cases—to the complete catalyst deactivation. Reduction of BTX concentration in acid gas streams is therefore a mandatory step before that gaseous streams reach the catalysts surface, thus increasing the process complexity as well as that of the reactor setup. Complementary and equally valuable technological options such as the use of either amine or solvent enrichment units (Acid Gas Enrichment units—AGE) [8] or the use of activated carbon beds [9] housed upstream of the catalytic desulfurization (SRU) unit, have successfully been implemented for BTX fractions removal.

Whatever the option at work, regeneration treatments of AGE or activated carbon units are periodically needed along with the downstream treatment of BTX wastes. These phases are costly and can cause the change of feed conditions in the desulfurization plant or even the complete feed shut down.

There are little doubts on the relevance of the selective H_2S oxidation to elemental sulfur from both an environmental and commercial viewpoint [10] (~75 Mt/year of elemental sulfur are globally produced from oil and gas processing units). However, the search for robust, highly effective and selective catalysts for the process remains a challenging area of research for the chemistry and engineering community engaged in the field. In particular, the development of robust and durable catalysts suitable to operate the selective H_2S oxidation under variable acid concentrations and showing an excellent resistance to aromatics deactivation is a highly challenging task and thus a widely investigated subject of research.

The last years have witnessed a growing interest of catalysis community towards the use of carbon nanomaterials as pure C-networks or light-heterodoped matrices (i.e., nanotubes (CNTs), nanofibers (CNFs) or thin-carbon film deposits, including N-doped counterparts) as metal-active phase supports or single-phase, metal-free catalysts for a wide series of chemical and electrochemical transformations [11,12]. Metal-free systems of this type have successfully been scrutinized as selective and efficient catalytic materials for a number of industrially relevant oxidation [13–18] and reduction [19–22] processes or as valuable promoters of other challenging catalytic transformations [23,24]. Nanocarbon-based materials, particularly in the form of N-doped networks, have already been exploited as effective metal-free systems for the selective hydrogen sulfide oxidation to elemental sulfur from natural gas tails [16,25–33]. The unique features of these single-phase systems (e.g., the absence of a metal active-phase, the prevalent basic surface character of the N-doped samples together with reduced production cost and environmental impact) have largely contributed to overcome the drawbacks classically encountered with metal nanoparticles-based catalysts. The absence of metal nanoparticles rules out unwanted sintering and leaching phenomena of the active phase occurring on heterogeneous catalysts typically operating under harsh reaction conditions. Moreover, their basic surface character can largely prevent the occurrence of cracking side-processes responsible for the rapid catalyst fouling with subsequent permanent alteration of its performance. This aspect is of particular relevance for catalytic materials operating in the presence of variable concentrations of aromatics whose ultimate and detrimental effects are well known for metal-based catalysts of the state-of-the-art.

In recent years, our group has proposed a versatile technology for the coating of 3D-shaped materials (including open-cell foam structures with variable void fractions), with thin and highly N-enriched mesoporous carbon deposits [15,26,34]. Accordingly, selected 3D hosting matrices underwent successive soaking/impregnation cycles using an aqueous solution of cheap and food-grade components, followed by controlled drying/calcination/annealing steps. Such an approach to the N-C coating of macroscopically shaped networks, besides reducing the formation of toxic by-products typically encountered with more conventional Chemical-Vapor-Deposition (CVD)-based synthetic schemes, offers a versatile and straightforward tool to the easy upscale of challenging metal-free catalytic materials.

The combination of this coating technology with a non-oxide ceramic support (i.e., silicon carbide—SiC) has generated ideal catalyst candidates suitable to operate the selective H_2S desulfurization efficiently under relatively high acid gas concentrations and in the presence of high contents of aromatics as contaminants (0.5 vol.% $\leq$ (H_2S) $\leq$ 2 vol.%; 1000 ppm $\leq$ (tol) $\leq$ 20,000 ppm) [35,36]. SiC as a support for the N-C active phase was selected in light of its high chemical inertness and stability under acidic/oxidative or basic environments along with its excellent mechanical resistance that made it the ideal choice for the target process. Moreover, its medium thermal conductivity [37] prevents or reduces the generation of local temperature gradients (hot spots) while operating highly exothermic transformations [27,38,39]. This additional feature, not available with more classical oxide-based ceramics, contributes to the ultimate catalyst stability and its lifetime on stream.

This contribution takes advantage from our recent findings in the area of metal-free catalysts for the highly efficient and selective H_2S desulfurization and it aims at pointing out the importance of controlling the morphological and chemical properties of the N-C phase to make a step forward in the direction of catalytic materials featured by improved desulfurization performance and resistance towards deactivation/fouling phenomena. We have recently demonstrated the excellent performance of a N-C/SiC catalyst in the presence of relatively high concentrations of toluene (up to 20,000 ppm, 2 vol.%) as contaminant in a desulfurization process operated under relatively harsh reaction conditions (GHSV up to 2400 h^{-1}). We have also demonstrated the existence of a beneficial "solvent effect" played by toluene on the process selectivity as a distinctive feature of these metal-free catalytic materials. With these catalysts, the toluene present in the gas stream was found to facilitate the desorption and removal of elemental sulfur from the catalyst surface thus reducing the sulfur residence time in contact with the N-C network and preserving the process from the occurrence of undesired over-oxidation paths at the catalyst surface. Hereafter we demonstrate how a more appropriate control of the soaking/thermal treatment cycles in the N-C phase growth holds beneficial effects on the chemical and morphological properties of the latter as well as on the ultimate materials performance in the direct H_2S oxidation from sour gases [H_2S = 0.3 vol.%; O_2/H_2S = 2.5; GHSV up to 3200 h^{-1}) containing aromatic contaminants (i.e., toluene) at concentrations as high as 40,000 ppm (4 vol.%). It should be pointed out that such a toluene concentration is markedly higher compared to that traditionally encountered in sour gas stream (i.e., 1200–2100 ppm) and it has been deliberately employed to validate the remarkable resistance of our metal-free catalyst towards BTX deactivation phenomena. To address these goals a new impregnation/thermal sequence for the N-C phase growth on SiC extrudates is proposed and a comparison with that previously reported by us has been used to shed light on the ideal chemical and morphological properties of the N-C phase for the desulfurization process to occur. In addition, N-C/SiC performance has been compared with that of the benchmark Fe_2O_3/SiO_2 catalyst [40] for the sake of completeness.

2. Results and Discussion

2.1. Catalysts Characterization and Properties of N-C^2/SiC and N-C^4/SiC

The SiC coating with the N-C phase was accomplished following two alternative sequences of the solid support soaking in a standardized impregnation solution and successive thermal treatments (see Section 3 for details). Figure 1 provides a visual sketch of the operative sequences employed for the preparation of N-C^2/SiC and N-C^4/SiC composites, respectively. At odds with N-C^4/SiC, it can be inferred that preparation of N-C^2/SiC follows a simplified synthetic path that includes only two impregnation/drying steps (green cycles) before undergoing annealing treatment at 900 °C for 2 h under inert atmosphere (red cycle) without passing any intermediate calcination step (orange cycle in N-C^4/SiC sequence).

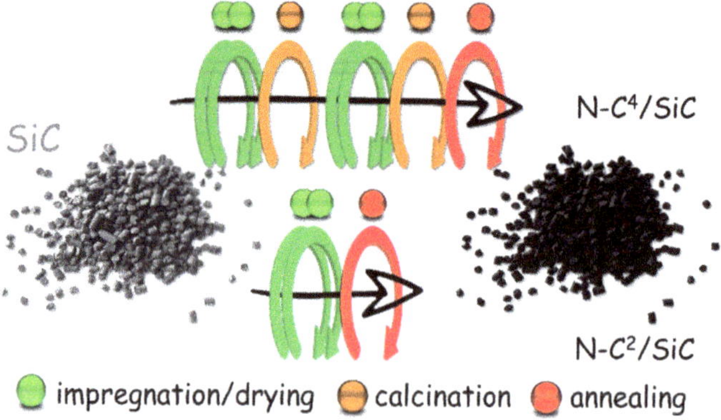

Figure 1. Quick visual representation of the two alternative synthetic paths used for the preparation of N-C^4/SiC (upper black arrow) and N-C^2/SiC (down black arrow) composites, respectively. Green cycles refer to the impregnation/drying steps, orange cycles refer to the material calcination at 400 °C for 2 h and red cycles refer to the material annealing at 900 °C for 2 h under Argon atmosphere.

Although a complete N-C^4/SiC characterization and its H$_2$S desulfurization properties have previously been detailed by us elsewhere, [26,34,36] their comparison with those of N-C^2/SiC is necessary to better highlight the improved properties and catalytic performance of the latter. Both composites (Figure 2A) were analyzed in terms of specific surface area (SSA), total pore volume and pore-size distribution by N$_2$ physisorption at the liquid N$_2$ temperature (77 K).

As Table 1 shows, the specific surface areas (SSA) of the two composites are very close each other and they are almost twice than that measured on the plain SiC support. Both composites present Type IV isothermal profiles (Figure 2A) with distinctive H$_2$ hysteresis loops in the 0.45–1.0 P/P$_0$ range, typical of mesoporous networks featured by complex pore structures of ill-defined shape [41]. Sample N-C^4/SiC presents a more pronounced hysteresis loop that is ascribed to the presence of a large extent of smaller mesopores and a lower mean pore-size distribution (Table 1) that facilitate the occurrence of capillary condensation phenomena. This datum is in line with the pore-volume distribution curves recorded on the three samples at comparison (Figure 2B). From the inspection of these profiles, it can be deduced a reduced content of small mesopores (in the 2–5 nm range) in N-C^2/SiC compared to its counterpart N-C^4/SiC, together with the presence of larger mesopores (prevalently in the 20-60 nm range) available on the former only. As far as N-C phase mass content is concerned, thermogravimetric analysis (TGA) on air (50 mL min^{-1}) on the two samples (Figure 2D) showed only negligible differences on the content of organic deposit (6.7 wt.% on N-C^2/SiC and 6.9 wt.% on N-C^4/SiC) whatever the catalyst preparation sequence adopted (Figure 1). Both profiles evidence a distinctive weight loss at 622 and 613 °C for N-C^2/SiC and N-C^4/SiC, respectively, where the derivative of the thermogravimetric curves (DTG) holds its maximum value.

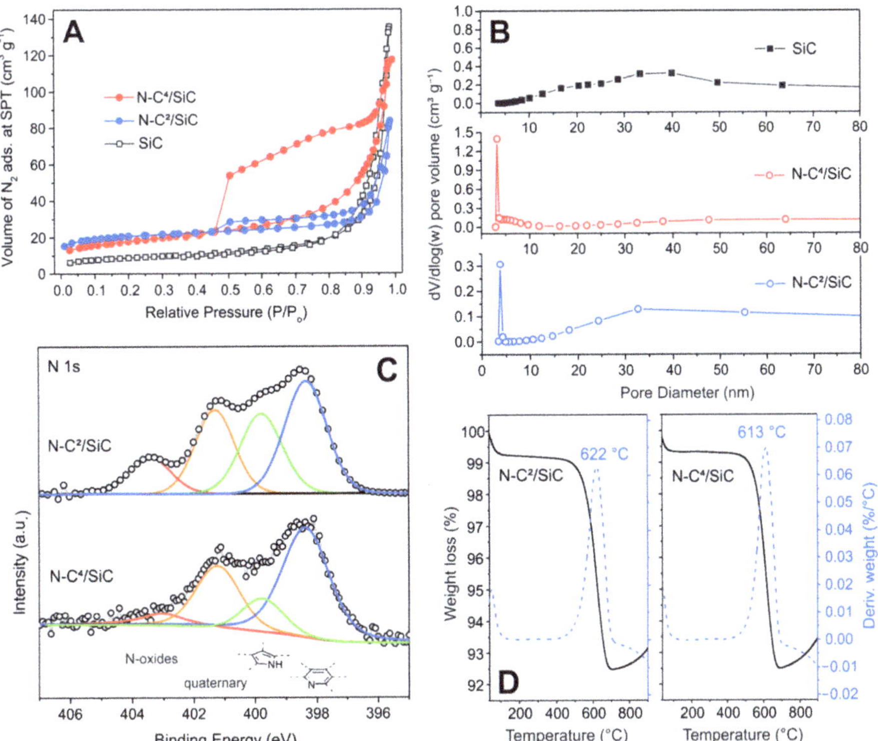

Figure 2. (**A**) N_2 adsorption-desorption isotherm linear plot of SiC (light grey curve), N-C^2/SiC (blue curve) and N-C^4/SiC samples (red curve) recorded at 77 K along with (**B**) the respective pore-size distributions measured (BJH method). (**C**) High-resolution XPS N 1s core level region of N-C^2/SiC (up) and N-C^4/SiC (down) at comparison, along with the relative curves fittings. (**D**) Thermogravimetric/derivative of the thermogravimetric curves (TG/DTG) profiles of N-C^2/SiC (left-side hand) and N-C^4/SiC (right-side hand) at comparison. Weight loss is measured arbitrarily in the 200–700 °C temperature range. Operative conditions: Air, 50 mL/min; heating rate: From 40 to 900 °C at 10 °C/min.

The XPS analysis of the materials at comparison confirmed the same composition providing largely superimposable survey profiles (Figure S3). The high-resolution N 1s analysis of the two samples has pointed out a slightly changed composition of the N-configurations available. Table 1 lists the relative % of the different N-species in the samples. N 1s peaks deconvolution (Figure 2C) accounts for three main components centered at 398.4 ± 0.2 eV (N-pyridinic—blue line), 399.7 ± 0.2 eV (N-pyrrolic—green line) and 401.2 ± 0.2 eV (N-quaternary—orange line), along with an additional shoulder (more pronounced in N-C^2/SiC) at higher binding energies (403.0 ± 0.3 eV—red line) and ascribed to the presence of N-oxidized species.

As Table 1 shows, the relative % of N-species obtained by the two alternative impregnation/thermal sequences (Figure 1) give rise to a redistribution of the N-configurations available at the materials surface. In particular, reducing the number of impregnation/drying steps and omitting the material calcination at 400 °C for 2 h, the percentage of pyrrole moieties is nearly doubled, that of N-oxide species increases, while that of basic N-pyridine sites decreases appreciably.

Table 1. Selected chemico-physical and morphological properties of catalysts and precursors.

Entry	Sample	SSA [a] $(m^2 g^{-1})$	Total Pore Volume $(cm^3 g^{-1})$ [b]	Average Pore Size (nm) [c]	Surface Basic Sites $(mmol g^{-1})$ [d]	N-C wt.% (from TGA)	N wt.% (from EA)	XPS Data, N-Species (%) [f]				
								N at.% [e]	Pyridinic	Pyrrolic	Graphitic	Oxidized
1	SiC	30	0.21	27.3	- [g]	-	-	-	-	-	-	-
2	N-C^2/SiC	69	0.11	14.8	0.63	6.7	2.1	5.1	36.8	26.0	25.6	11.6
3	N-C^4/SiC	61	0.18	5.6	0.45	6.9	1.5	4.5	53.1	13.1	26.8	7.0
4	Fe$_2$O$_3$/SiO$_2$	160	0.40	10.3	n.d.	-	-	-	-	-	-	-

[a] Brunauer-Emmett-Teller (BET) specific surface area (SSA) measured at T = 77 K. [b] Total pore volume determined using the adsorption branch of N_2 isotherm at P/P_0 = 0.98. [c] Determined by BJH desorption average pore width (4V/A). [d] measured by acid-base titration. [e] Determined by XPS analysis. [f] Determined by high resolution XPS N 1s core region and its relative peak deconvolution. [g] The pH value of an aqueous SiC dispersion lies close to pH 6.6 whereas the pH value of aqueous N-C^x/SiC dispersions (x = 2, 4) ranks close to 9.4. Data reported for elemental analysis (EA) and acid-base titration are calculated as the mean values over three independent runs. *n.d.* = not determined.

At odds with this trend, a quantitative estimation of basic sites available at the surface of both catalysts carried out by acid-base titration (see Section 3) has unveiled the higher basic character of N-C^2/SiC (0.63 mmol g^{-1}) respect to N-C^4/SiC (0.45 mmol g^{-1}). This discrepancy can be justified by the higher N-content of the N-C phase in N-C^2/SiC. Indeed, according to the N-C wt.% measured by TGA on the two samples (see Table 1 and Figure 2D) and the N wt.% measured by EA (Table 1), the N wt.% content normalized to the weight of N-C coating was calculated in 22 N wt.% and 31 N wt.% for N-C^4/SiC and N-C^2/SiC, respectively. In spite of a reduction in the percentage of basic pyridine sites for N-C^2/SiC, its markedly higher N-content reasonably accounts for its higher basic surface character [26,42].

2.2. Desulfurization Performance of N-C^2/SiC and N-C^4/SiC in the Presence of a Relatively High Vol.% of Toluene as Acid Gas Contaminant

The catalysts screening in the sour gas desulfurization starts from the awareness that aromatic contaminants like toluene hold a positive "solvent effect" on the catalytic outcomes of these metal-free systems [36]. Indeed, they favor a faster removal of sulfur deposits from the material mesopores, thus preventing the occurrence of undesired over-oxidation paths. In a recent desulfurization report with N-C^4/SiC, we have already shown the remarkably high resistance of this metal-free catalyst towards deactivation/fouling in the presence of toluene as the acid gas stream contaminant up to 5000 ppm. We also claimed an increase of the elemental sulfur rate up to 30% compared to selectivity values recorded for the same metal-free catalyst operated under identical—but toluene-free—conditions [36]. The comparative study between N-C^4/SiC and N-C^2/SiC points out on the importance of controlling the morphology and chemical composition of the N-C phase in order to get more robust, selective and efficient desulfurization catalysts suitable to operate the process under severe operative conditions. In this study, toluene was selected again as a model aromatic contaminant in sour gases [43] and the performance of two metal-free systems were compared for the sake of completeness with that of the benchmark Fe$_2$O$_3$/SiO$_2$ catalyst under the same conditions.

Aimed at stressing the relevance of morphological and chemical surface properties of N-C active phase in the process, we deliberately selected harsh operative conditions since the beginning of the desulfurization reaction. The process was then followed for a relatively long time (>200 h) and until the three samples clearly followed distinct catalytic paths. As a first trial, a mixture of H$_2$S (0.3 vol.%; O$_2$-to-H$_2$S ratio = 2.5; steam: 10 vol.%) and toluene (40,000 ppm; 4 vol.%) was passed downward the catalyst bed (6 g, V$_{cat}$ ~7.5 cm^3 for N-C^4/SiC and N-C^2/SiC and 6 g, V$_{cat}$ ~5.8 cm^3 for Fe$_2$O$_3$/SiO$_2$) heated at 210 °C and with a GHSV of 3200 h^{-1} (STP) (Figure 3A).

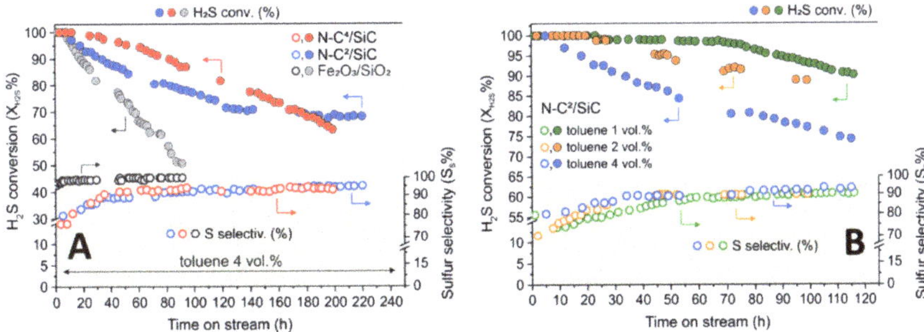

Figure 3. (**A**) Desulfurization performance on N-C^2/SiC, N-C^4/SiC and Fe$_2$O$_3$/SiO$_2$ catalysts of an acid gas stream ([H$_2$S] = 0.3 vol.%) in the presence of 40,000 ppm of toluene (4 vol.%) as contaminant in the stream. Catalysis details: 6 g (V$_{cat}$ ~ 7.5 cm^3 for N-C^4/SiC and N-C^2/SiC) or 6 g (V$_{cat}$ ~ 5.8 cm^3 for Fe$_2$O$_3$/SiO$_2$); O$_2$-to-H$_2$S ratio = 2.5, [H$_2$O] = 10 vol.%, He (balance); reaction temperature = 210 °C, GHSV (STP) = 3200 h^{-1}. (**B**) Desulfurization performance on N-C^2/SiC at variable toluene concentrations (10,000, 20,000 and 40,000 ppm or 1, 2 and 4 vol.%).

The long-term desulfurization process (up to 220 h) in the presence of 40,000 ppm of toluene in the stream served to highlight the excellent sulfur selectivity (S$_S$ up to ~94% with N-C^2/SiC at the steady-state-conditions) as well as the remarkably high coke resistance of both metal-free catalysts under severe and prolonged operative conditions. As Figure 3A shows, all catalysts ensure a quantitative H$_2$S conversion (100%) within the first 10 h on stream. Afterwards, H$_2$S conversion (X$_{H2S}$) decreases appreciably whatever the nature of the catalyst employed, with the benchmark Fe$_2$O$_3$/SiO$_2$ showing the much faster deactivation rate compared to its metal-free counterparts. Under these conditions, the iron-based catalyst shows a quantitative selectivity towards elemental sulfur although the high toluene content in the gas stream rapidly compromises its H$_2$S conversion capacity that falls below 50% just after 85 h on reaction. In spite of a slightly lower sulfur selectivity (S$_S$ in the 93–95% range), N-C^2/SiC and N-C^4/SiC show a markedly higher deactivation resistance and follow distinct deactivation paths. The N-C^4/SiC starts an appreciable deactivation only after 25 h on stream. Afterwards, X$_{H2S}$ gradually but constantly decreases down to 65% (after ~200 h on reaction). Noteworthily, N-C^2/SiC shows a more rapid deactivation in the first hours on stream that however reaches a H$_2$S conversion plateau around 160 h that is almost constantly maintained even after 220 h. As far as sulfur selectivity is concerned, both metal-free catalysts constantly rank above 90%. According to our previous report, the positive S$_S$ increase is ascribable to a co-solvent action played by toluene. Indeed, it facilitates the dissolution of sulfur deposits thus reducing their contact time with the N-C active phase and hence limiting the occurrence of undesired over-oxidation paths. The higher stability of N-C^2 active phase must be searched in the minor but critical chemical and morphological differences with N-C^4 phase and hence within the simplified impregnation/thermal sequence for the synthesis of N-C^2/SiC compared to N-C^4/SiC. It can be inferred that a larger mean pore size distribution in N-C^2 (Table 1, entry 2 vs. 3), a lower percentage of small mesopores in favor of larger ones (Figure 2B) together with a higher basic surface character of the sample (Table 1) account for its improved catalytic performance. Indeed, larger mesopores reduce the occurrence of pore clogging phenomena, ensure a more effective reagents access to the catalyst active phase and allow a more effective toluene scrubbing action towards the formed sulfur deposits. At the same time, a higher basic surface character creates the ideal microenvironment for the generation of local H$_2$S gradients and reduces the occurrence of cracking side-processes responsible for undesired "catalyst coking" [15,44–46]. The ability of N-C^2/SiC to stabilize on a relatively high H$_2$S conversion values is attributed to the simultaneous occurrence of all these phenomena. Selectivity values up to 94% at the steady-state conditions together with a X$_{H2S}$ constantly lying on 68% denote an efficient and selective desulfurization process where the toluene constant rate in the stream no longer affects the catalyst performance. On the other hand, it

This positive trend in the absence of harsher oxidative conditions, led us to consider the action of toluene on the performance of our metal-free catalyst as that of an "interfering solvent" rather than a source of carbon for the growth of coke deposits. Indeed, the toluene confinement into the pores of the catalyst active phase alters the performance of the latter reversibly without causing any real catalyst coking and thus any irreversible deactivation. If the use of steam is known to be functional to the process by creating a thin water film on N-C surface that favors the diffusion of hydrophilic H_2S molecules into the catalyst pores, [28] the co-existence with a hydrophobic co-solvent (e.g., toluene) will translate into a depletion of reagents uptake and their diffusion towards the catalyst active sites. Anyhow, once toluene molecules are gradually desorbed from the pores and channels of the N-C network by using a toluene-free sour gas stream, the catalyst recovers its original performance.

The excellent coke resistance of $N-C^2/SiC$ catalyst in the presence of relatively high concentrations of aromatics in the gas stream (40,000 ppm), has been confirmed by the analysis of the recovered $N-C^2/SiC$ (spent catalyst) after 720 h on reaction. Figure 5A refers to the TGA analysis of the freshly prepared $N-C^2/SiC$ (left-side hand) with its spent (right-side hand) counterpart put at comparison.

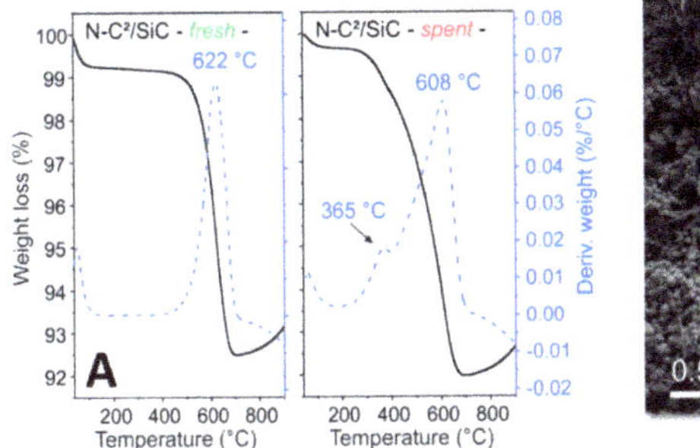

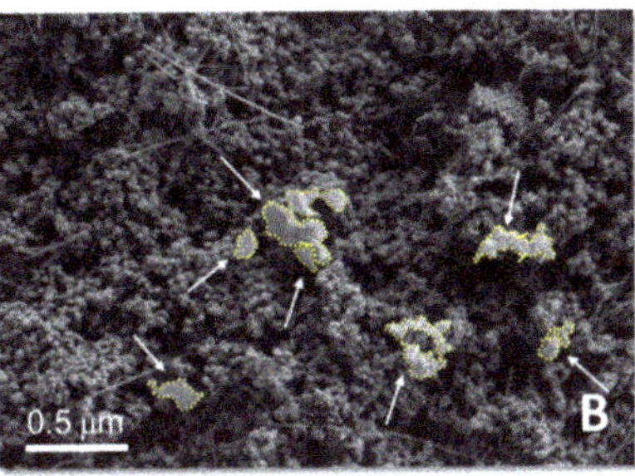

Figure 5. (**A**) TG/DTG profiles of the freshly prepared $N-C^2/SiC$ (left-side hand) and its exhaust counterpart (right-side hand) at comparison. Weight loss is measured arbitrarily in the 200–700 °C temperature range and it corresponded to: 6.7 wt. loss % on the fresh $N-C^2/SiC$ and 7.6 wt. loss % on the spent $N-C^2/SiC$. Operative conditions: Air, 50 mL/min; heating rate: from 40 to 900 °C at 10 °C/min. (**B**) SEM micrograph of $N-C^2/SiC$ after its recovery at the end of a long-term catalytic run of 720 h. White arrows indicate residual sulfur deposits marked all around by yellow dashed lines.

The spent sample presents a minor shoulder featured by a maximum weight loss centered around 365 °C that accounts for about 0.9% of the overall weigh loss after the complete N-C active phase burning. Although we cannot definitively rule out the generation to a certain extent of low-melting coke deposits, we believe that such a little shoulder in the TGA profile of the exhaust sample is reasonably ascribable to the oxidation of sulfur residues. In spite of a temperature in the last part of the catalytic run (250 °C) that is higher than the sulfur dewpoint, some residues still remain available on the catalyst surface. Indeed, the SEM analysis of the recovered $N-C^2/SiC$ (Figure 5B) clearly shows their presence in the form of patchy islands, whose generation is attributed to the residual catalyst activity during its cooling phase.

3. Experimental Section

3.1. Materials and Methods

β-SiC supports [extrudates (3 × 1 mm; $h × \oslash$), V = ~0.002 cm^3, SSA measured by N_2 physisorption (at 77 K) of 30 ± 1 m^2 g^{-1}] were provided by SICAT SARL (www.sicatcatalyst.com), thoroughly washed with deionized water in order to remove powdery fractions, hence dried at 130 °C overnight before use. Ammonium carbonate (($NH_4)_2CO_3$, MW: 96.09 g mol^{-1}; Lot: A0356079), D-glucose 100% ($C_6H_{12}O_6$, MW: 180.16 g mol^{-1}) and citric acid ($C_6H_8O_7$ anhydrous, ≥99.5%, MW: 192.12 g mol^{-1}) were provided by ACROS Organic[TM], MYPROTEINTM and VWR Chemicals, respectively. Unless otherwise stated, all reagents and solvents were used as provided by commercial suppliers without any further purification/treatment. N-C^4/SiC sample was prepared following the literature procedure previously reported by us [26,34,36] (and described in brief hereafter for the sake of completeness). N-C^2/SiC composite was prepared from the same soaking water solution of food-grade components, using a simplified impregnation/thermal sequence (vide infra). Scanning Electron Microscopy (SEM) was carried out on an UHR-SEM Gaia 3 FIB/SEM (TESCAN, Brno-Kohoutovice, Czech Republic). A 10 kV electron beam was used for SEM imaging operated in high-vacuum mode, using BSE and SE detectors. N_2 adsorption-desorption measurements were carried out on a Micromeritics$^®$ (Milan, Italy) sorptometer at the liquid N_2 temperature and relative pressures between 0.06 and 0.99 P/P_0. Each sample was outgassed at 250 °C under ultra-high vacuum for 8 h prior analysis in order to desorb moisture and adsorbed volatile species. The X-ray Photoelectron Spectroscopy (XPS) was carried out in an ultrahigh vacuum (UHV) spectrometer (Prevac, Rogów, Poland) equipped with a CLAM4 (MCD) hemispherical electron analyzer. The Al Kα line (1486.6 eV) of a dual anode X-ray source was used as incident radiation. Survey and high-resolution spectra were recorded in constant pass energy mode (100 and 20 eV, respectively). The CASA XPS program with a Gaussian-Lorentzian mix function and Shirley background subtraction was employed to deconvolute XPS spectra. Elemental analyses were performed on a Thermo FlashEA 1112 Series CHNS-O analyzer (Thermo Fisher Scientific, Waltham, MA, USA) and elemental average values were calculated over three independent runs. Powder Diffraction (PXRD) measurements were carried out on a Bruker D-8 Advance diffractometer (Bruker, Billerica, MA, USA) quipped with a Vantec detector (Cu Kα radiation) working at 40 kV and 40 mA. X-ray diffractogram was recorded in the 10–80° 2θ region at room temperature in air. Iron loading for the benchmark Fe_2O_3/SiO_2 was fixed by Inductively Coupled Plasma Atomic Emission spectrophotometry (ICP-AES) after sample acidic mineralization, using an Optima 2000 Perkin Elmer Inductively Coupled Plasma (ICP) Dual Vision instrument (Perkin Elmer Italia, Milan, Italy). Thermogravimetric analyses were performed on air (50 mL min^{-1}) from 40 to 900 °C (heating rate: 10 °C min^{-1}) on an EXSTAR Seiko 6200 analyser (Riga, Latvia). Acid-base titration was accomplished using the following procedure [47–49]: 10 mg of N-C^x/SiC (x = 2 or 4) were suspended in 7 mL of a standardized HCl solution (3×10^{-3} M, standardized with Na_2CO_3 as primary standard) and stirred at room temperature for 48 h. After that, three aliquots of the solution were titrated with a standardized NaOH solution (2.5×10^{-3} M). The basic sites loading was finally calculated as the average value over the three independent titration runs.

3.2. General Procedure for the Preparation of N-C^2/SiC and N-C^4/SiC Catalysts

N-C^2/SiC and N-C^4/SiC catalysts were prepared from the same water soaking solution for the SiC support but following different impregnation/thermal treatment sequences. The impregnation solution was prepared by dissolving at room temperature 3 g of D-glucose and 4.5 g of citric acid in 20 mL of ultrapure Milli-Q water. Afterwards, 3.46 g of ammonium carbonate were added to the stirred solution during which an effervescence due to CO_2 evolution starts (CAUTION! CO_2 effervescence needs to be carefully controlled during this phase by portioning the amount of $(NH_4)_2CO_3$ added over time). The as-prepared solution was used for the soaking/impregnation of 20 g of SiC extrudates whatever the nature of the target composite prepared (N-C^2/SiC or N-C^4/SiC) [34]. N-C^4/SiC was obtained following previously reported procedures [36]. Accordingly, SiC was soaked twice in the above water solution and excess of water—remaining after each impregnation step—was gently evaporated at 40 °C for 3 h. Afterwards, the solid was dried at 130 °C overnight before being calcined in air at 400 °C for 2 h (heating rate 2 °C min^{-1}). The as-obtained composite underwent identical impregnation/thermal treatment sequence at the end of which the sample was annealed at 900 °C (heating rate: 10 °C min^{-1}) for 2 h under inert (Ar) atmosphere. As a result, a N-doped and mesoporous C-graphitic coat at the SiC outer surface was formed.

As far as N-C^2/SiC is concerned, it was obtained by soaking SiC twice in the impregnation solution, evaporating the excess of water at 40 °C for 2 h, drying the sample at 130 °C overnight before moving it directly to the annealing phase at 900 °C under inert atmosphere.

Fe$_2$O$_3$/SiO$_2$ (2.6 wt.% Fe) was prepared according to literature data [40]. To this aim, 10 g of SiO$_2$ powder were treated by incipient wetness impregnation with an aqueous solution of iron nitrate (Fe(NO$_3$)$_3$·9H$_2$O, 2.23 g; MW: 404,00 g mol^{-1}) in 10 mL of ultrapure Milli-Q water. The resulting solid was dried at 130 °C overnight before being calcined in air at 350 °C (heating rate: 5 °C min^{-1}) and maintained at the target temperature for additional 2 h before being used as such in catalysis. The final iron loading was measured by ICP-AES analysis and it was fixed to 2.6 wt.%. The XRD spectrum of the iron catalyst (Figure S1) was in accord with related literature reports [50].

Although the comparison of a metal-based catalyst with a metal-free one might appear as a meaningless exercise, Fe$_2$O$_3$/SiO$_2$ is a common benchmark system for the H$_2$S desulfurization, and its employment under conditions identical to those (hard) operated with the metal-free composites provides a clear-cut evidence of the superior performance and stability of the latter. Moreover, the direct H$_2$S oxidation to elemental sulfur in the presence of aromatics as contaminants in the gaseous stream is almost absent in the literature. In addition, SiO$_2$ as the metal active phase support was properly selected and compared with SiC, the latter being naturally coated by a thin layer of SiO$_x$C$_y$/SiO$_2$ once exposed to air at room temperature [27].

3.3. Selective H$_2$S Desulfurization of Sour Gases to Elemental Sulfur

The H$_2$S oxidation process can be described by Equations (1)–(3) reported below [51,52]. For catalytic trials, 6 g of N-C^2/SiC or N-C^4/SiC (V$_{cat}$ ~7.5 cm^3) were loaded on a silica wool pad, housed in a Pyrex tubular (⌀$_{ID}$: 16 mm) reactor housed in a vertical electrical furnace, and the catalytic reactions were operated isothermally under atmospheric pressure. A graphical representation of the desulfurization scheme is provided on Figure S2.

$$2H_2S + O_2 \rightarrow 2S + 2H_2O \qquad \Delta H = -187\ \text{kJ/mol} \qquad (1)$$

$$S + O_2 \rightarrow SO_2 \qquad \Delta H = -297\ \text{kJ/mol} \qquad (2)$$

$$2H_2S + 3O_2 \rightarrow 2SO_2 + 2H_2O \qquad \Delta H = -518\ \text{kJ/mol} \qquad (3)$$

The temperature of the furnace was controlled by a K-type thermocouple and a Minicor regulator. The reactants gas mixture [H$_2$S (0.3 vol.%), O$_2$/H$_2$S = 2.5, H$_2$O (10 vol.%) in inert He as carrier (balance)] was passed downward through the catalyst bed, being gas flow rates monitored through Brooks 5850TR mass flow controllers. Steam (10 vol.%) was ensured by bubbling the inert carrier in a saturator containing hot water at 61 °C. *CAUTION! H$_2$S is a colourless, flammable, highly toxic gas. It must be handled—including its solutions—rigorously under a fume-hood and with all necessary precautions, especially a specific leak detector installed close to the operating setup.* The gas hourly space velocity (GHSV) was fixed at 3200 h^{-1} (corresponding to 400 mL min^{-1} or 4000 mL g$_{cat}$$^{-1}$ h^{-1}) and the O$_2$-to-H$_2$S molar ratio was kept constant to 2.5. The influence of toluene on the catalyst performance was investigated by feeding the reactants stream with toluene vapors at concentrations comprised between 1 vol.% (10,000 ppm) up to 4.0 vol.% (40,000 ppm). Toluene was fed-up in the stream by flowing He through a saturator containing pure toluene constantly maintained at 40 °C. To this purpose an independent line of He (Figure S2) was used to feed-up toluene in the reagents stream and its target concentration was adjusted by regulating the flow of the carrier. The amount of toluene passing through the catalyst was double checked by measuring the real liquid volume of toluene vaporized per day of experiment. All catalytic runs were carried out in continuous mode. Hence, most of the formed elemental sulfur was vaporized (because of the high partial pressure of sulfur at the target reaction temperatures) and condensed alongside with steam at the reactor outlet in a trap maintained at room temperature. The analysis of the inlet and outlet gases was performed on-line using a Varian CP-3800 gas chromatograph (GC) equipped with a Chrompack CP SilicaPLOT capillary column and a thermal catharometer detector (TCD) for the detection of O_2, H_2S, H_2O, and SO_2 (down to 30 ppm). H_2S and SO_2 concentrations were recalculated on the basis of the corrected flow after steam condensation in a trap (Figure S2). All connecting lines were wrapped with thermal tapes maintained at 140 °C in order to prevent any condensation phenomena.

4. Conclusions

To summarize, the optimization of the SiC impregnation/thermal treatment sequences for the control of the surface chemistry and morphology of a highly N-rich carbon phase coating has been proposed. The new sequence for the N-C^2/SiC preparation has pointed out the importance of controlling the chemico-physical properties of the N-C phase as to get more efficient, selective and stable metal-free catalysts to be employed in the selective H_2S oxidation of sour gas streams and in the presence of aromatic contaminants concentrations as high as 40,000 ppm (4 vol.%). In the study, we have demonstrated how larger mesopores at the N-C active phase along with its higher basic surface character hold largely beneficial effects on the catalyst performance and its stability on stream. While the former reduces the occurrence of pore clogging phenomena, ensures a more effective reagents access to the catalyst active phase and allows a more effective scrubbing/removal action of the sulfur deposits by the toluene, the latter creates the ideal microenvironment for the generation of local H_2S gradients and reduces the occurrence of cracking side-processes responsible for the "catalyst coking". Most importantly, harsh H_2S desulfurization conditions in the presence of an intermittent toluene rate (from 0 to 4 vol.% and again down to 0 vol.%) in the sour gas stream has allowed to better elucidate the effect of aromatics on the performance and long-term stability of these metal-free desulfurization catalysts. Our results have pointed out that metal-free catalysts of this type suffer only marginally of irreversible deactivation caused by the generation of coke deposits. On the other hand, the reduced X_{H2S} efficiency in the presence of toluene can be reasonably ascribed to a competitive pore-filling by the toluene as the steam co-solvent. While steam facilitates H_2S diffusion into the pores and channels of the N-C active phase, the hydrophobic toluene can detrimentally compete with the reagent uptake on the catalyst active phase. However, once toluene molecules are gradually desorbed from the pores and channels of the N-C network (i.e., purging the catalyst under a toluene-free sour gas stream), the latter recovers its original performance.

Overall, three catalytic system at comparison (N-C^2/SiC, N-C^4/SiC and Fe_2O_3/SiO_2) have served to highlight the role of metal-free catalysts and their surface chemico-physical properties on their H_2S desulfurization performance in the presence of aromatics as contaminants. As shown in Figure 3A, while the iron-based catalyst rapidly deactivates because of the fouling of its active-phase (catalyst coking), the two metal-free systems behave differently as a function of their chemical and morphological properties. Under these conditions, the higher the basic surface properties and the higher the density of larger mesopores in the material, the higher the catalyst stability and durability on run.

Supplementary Materials: The following are available online at https://www.mdpi.com/2073-434 4/11/2/226/s1, Figure S1: XRD profile of Fe_2O_3/SiO_2 catalyst (Fe 2.6 wt.%), Figure S2: Schematic representation of a desulfurization apparatus, Figure S3: XPS survey spectra of N-C^2/SiC and N-C^4/SiC.

Author Contributions: Conceptualization, C.P.-H. and G.G.; methodology, C.D.-V., C.P.H., J.-M.N., C.P. and G.G.; validation, C.D.-V., C.P.H., J.-M.N. and G.G.; formal analysis, C.D.-V., T.T.-H, G.T. and L.N.-D.; investigation, C.D.-V., C.P.H., J.-M.N., G.T. and G.G.; resources, C.P.-H., G.G., C.P. and L.N.-D.; data curation, C.D.-V., C.P.-H., G.T. and G.G.; writing—original draft preparation, G.G. and C.P.-H.; writing—review and editing, G.G., G.T., C.P., L.N.-D. and C.P.-H.; supervision, C.P.-H., J.-M.N., G.G. and L.N.-D.; funding acquisition, C.P.-H., G.G. and L.N.-D. All authors have read and agreed to the published version of the manuscript.

Funding: This research was funded by the TRAINER project (Catalysts for Transition to Renewable Energy Future) of the "Make our Planet Great Again" program (Ref. ANR-17-MPGA-0017), the PRIN 2017 Project Multi-e (20179337R7) "Multielectron transfer for the conversion of small molecules: an enabling technology for the chemical use of renewable energy" and the Vietnam National Foundation for Science and Technology Development (NAFOSTED; grant number 104.05-2017.336). This research was also supported by SATT-Conectus through the DECORATE project.

Data Availability Statement: Data available on request.

Acknowledgments: SICAT SARL (www.sicatcatalyst.com) is gratefully acknowledged for providing SiC pellet samples.

Conflicts of Interest: The authors declare no conflict of interest.

References

1. Comparison Against Other Fossil Fuels. Available online: https://www.swarthmore.edu/environmental-studies-capstone/comparison-against-other-fossil-fuels (accessed on 7 December 2020).
2. Faramawy, S.; Zaki, T.; Sakr, A.A.-E. Natural gas origin, composition, and processing: A review. *J. Nat. Gas Sci. Eng.* **2016**, *34*, 34–54. [CrossRef]
3. Wieckowska, J. Catalytic and adsorptive desulphurization of gases. *Catal. Today* **1995**, *24*, 405–465. [CrossRef]
4. Eow, J.S. Recovery of Sulfur from Sour Acid Gas: A Review of the Technology. *Environ. Prog.* **2002**, *21*, 143–162. [CrossRef]
5. Piéplu, A.; Saur, O.; Lavalley, J.C.; Legendre, O.; Nédez, C. Claus Catalysis and H_2S Selective Oxidation. *Catal. Rev. Sci. Eng.* **1998**, *40*, 409–450. [CrossRef]
6. Zhang, X.; Tang, Y.; Qu, S.; Da, J.; Hao, Z. H_2S-Selective Catalytic Oxidation: Catalysts and Processes. *ACS Catal.* **2015**, *5*, 1053–1067. [CrossRef]
7. Ballaguet, J.-P.R.; Vaidya, M.M.; Duval, S.A.; Harale, A.; Khawajah, A.H.; Tammana, V.V.R. Sulfur Recovery Process for Treating Low to Medium Mole Percent Hydrogen Sulfide Gas Feeds with BTEX in a Claus Unit. U.S. Patent 9981848-B2, 29 May 2018.
8. Bahadori, A. *Natural Gas Processing Technology and Engineering Design*, 1st ed.; Gulf Professional Publishing: Houston, TX, USA, 2014.
9. Jahangiri, M.; Shahtaheri, S.J.; Adl, J.; Rashidi, A.; Kakooei, H.; Forushani, A.R.; Ganjali, M.R.; Ghorbanali, A. The adsorption of benzene, toluene and xylenes (BTX) on the carbon nanostructures: The study of different parameters. *Fresenius Environ. Bull.* **2011**, *20*, 1036–1045.
10. Sulphur Prices, Markets & Analysis. Available online: http://Www.Icis.Com/Fertilizers/Sulphur/ (accessed on 9 December 2020).
11. Serp, P.; Corrias, M.; Kalck, P. Carbon nanotubes and nanofibers in catalysis. *Appl. Catal. A Gen.* **2003**, *253*, 337–358. [CrossRef]
12. Su, D.S.; Perathoner, S.; Centi, G. Nanocarbons for the Development of Advanced Catalysts. *Chem. Rev.* **2013**, *113*, 5782–5816. [CrossRef]
13. Ba, H.; Luo, J.; Liu, Y.; Duong-Viet, C.; Tuci, G.; Giambastiani, G.; Nhut, J.-M.; Nguyen-Dinh, L.; Ersen, O.; Su, D.S.; et al. Macroscopically shaped monolith of nanodiamonds @nitrogen-enriched mesoporous carbon decorated SiC as a superiormetal-free catalyst for the styrene production. *Appl. Catal. B Environ.* **2017**, *200*, 343–350. [CrossRef]
14. Diao, J.; Feng, Z.; Huang, R.; Liu, H.; Hamid, S.B.A.; Su, D.S. Selective and Stable Ethylbenzene Dehydrogenation to Styrene over Nanodiamonds under Oxygen-lean Conditions. *ChemSusChem* **2016**, *9*, 662–669. [CrossRef]
15. Ba, H.; Liu, Y.; Truong-Phuoc, L.; Duong-Viet, C.; Nhut, J.-M.; Nguyen-Dinh, L.; Ersen, O.; Tuci, G.; Giambastiani, G.; Pham-Huu, C. N-Doped Food-Grade-Derived 3D Mesoporous Foams as Metal-Free Systems for Catalysis. *ACS Catal.* **2016**, *6*, 1408–1419. [CrossRef]
16. Chizari, K.; Deneuve, A.; Ersen, O.; Florea, I.; Liu, Y.; Edouard, D.; Janowska, I.; Begin, D.; Pham-Huu, C. Nitrogen-Doped Carbon Nanotubes as a Highly Active Metal-Free Catalyst for Selective Oxidation. *ChemSusChem* **2012**, *5*, 102–108. [CrossRef]
17. Li, M.; Xu, F.; Li, H.; Wang, Y. Nitrogen-doped porous carbon materials: Promising catalysts or catalyst supports for heterogeneous hydrogenation and oxidation. *Catal. Sci. Technol.* **2016**, *6*, 3670–3693. [CrossRef]
18. Luo, J.; Peng, F.; Wang, H.; Yu, H. Enhancing the catalytic activity of carbon nanotubes by nitrogen doping in the selective liquid phase oxidation of benzyl alcohol. *Catal. Commun.* **2013**, *39*, 44–49. [CrossRef]
19. Tang, C.; Zhang, Q. Nanocarbon for Oxygen Reduction Electrocatalysis: Dopants, Edges, and Defects. *Adv. Mater.* **2017**, *29*, 1604103. [CrossRef]
20. Zhang, L.; Jia, Y.; Yan, X.; Yao, X. Activity Origins in Nanocarbons for the Electrocatalytic Hydrogen Evolution Reaction. *Small* **2018**, *14*, e1800235. [CrossRef]
21. Zhao, S.; Wang, D.-W.; Amal, R.; Dai, L. Carbon-Based Metal-Free Catalysts for Key Reactions Involved in Energy Conversion and Storage. *Adv. Mater.* **2018**, *31*, e1801526. [CrossRef]
22. Tuci, G.; Filippi, J.; Ba, H.; Rossin, A.; Luconi, L.; Pham-Huu, C.; Vizza, F.; Giambastiani, G. How to Teach an Old Dog New (Electrochemical) Tricks: Aziridine-Functionalized CNTs as Efficient Electrocatalysts for the Selective CO_2 Reduction to CO. *J. Mater. Chem. A* **2018**, *6*, 16382–16389. [CrossRef]

23. Duan, X.; Ao, Z.; Sun, H.; Indrawirawan, S.; Wang, Y.; Kang, J.; Liang, F.; Zhu, Z.H.; Wang, S. Nitrogen-Doped Graphene for Generation and Evolution of Reactive Radicals by Metal-Free Catalysis. *ACS Appl. Mater. Interfaces* **2015**, *7*, 4169–4178. [CrossRef]
24. Yang, Y.; Zhang, W.; Ma, X.; Zhao, H.; Zhang, X. Facile Construction of Mesoporous N-Doped Carbons as Highly Efficient 4-Nitrophenol Reduction Catalysts. *ChemCatChem* **2015**, *7*, 3454–3459. [CrossRef]
25. Liu, Y.; Duong-Viet, C.; Luo, J.; Hébraud, A.; Schlatter, G.; Ersen, O.; Nhut, J.-N.; Pham-Huu, C. One-Pot Synthesis of a Nitrogen-Doped Carbon Composite by Electrospinning as a Metal-Free Catalyst for Oxidation of H_2S to Sulfur. *ChemCatChem* **2015**, *7*, 2957–2964. [CrossRef]
26. Ba, H.; Liu, Y.; Truong-Phuoc, L.; Duong-Viet, C.; Mu, X.; Doh, W.H.; Tran-Thanh, T.; Baaziz, W.; Nguyen-Dinh, L.; Nhut, J.-M.; et al. A highly N-doped carbon phase "dressing" of macroscopic supports for catalytic applications. *Chem. Commun.* **2015**, *51*, 14393–14396. [CrossRef]
27. Duong-Viet, C.; Ba, H.; El-Berrichi, Z.; Nhut, J.-M.; Ledoux, M.J.; Liu, Y.; Pham-Huu, C. Silicon carbide foam as a porous support platform for catalytic applications. *New J. Chem.* **2016**, *40*, 4285–4299. [CrossRef]
28. Sun, F.; Liu, J.; Chen, H.; Zhang, Z.; Qiao, W.; Long, D.; Ling, L. Nitrogen-Rich Mesoporous Carbons: Highly Efficient, Regenerable Metal-Free Catalysts for Low-Temperature Oxidation of H_2S. *ACS Catal.* **2013**, *3*, 862–870. [CrossRef]
29. Duong-Viet, C.; Truong-Phuoc, L.; Tran-Thanh, T.; Nhut, J.-M.; Nguyen-Dinh, L.; Janowska, I.; Begin, D.; Pham-Huu, C. Nitrogen-doped carbon nanotubes decorated silicon carbide as a metal-free catalyst for partial oxidation of H_2S. *Appl. Catal. A Gen.* **2014**, *482*, 397–406.
30. Ba, H.; Duong-Viet, C.; Liu, Y.; Nhut, J.-M.; Granger, P.; Ledoux, M.J.; Pham-Huu, C. Nitrogen-doped carbon nanotube spheres as metal-free catalysts for the partial oxidation of H_2S. *C. R. Chim.* **2016**, *19*, 1303–1309. [CrossRef]
31. Yu, Z.; Wang, X.; Hou, Y.-N.; Pan, X.; Zhao, Z.; Qiu, J. Nitrogen-doped mesoporous carbon nanosheets derived from metal-organic frameworks in a molten salt medium for efficient desulfurization. *Carbon* **2017**, *117*, 376–382. [CrossRef]
32. Shen, L.; Lei, G.; Fang, Y.; Cao, Y.; Wang, X.; Jiang, L. Polymeric carbon nitride nanomesh as an efficient and durable metal-free catalyst for oxidative desulfurization. *Chem. Commun.* **2018**, *54*, 2475–2478. [CrossRef]
33. Duong-Viet, C.; Ba, H.; Liu, Y.; Truong-Phuoc, L.; Nhut, J.M.; Pham-Huu, C. Nitrogen-doped carbon nanotubes on silicon carbide as a metal-free catalyst. *Chin. J. Catal.* **2014**, *35*, 906–913. [CrossRef]
34. Pham-Huu, C.; Giambastiani, G.; Liu, Y.; Ba, H.; Nguyen-Dinh, L.; Nhut, J.-M.; Duong-Viet, C. Method for preparing highly nitrogen-doped mesoporous carbon composites. EP Patent 3047905 A1, 29 April 2020.
35. Mokhatab, S.; Poe, W.A. (Eds.) Typical [H_2S] and [BTX] in sour gas are assumed close to 0.25 vol.% and about 2000 ppm, respectively, with the latter being tentatively composed by: Benzene = c.a. 900 ppm, Toluene = c.a. 750 ppm and Xylene = c.a. 400 ppm. In *Handbook of Natural Gas Transmission and Processing*; Gulf Professional Publishing: Houston, TX, USA, 2012.
36. Duong-Viet, C.; Nhut, J.-M.; Truong-Huu, T.; Tuci, G.; Nguyen-Dinh, L.; Liu, Y.; Pham, C.; Giambastiani, G.; Pham-Huu, C. A Nitrogen-Doped Carbon Coated Silicon Carbide as a Robust and Highly Efficient Metal-Free Catalyst for Sour Gases Desulfurization in the Presence of Aromatics as Contaminants. *Catal. Sci. Technol.* **2020**, *10*, 5487–5500. [CrossRef]
37. Mukasyan, A.S. Silicon Carbide. In *Concise Encyclopedia of Self-Propagating High-Temperature Synthesis*; Borovinskaya, I.P., Gromov, A.A., Levashov, E.A., Maksimov, Y.M., Mukasyan, A.S., Rogachev, A.S., Eds.; Elsevier Inc.: Amsterdam, The Netherlands, 2017; pp. 336–338.
38. Keller, N.; Pham-Huu, C.; Estournès, C.; Ledoux, M.J. Low temperature use of SiC-supported NiS_2-based catalysts for selective H_2S oxidation Role of SiC surface heterogeneity and nature of the active phase. *Appl. Catal. A Gen.* **2002**, *234*, 191–205. [CrossRef]
39. Nguyen, P.; Edouard, D.; Nhut, J.M.; Ledoux, M.J.; Pham, C.; Pham-Huu, C. High thermal conductive b-SiC for selective oxidation of H_2S: A new support for exothermal reactions. *Appl. Catal. B Environ.* **2007**, *76*, 300–310. [CrossRef]
40. Terörde, R.J.A.M.; van den Brink, P.J.; Visser, L.M.; van Dillen, A.J.; Geus, J.W. Selective oxidation of hydrogen sulfide to elemental sulfur using iron oxide catalysts on various supports. *Catal. Today* **1993**, *17*, 217–224. [CrossRef]
41. Sing, K.S.W.; Everett, D.H.; Haul, R.A.W.; Moscou, L.; Pierotti, R.A.; Rouquérol, J. Reporting Physisorption Data for Gas/Solid Systems with Special Reference to the Determination of Surface Area and Porosity. *Pure Appl. Chem.* **1985**, *57*, 603–619. [CrossRef]
42. Arrigo, R.; Hävecker, M.; Wrabetz, S.; Blume, R.; Lerch, M.; McGregor, J.; Parrott, E.P.J.; Zeitler, J.A.; Gladden, L.F.; Knop-Gericke, A.; et al. Tuning the Acid/Base Properties of Nanocarbons by Functionalization via Amination. *J. Am. Chem. Soc.* **2010**, *132*, 9616–9630. [CrossRef]
43. Toluene was also Selected on the Basis of Its Intermediate Toxicity Degree among BTX Contaminants, i.e., Xylene > Toluene > Benzene. Available online: http://universulphur.com/mespon/2015_presentations/session_b/5.%20Dealing%20with%20Aromatics%20in%20the%20Sulfur%20Recovery%20Unit%20--%20Eric%20Roisin%20--%20Axens.pdf (accessed on 21 December 2020).
44. Zhao, Z.; Dai, Y.; Ge, G.; Guo, X.; Wang, G. Facile simultaneous defect production and O,N-doping of carbon nanotubes with unexpected catalytic performance for clean and energy-saving production of styrene. *Green Chem.* **2015**, *17*, 3723–3727. [CrossRef]
45. Jin, X.; Balasubramanian, V.V.; Selvan, S.T.; Sawant, D.P.; Chari, M.A.; Lu, G.Q.; Vinu, A. Highly Ordered Mesoporous Carbon Nitride Nanoparticles with High Nitrogen Content: A Metal-Free Basic Catalyst. *Angew. Chem. Int. Ed.* **2009**, *48*, 7884–7887. [CrossRef]
46. Gounder, R.; Iglesia, E. Catalytic Consequences of Spatial Constraints and Acid Site Location for Monomolecular Alkane Activation on Zeolites. *J. Am. Chem. Soc.* **2009**, *131*, 1958–1971. [CrossRef]

47. Tuci, G.; Zafferoni, C.; Rossin, A.; Milella, A.; Luconi, L.; Innocenti, M.; Truong Phuoc, L.; Duong-Viet, C.; Pham-Huu, C.; Giambastiani, G. Chemically Functionalized Carbon Nanotubes with Pyridine Groups as Easily Tunable N-Decorated Nanomaterials for the Oxygen Reduction Reaction in Alkaline Medium. *Chem. Mater.* **2014**, *26*, 3460–3470. [CrossRef]
48. Tuci, G.; Luconi, L.; Rossin, A.; Berretti, E.; Ba, H.; Innocenti, M.; Yakhvarov, D.; Caporali, S.; Pham-Huu, C.; Giambastiani, G. Aziridine-Functionalized Multiwalled Carbon Nanotubes: Robust and Versatile Catalysts for the Oxygen Reduction Reaction and Knoevenagel Condensation. *ACS Appl. Mater. Interfaces* **2016**, *8*, 30099–30106. [CrossRef]
49. Moaseri, E.; Baniadam, M.; Maghrebi, M.; Karimi, M. A Simple Recoverable Titration Method for Quantitative Characterization of Amine-Functionalized Carbon Nanotubes. *Chem. Phys. Lett.* **2013**, *555*, 164–167. [CrossRef]
50. Cai, Z.; Leong, E.; Wang, Z.; Niu, W.; Zhang, W.; Ravaine, S.; Yakovlev, N.; Liu, Y.; Teng, J.; Lu, X. Sandwich-structured $Fe_2O_3@SiO_2@Au$ nanoparticles with magnetoplasmonic responses. *J. Mater. Chem. C* **2015**, *3*, 11645–11652. [CrossRef]
51. Fahim, M.A.; Alsahhaf, T.A.; Elkilani, A. Fundamentals of Petroleum Refining. Elsevier, B.V.: Amsterdam, The Netherlands, 2010; pp. 377–402.
52. Ebbing, D.D.; Gammon, S.D. *General Chemistry*, 9th ed.; Houghton Mifflin Company: Boston, MA, USA, 2009.

 *catalysts*

Article

Reaction Mechanism of Simultaneous Removal of H$_2$S and PH$_3$ Using Modified Manganese Slag Slurry

Jiacheng Bao [1], Xialing Wang [1], Kai Li [1], Fei Wang [1], Chi Wang [2], Xin Song [1], Xin Sun [1,*] and Ping Ning [1,*]

[1] Faculty of Environmental Science and Engineering, Kunming University of Science and Technology, Kunming 650500, China; jcBao@stu.kust.edu.cn (J.B.); wangxialing@stu.kust.edu.cn (X.W.); 20130203@kust.edu.cn (K.L.); 20192016@kust.edu.cn (F.W.); songxin@kust.edu.cn (X.S.)

[2] Faculty of Chemical Engineering, Kunming University of Science and Technology, Kunming 650500, China; 20150089@kust.edu.cn

* Correspondence: sunxin@kust.edu.cn (X.S.); ningping@kust.edu.cn (P.N.)

Received: 25 October 2020; Accepted: 26 November 2020; Published: 27 November 2020

Abstract: The presence of phosphine (PH$_3$) and hydrogen sulfide (H$_2$S) in industrial tail gas results in the difficulty of secondary utilization. Using waste solid as a wet absorbent to purify the H$_2$S and PH$_3$ is an attractive strategy with the achievement of "waste controlled by waste". In this study, the reaction mechanism of simultaneously removing H$_2$S and PH$_3$ by modified manganese slag slurry was investigated. Through the acid leaching method for raw manganese slag and the solid–liquid separation subsequently, the liquid-phase part has a critical influence on removing H$_2$S and PH$_3$. Furthermore, simulation experiments using metal ions for modified manganese slag slurry were carried out to investigate the effect of varied metal ions on the removal of H$_2$S and PH$_3$. The results showed that Cu^{2+} and Al^{3+} have a promoting effect on H$_2$S and PH$_3$ conversion. In addition, the Cu^{2+} has liquid-phase catalytic oxidation for H$_2$S and PH$_3$ through the conversion of Cu(II) to Cu(I).

Keywords: phosphine; hydrogen sulfide; manganese slag; metal ions; reaction mechanism

---

1. Introduction

The presence of phosphine (PH$_3$) and hydrogen sulfide (H$_2$S) in industrial gases can result in reduced feed gas quality, excessive equipment corrosion, and catalyst deactivation and poisoning, limiting industrial gas recovery and utilization [1], especially yellow phosphorus off-gas. In addition, H$_2$S, as a highly toxic, corrosive gas, can not only cause air pollution but also eye irritation and breathing problems. Even exposure to small amounts of H$_2$S will pose a serious threat to humans [2]. Additionally, PH$_3$ may cause immediate death if one is exposed to a concentration level of 50 ppm, according to the National Institute for Occupational Safety and Health (NIOSH) [3]. Therefore, it is desirable to remove them from the point of view of the highly efficient use of industrial gases and human health.

Currently, the wet process has more advantages for simultaneously removing H$_2$S and PH$_3$ compared to the dry process, due to lower cost and easier preparation process [4,5]. However, the wet process for the removal of H$_2$S and PH$_3$ could rapidly consume oxidant, thus leading to a decrease in removal efficiency [4]. Hence, there is an urgent need to modify the traditional wet method to meet stricter environmental laws. In recent years, the use of metal ore, tailings, and metal smelting slag to remove industrial waste gas has attracted the attention of researchers [6–8]. Smelting slag contains a large number of transition metals such as Fe, Mn, and basic oxides, which could have been favorable for absorption of PH$_3$ and H$_2$S in the wet process due to the liquid phase catalytic oxidation ability of transition metals and higher alkalinity of basic oxides such as CaO, MgO, etc. The liquid catalytic oxidation by transition metals can be favorable for the oxidation of gas pollutants, thus being conducive to absorbing H_2S or/and PH_3 [9–11]. At present, China has become the largest manganese producer, consumer, and exporter, and has emitted more than 2×10^6 tons of manganese slag (MS) every year, which mainly originate from the manganese metallurgy process. Piling manganese slag waste has caused color stain and pollution in the composition of soil and surface water and contaminated groundwater and rivers due to surface runoff [12]. Therefore, there is a pressing need to dispose of MS in a green and highly efficient manner. The chemical compositions of MS are rich in Mn-based oxides and contain a number of basic oxides, which make it possible to become a highly efficient absorbent for the removal of H_2S and PH_3.

According to our previous studies [13], modified manganese slag (MS) slurry was used to remove H_2S and PH_3 and proved a great absorbent. However, a detailed investigation of the reaction mechanism was still lacking. Thus, our research extends the knowledge into the reaction mechanism of H_2S and PH_3 using modified MS slurry. In this paper, the emphasis was laid on the role of modified MS slurry in removing H_2S and PH_3. In addition, to understand the liquid phase catalytic oxidation ability of modified MS slurry, we simulated the composition of MS slurry in the form of different metal salt solutions based on the actual composition of electrolytic MS. In addition, XPS (X-ray photoelectron spectrometer), IC (ion chromatography), and XRF (X-ray fluorescence) and XRD (X-ray diffraction) techniques were used to investigate the samples' surface information. Thus, a mechanism of simultaneous removal of PH_3 and H_2S using modified MS slurry was proposed.

2. Results and Discussion

2.1. Effect of Different Component Slurry after Acid Leaching on Simultaneous Removal of H_2S and PH_3

In order to investigate the mechanism of simultaneous removal of H_2S and PH_3 using modified MS slurry, the acid leaching method was used to treat the raw MS. Thus, the majority of soluble metal ions can be leached and transferred from the solid phase of raw MS to the aqueous solution. It can be seen from Figure 1 that XRD results show that the main mineral phases of raw MS were $3CaO \cdot SiO_2$ (C_3A), $2CaO \cdot SiO_2$ (C_2A), CaO, Al_2O_3, and merwinite. These components were acid-soluble, which can be decomposed by the acid solution. Table 1 also shows that the main elements in raw MS were Si, Ca, and Mn. Thus, through acid leaching and then filtering, the active components mainly including metal ions could be removed from the raw MS.

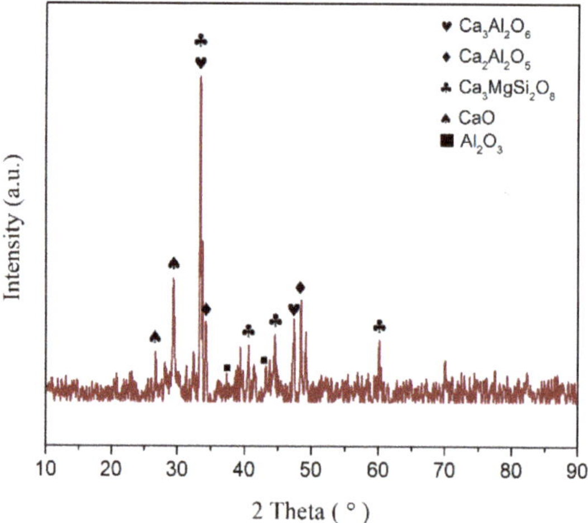

Figure 1. XRD pattern of raw manganese slag.

Table 1. XRF analysis of electrolytic manganese slag (wt.).

Element	Ca	Si	Mn	Al	Mg	S	O
Raw electrolytic manganese slag	25.54	13.09	11.33	4.80	3.54	1.87	37.48

As shown in Figure 2, the 100% H_2S removal efficiency of modified MS slurry (MS + $CuSO_4$ group) can maintain 7 h while the 65% PH_3 removal efficiency of modified MS slurry can maintain 5 h. In addition, the acid leaching residues and $CuSO_4$ slurry can only obtain around 7% H_2S removal efficiency and approximately 9% PH_3 removal efficiency. Compared with the simultaneous removal of H_2S and PH_3 by modified MS slurry, the acid leaching residue and $CuSO_4$ slurry obtained a poorer removal efficiency due to the consumption of active components after the acid leaching method. Additionally, the raw MS and Cu^{2+} have a synergistic effect on the removal of H_2S and PH_3. Thus, leached metal ions in slurry have a leading role in removing H_2S and PH_3. The mixture of MS and $CuSO_4$ slurry obtained the best PH_3 conversion efficiency while the effect of $CuSO_4$ solution was not obvious. According to our previous report [13], the increase in PH_3 conversion efficiency can be ascribed to the oxidation ability of Cu^{2+} and Mn^{2+} from MS, thus inhibiting the formation of CuS/Cu_2S.

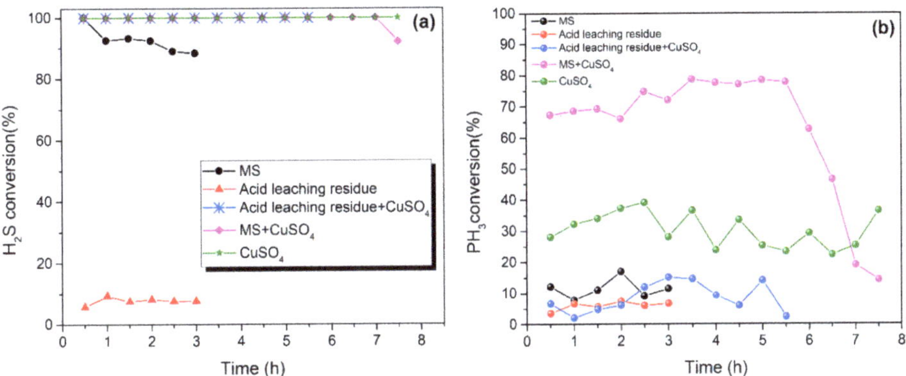

Figure 2. Effects of different component slurry after acid leaching on simultaneous removal of (a) H_2S and (b) PH_3 by manganese slag before and after acid leaching. Experimental conditions: H_2S concentration = 800 ppm; PH_3 concentration = 400 ppm; gas flow rate = 110 mL min^{-1}; reaction temperature = 35 °C; Oxygen content = 1 vol %; stirring rate = 800 r/min.

2.2. Effect of Simulated Modified MS Slurry on Simultaneous Removal of H_2S and PH_3

The metal ions leached from MS during the reaction period played a leading role in removing H_2S and PH_3. Thus, to gain more insight, the modified MS slurry was simulated in the form of metal salts based on the real component composition. The various simulated metal salts species (including metal nitrates, metal chlorides, and metal sulfates) were investigated and the simulated contents were listed in Table 2. The H_2S and PH_3 removal efficiency by simulated modified MS slurry is shown in Figure 3. The results showed that the three groups obtained a 100% H_2S conversion efficiency; whereas Group 3 (metal chlorides) obtained the higher PH_3 removal efficiency relative to those of the other groups, with the order being metal chlorides > metal nitrate > metal sulfates. In order to achieve a better understanding of the variation in PH_3 and H_2S conversion by different metal salts, the reaction products in aqueous solutions were detected by the IC method. The IC results, as shown in Figure 4, showed that the generated PO_3^- and SO_4^{2-} concentrations in Group 3 increased with the reaction proceeding, which indicated that the H_2S and PH_3 could be converted to PO_3^- and SO_4^{2-}, respectively. Thus, the addition of metal sulfates inhibited higher H_2S and PH_3 conversion efficiency because the generation of SO_4^{2-} was accelerated and inhibited the oxidation of PH_3 when the copper concentration was at relatively high concentration, according to our previous study [13]. In other words, more sulfates existing in the solution led to lower PH_3 conversion efficiency. Hence, the metal chloride was chosen for the following experiments.

Table 2. Composition of simulated modified manganese slag slurry.

Samples	Composition (wt.%), Total Mass = 3 g, Balanced in SiO_2				
	Ca^{2+}	Mg^{2+}	Mn^{2+}	Al^{3+}	Extra Added Cu^{2+}
Group 1 (metal nitrates)	25.54	3.54	11.33	4.80	0.01 mol
Group 2 (metal chlorides)	25.54	3.54	11.33	4.80	0.01 mol
Group 3 (metal sulfates)	25.54	3.54	11.33	4.80	0.01 mol

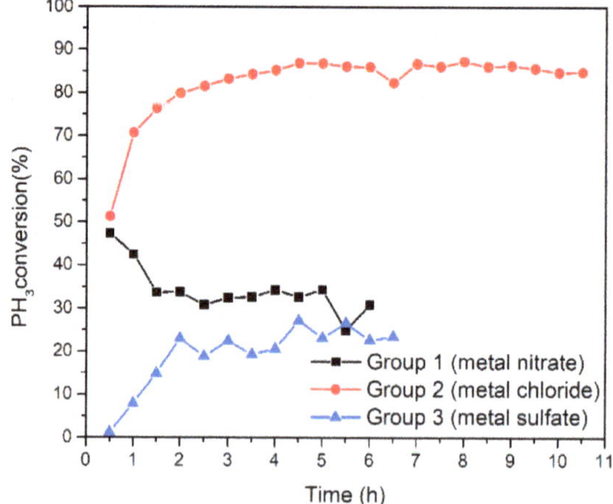

Figure 3. Effect of different simulated modified manganese slag (MS) slurry on simultaneous removal of H_2S and PH_3 by manganese slag before and after acid leaching. Experimental conditions: H_2S concentration = 800 ppm; PH_3 concentration = 400 ppm; gas flow rate = 110 mL min^{-1}; reaction temperature = 35 °C; Oxygen content = 1 vol %; stirring rate = 800 r/min.

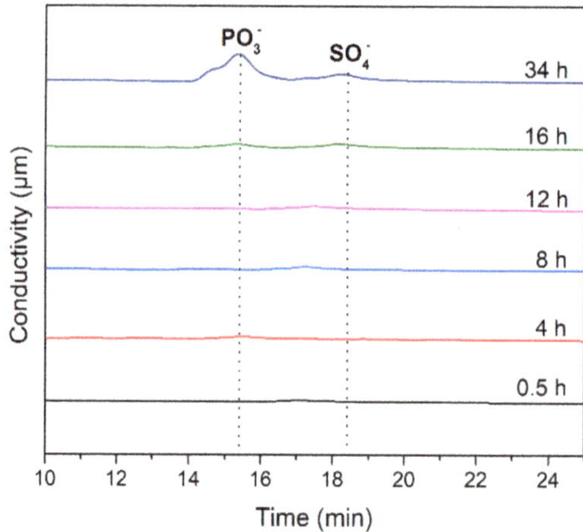

Figure 4. Ion concentration in Group 3 (metal chloride) as a function of time (50× dilution).

2.3. Effect of Single and Multi-Metal Ions on Simultaneous Removal of H_2S and PH_3

To further explore the role of metal ions in removing H_2S and PH_3, we provide a series of experiments to examine the effect of metal ions on H_2S and PH_3 removal. It can be seen from Figure 5 that all groups can achieve 100% H_2S removal efficiency. The highest PH_3 conversion efficiency of $Cu^{2+} + Al^{3+}$, $Cu^{2+} + Ca^{2+}$, $Cu^{2+} + Mg^{2+}$, and $Cu^{2+} + Mn^{2+}$ were 95.52%, 89.45%, 87.43%, and 81.63%, respectively. The $Al^{3+} + Cu^{2+}$ group obtained higher PH_3 conversion efficiency relative to that of Cu^{2+} alone, which indicated that the Al^{3+} and Cu^{2+} have a synergistic effect on PH_3 removal. The addition of Mg^{2+}, Ca^{2+}, and Mn^{2+} slightly reduced the PH_3 conversion efficiency compared to those of the Cu^{2+} group and $Al^{3+} + Cu^{2+}$ group. Thus, the analysis emphasis was laid on the $Al^{3+} + Cu^{2+}$ group, and the reaction products in long-term experiments are analyzed in the next section (Section 2.4).

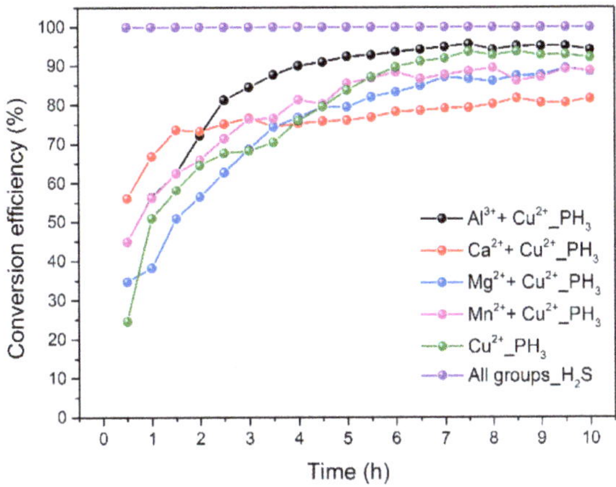

Figure 5. Effect of different single and multi-metal ions (metal chlorides) on simultaneous removal of H_2S and PH_3. Experimental conditions: H_2S concentration = 800 ppm; PH_3 concentration = 400 ppm; gas flow rate = 110 mL min^{-1}; reaction temperature = 35 °C; Oxygen content = 1 vol %; stirring rate = 800 r/min.

2.4. Reaction Mechanism of Metal Ions to Simultaneous Removal of H_2S and PH_3

To understand the effect of Al^{3+} combined with Cu^{2+} on simultaneous removal of H_2S and PH_3, the long-term experiments were carried out and the variation in the pH value of solution, solid or aqueous products is analyzed in this section. For the Cu^{2+} group, Figure 6a shows that the PH_3 conversion efficiency increased initially and then decreased with the reaction proceeding, accompanied by a gradual decrease in the pH value. The XRD results in Figure 6b indicate that the main reaction product was CuS, Cu_8S_5, and CuCl, which resulted from the reaction between H_2S and Cu^{2+} (Equation (1)). Meanwhile, the H^+ ion was increased, thereby leading to a decrease in pH value. The $Al^{3+} + Cu^{2+}$ group showed a similar conversion trend for H_2S and PH_3. However, the reaction products of the $Al^{3+} + Cu^{2+}$ group became different. It can be seen from Figure 6d that the $Al_2(SO_4)_3$, $CuSO_4$, $Cu_4(SO_4)(OH)_6 \cdot 2H_2O$, and CuS were the main reaction products for removing H_2S while $AlPO_3$, AlP, and $Cu_5O_2(PO_4)_2$ were the main reaction products for removing PH_3. When Al^{3+} was added into the solution, the pH value slowly decreased during 5 h of initial reaction time, which indicated that Al^{3+} could inhibit the rapid decline of pH value, although the pH value subsequently decreased dramatically.

$$Cu^{2+} + H_2S \rightarrow CuS + 2H^+ \tag{1}$$

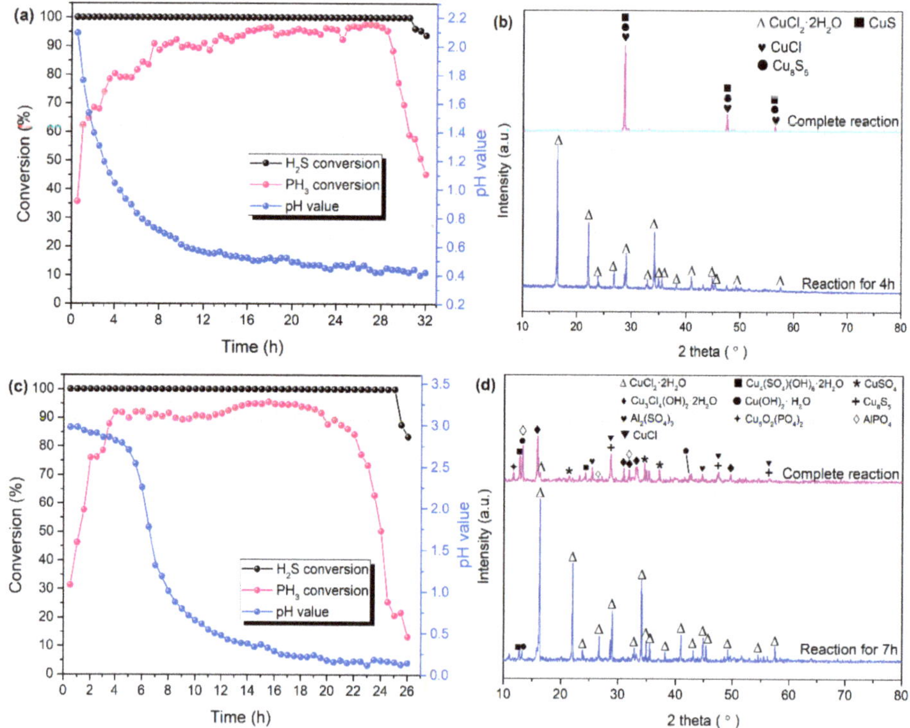

Figure 6. Variation in pH value of (**a**) Cu^{2+} group and (**c**) $Al^{3+} + Cu^{2+}$ group as a function of time. XRD patterns of reaction process of (**b**) Cu^{2+} group and (**d**) $Al^{3+} + Cu^{2+}$ group. Experimental conditions: H_2S concentration = 800 ppm; PH_3 concentration = 400 ppm; gas flow rate = 110 mL min^{-1}; reaction temperature = 35 °C; Oxygen content = 1 vol %; stirring rate = 800 r/min.

To gain more insight, we conducted several XPS studies for the reaction products of the Cu^{2+} group and $Al^{3+} + Cu^{2+}$ groups with the reaction proceeding. For the $Al^{3+} + Cu^{2+}$ group, as shown in Figure 7a, the binding energy (B. E.) centered at about 159.4 eV for S $2p_{3/2}$ could be attributed to S^{2-} [14], which indicated that CuS was generated in the $Al^{3+} + Cu^{2+}$ group. In addition, the B. E. located at 164.2 eV may be ascribed to polysulfide species [15], which indicated that CuS/Cu_8S_5 was oxidized in the existence of oxygen. With the reaction further proceeding, the sulfate was formed as evidence that the B. E. of 169.8 eV appeared [16] and the sulfate content further increased from 7 h to complete reaction, which indicated that CuS was further oxidized to $CuSO_4$ by oxygen in the presence of the water environment [17]. This was consistent with the XRD results as shown in Figure 6d. For the Cu^{2+} group, the variation in S valence state in the initial reaction period was similar to the $Al^{3+} + Cu^{2+}$ group. However, with further increase of the reaction time, the content of the polysulfide species (53.2%, calculated by XPS data as shown in Table 3) was the same as the sulfate content (46.8%); whereas the sulfate content (91.4%) was dominant in the $Al^{3+} + Cu^{2+}$ group. This can be explained by noting that the addition of Al^{3+} could effectively accelerate the CuS oxidation, thus generating more sulfate species.

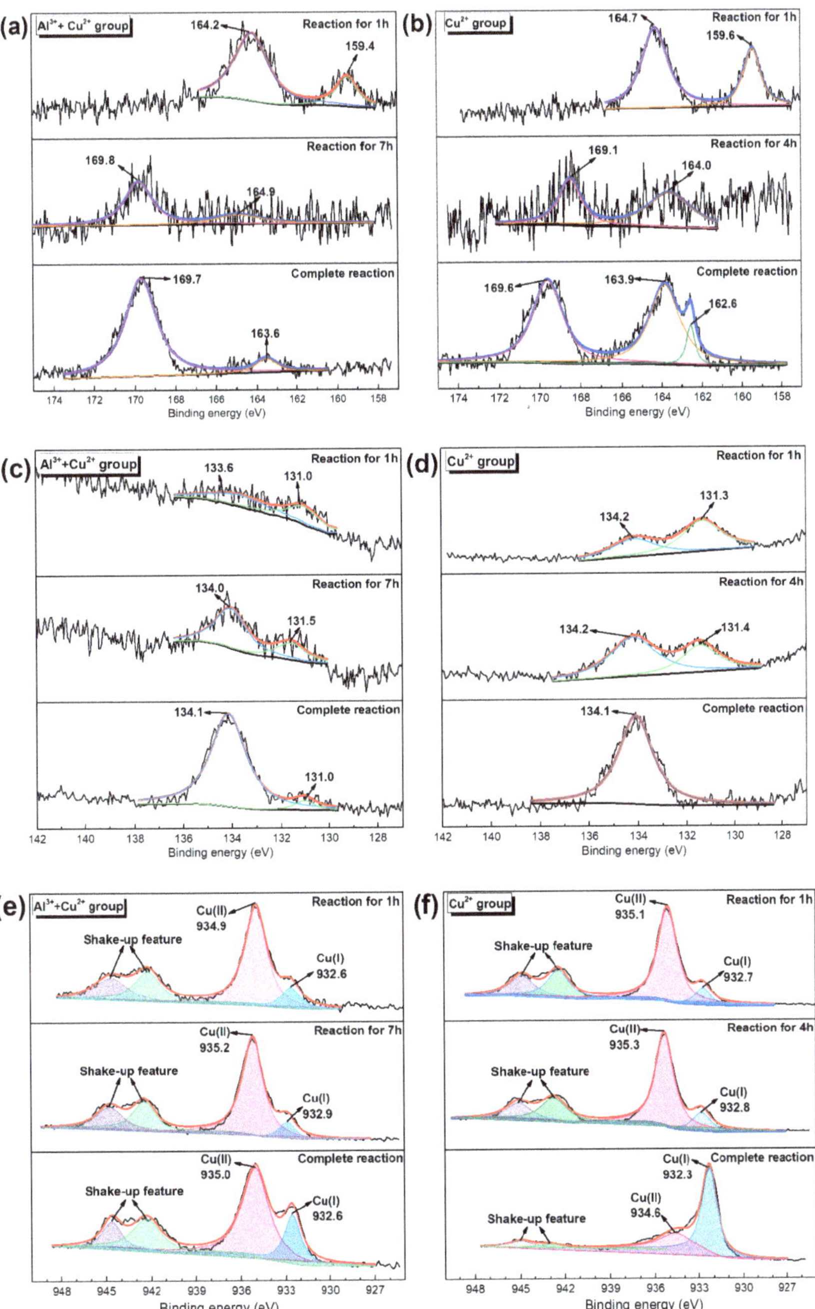

Figure 7. XPS spectra of Al^{3+} + Cu^{2+} group and Cu^{2+} group with different reaction time for surveys of (**a**,**b**) S 2p, (**c**,**d**) P 2p, and (**e**,**f**) Cu 2p.

Table 3. XPS data of $Al^{3+} + Cu^{2+}$ group and Cu^{2+} group with different reaction time for surveys of S 2p; P 2p, and Cu 2p.

Sample	Element	Parameter	1 h		4 h		Complete Reaction		
		Position (eV)	164.7	159.6	169.1	164.0	169.6	163.9	162.6
	S	Atomic ratio (%)	68.2	31.9	40.4	59.6	46.8	46.5	6.7
		Substance	S_n^{1}	S^{2-}	SO_4^{2-}	S	SO_4^{2-}	S_n	S_n
		Position (eV)	134.2	131.3	134.2	131.4	134.1		
Cu^{2+} Group	P	Atomic ratio (%)	39.9	60.1	64.7	35.3	100.0		
		Substance	PO_4^{3-}	P^{3-}	PO_4^{3-}	P^{3-}	PO_4^{3-}		
		Position (eV)	935.1	932.7	932.8	935.3	932.3	934.6	
	Cu	Atomic ratio (%)	87.9	12.1	15.2	84.8	70.2	29.8	
		Substance	$CuSO_4$	CuS	CuS	$CuSO_4$	CuS	$CuSO_4$	

Sample	Element	Parameter	1 h		4 h		Complete Reaction	
		Position (eV)	164.2	159.4	169.8	164.9	169.7	163.6
	S	Atomic ratio (%)	78.4	21.6	74.2	25.8	91.4	8.6
		Substance	S_n	S^{2-}	SO_4^{2-}	S	S_n	S_n
		Position (eV)	133.6	131.0	134.0	131.5	134.1	131.0
$Al^{3+} + Cu^{2+}$ Group	P	Atomic ratio (%)	51.7	48.3	69.8	30.2	94.4	5.6
		Substance	PO_4^{3-}	P^{3-}	PO_4^{3-}	P^{3-}	PO_4^{3-}	P^{3-}
		Position (eV)	934.9	932.6	935.2	932.9	935.0	932.6
	Cu	Atomic ratio (%)	83.5	16.5	86.5	13.5	73.6	26.4
		Substance	Cu(II)	Cu(I)	Cu(II)	Cu(I)	Cu(II)	Cu(I)

[1] S_n refers to the polysulfide species.

As shown in Figure 7c, when the $Al^{3+} + Cu^{2+}$ group reacted for 1 h, phosphate was formed, since the peak at 134.6 eV could be attributed to PO_4^{3-} while the B. E. of 131.0 eV may be ascribed to P^{3-}. With an increase in reaction time, the PO_4^{3} content increased from 51.7% to 69.8%, up to 94.4% in the final. The variation of the P valence state in the Cu^{2+} group as shown in Figure 7d was the same as for the $Al^{3+} + Cu^{2+}$ group; but the generation rate of phosphate in the Cu^{2+} group was slower than that of the $Al^{3+} + Cu^{2+}$ group. When the reaction time was for 1 h, the relative phosphate content in the Cu^{2+} group was 39.9%, thus leading to a slower increased PH_3 conversion efficiency than the Cu^{2+} group. The faster conversion of PH_3 to phosphate resulted in accelerated PH_3 absorption in the initial period of reaction.

It can be seen from Figure 7e that the B. E. located at around 934.9 and 932.6 eV in the $Al^{3+} + Cu^{2+}$ group could be attributed to Cu(II) and Cu(I), respectively [18]. The relative Cu^{2+} content in the $Al^{3+} + Cu^{2+}$ group was 83.5% for 1 h of reaction time and then decreased to 73.6% until the complete reaction, which indicated that part of Cu^{2+} was converted to Cu^+ as evidenced by the formation of CuCl and Cu_8S_5 by the XRD technique. From Figure 7f, it can be seen that the ratio of Cu(II)/Cu(I) in the Cu^{2+} group decreased dramatically relative to the $Al^{3+} + Cu^{2+}$ group, which indicated that Cu^{2+} ion had a liquid-phase catalytic oxidation effect on removing H_2S and PH_3 through variation in the valence state of Cu^{2+} to Cu^+.

Based on the above analysis, the metal ions played a leading role in removing H_2S and PH_3. The reaction process can be summarized as follows (Equations (2)–(13)):

$$H_2S \leftrightarrow HS^- + H^+ \leftrightarrow S^{2-} + 2H^+ \tag{2}$$

$$Me^{2+} + H_2S + 2H_2O \rightarrow MeS \downarrow + 2H^+ \quad \left(Me^{2+} = Cu^{2+}\right) \tag{3}$$

$$CuS(s) \rightarrow Cu^{2+}(aq) + S^{2-} \quad \left(Me^{2+} = Cu^{2+}\right) \tag{4}$$

$$8Cu^{2+} + PH_3 + 4H_2O \rightarrow 8Cu^+ + PO_4^{3-} + 11H^+ \tag{5}$$

$$4Cu^+ + O_2 + 2H_2O \rightarrow 4Cu^{2+} + 4OH^- \tag{6}$$

$$Cu^+ + O_2 \rightarrow Cu^{2+} + O_2^- \tag{7}$$

$$O_2 + 2H_2S \rightarrow 2S \downarrow + 2H_2O \tag{8}$$

$$S + 1.5O_2 + 2H_2O \rightarrow SO_4^{2-} + 2H^+ \tag{9}$$

$$(CuS)_n + O_2 \rightarrow (CuS)_{n-1}(CuS)^{\bullet+} + O_2^{\bullet-} \tag{10}$$

$$Cu^{2+} + H_2O \rightarrow Cu^+ + H^+ + OH^\bullet \tag{11}$$

$$2OH^\bullet \rightarrow H_2O_2 \tag{12}$$

$$4H_2O_2 + S^{2-} \rightarrow SO_4^{2-} + 4H_2O \tag{13}$$

3. Materials and Methods

3.1. Materials

The manganese slag samples were collected from Wenshan, China. The electrolytic manganese slag is firstly dried at 105 °C in the oven for 12 h, then mechanically ground by ball mill and sieved to 200 mesh (74 μm) for use. The standard gases include N_2 (≥99.99%), H_2S/N_2 (1.00% H_2S, *v/v*), PH_3/N_2 (1.00% PH_3, *v/v*), and O_2 (≥99.99%), all of which were purchased from Dalian Special Gases Co., Ltd., Dalian, China.

3.2. Acid Leaching Procedure and Preparation of Modified MS Slurry

First, the electrolytic manganese slag was weighed and transferred into a three-necked flask. Then, 2 mol L^{-1} hydrochloric acid solution was taken into the three-necked flask with the connecting condensing device. The solid-to-liquid ratio of the slag to the acid solution was 1:6. The acid leaching temperature was set at 100 °C for 60 min. After acid leaching, the acid-leached solution was filtered. Then, filter residue and filtrate were obtained. The obtained filter residue was named as acid leaching residue. The acid leaching residue was then mixed with deionized water to obtain a 40 mL acid leaching residue slurry. Furthermore, the acid leaching residue was mixed with 0.01 mol $CuSO_4$, named as acid leaching residue + $CuSO_4$. The mixture of acid leaching residue and $CuSO_4$ was added into 40 mL deionized water to prepare the acid leaching residue + $CuSO_4$ slurry. The modified MS slurry was obtained by a mixture of 3 g MS, 40 mL deionized water, and 0.01 mol $CuSO_4$, named MS + $CuSO_4$. In addition, the 3 g of raw MS was added to 40 mL deionized water to prepare the raw MS slurry, named MS. The $CuSO_4$ solution also was prepared by adding a mixture of 40 mL deionized water and 0.01 mol $CuSO_4$, named $CuSO_4$.

3.3. Analytical Method

The solid samples were characterized by XPS technology to obtain the information of S 2p, P 2p, and Cu 2p that were measured by the ESCALAB 250 X-ray photoelectron spectrometer (Thermo Fisher Scientific, Waltham, MA, USA) with a resolution of 0.45 eV (Ag) and 0.82 eV (PET). The sensitivity of the instrument is 180 kbps (200 μm, 0.5 eV). The anions in the solutions were detected by ICS-600 (Thermo Fisher Scientific, Waltham, MA, USA). The ion chromatography is equipped with an AS12A anion separation column consisting of sodium carbonate and sodium bicarbonate. The instrument is turned on for 1–2 h, and the background conductance is less than 30 μs to start measurement. XRD method was used to determine the phase structure of the solid samples prior and after H_2S and PH_3 absorption experiments by a D/MAX-2200 X-ray diffractometer (Rigaku, Tokyo, Japan) with Cu Kα ray, at a voltage of 36 kV, at a current of 30 mA, with scanning range between 10 and 90°, and at a scanning speed of 5°/min.

3.4. Catalytic Activity Test

The experiment device for evaluating the activity of the remover is shown in Figure 8. The simulated flue gas consists of O_2, H_2S, PH_3, and N_2, all of which were supplied by gas cylinders (1a–1d), and their concentration were 1 vol %, 800 ppm (parts per million by volume), 400 ppm, respectively, with a total gas flow rate of 110 mL/min. The gas flow in each gas path is controlled by mass flow controllers (2) (Beijing Seven-Star Electronics Co., Ltd., Beijing, China) with a digital display (4) (Beijing Seven-star Electronics Co., Ltd., Beijing, China)., then all of the gases were mixed in the mixing tank (5) (J.K Fluid Technology, Jiaxing, China). The mixed gas will reach the desired concentration by adjusting the mass flow controllers, and be analyzed by gas chromatography (6) (FULI-9790II gas chromatography, FULI Instrument Co., Ltd., Taizhou, China). Then, the mixed gas passes through three-necked flask (10) combined with the constant temperature magnetic stirrer (DF-101S, INESA Scientific Instrument Co., Ltd., Shanghai, China), contacting the prepared modified manganese slag slurry in which mixed gas will react with the slurry. During the reaction process, the pH value of slurry was measured by pH meter (PHS-3C, INESA Scientific Instrument Co., Ltd., Shanghai, China). Moreover, the tail gas will first be analyzed by a gas chromatography and then be removed in the tail gas treatment system (9) (scrubbed by the 5 wt.% $KMnO_4$ solution with 5 wt.% H_2SO_4).

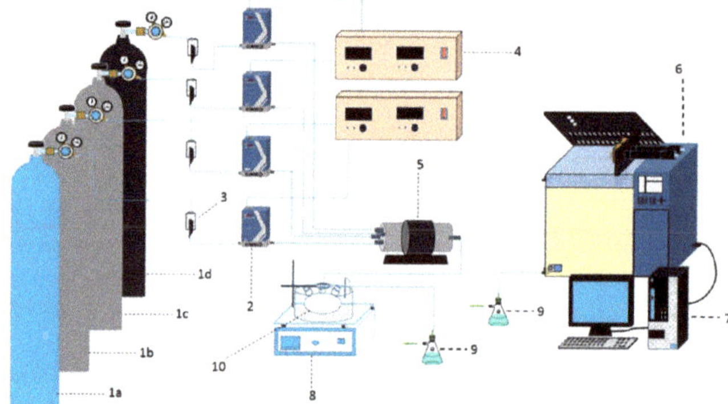

Figure 8. Experimental device diagram of extractant activity evaluation. 1a, 99.99% O_2 cylinder gas; 1b, 1 vol % H_2S cylinder gas; 1c, 1 vol % PH_3 cylinder gas; 1d, 99.99 vol % N_2 cylinder gas; 2, mass flow meter; 3, on-off valve; 4, digital display; 5, gas mixing tank; 6, FULI-9790II gas chromatography; 7, data analysis; 8, constant temperature magnetic stirrer; 9, exhaust gas absorption bottle; 10, three-necked flask.

The calculation method for the conversion efficiency of PH_3 and H_2S is shown in Equation (14).

$$PH_3(H_2S) \text{ conversion efficiency } \% = \frac{PH_3(H_2S)_{\text{inlet}} - PH_3(H_2S)_{\text{outlet}}}{PH_3(H_2S)_{\text{inlet}}} \times 100 \qquad (14)$$

where PH_3 inlet and H_2S inlet refer to the inlet concentrations of PH_3 and H_2S at the initial of the reactor, respectively, in mg/m^3; and PH_3 outlet and H_2S outlet refer to the outlet concentrations of PH_3 and H_2S at the end of the reactor, respectively, in mg/m^3.

4. Conclusions

In this study, modified MS slurry was systematically carried out toward simultaneously removing H_2S and PH_3. On the basis of different characterization techniques, we investigated the reaction mechanism of removing H_2S and PH_3 by modified MS slurry. The main findings of this study are as follows:

(1)   Through acid leaching experiments, the liquid-phase part after filtration has a leading role in removing H_2S and PH_3. The highest PH_3 conversion efficiency of leaching residue slurry + $CuSO_4$ group can only obtain 15.14%, which indicated that the main active components were consumed by the acid leaching method.

(2)   By simulation of the modified MS slurry with metal ions based on real chemical composition, the catalytic activity for H_2S and PH_3 is relative to the types of metal salts, with the order being metal chlorides > metal nitrates > metal sulfates. All of the metal salts can obtain 100% H_2S removal efficiency. In addition, the metal chlorides can maintain above 70% PH_3 conversion efficiency for 10.5 h and the highest PH_3 conversion efficiency was 86.85%; whereas the highest PH_3 removal efficiency of the metal nitrates and metal sulfates can only obtain 47.36% and 27.24%, respectively. Furthermore, Al^{3+} and Cu^{2+} has a synergistic effect on removing H_2S and PH_3 compared to Ca^{2+}, Mg^{2+}, and Mn^{2+} combined with Cu^{2+} groups.

(3) H_2S was oxidized to element S and sulfate due to the reaction between Cu^{2+} and H_2S and part of the H_2S oxidation by O_2, while the PH_3 was oxidized to PO_4^{3-} by liquid-phase catalytic oxidation of metal ions with the conversion of Cu^{2+} to Cu^+.

(4)   The best PH_3 and H_2S conversion efficiency was obtained by the modified MS slurry (MS + $CuSO_4$), and the maximum removal efficiency of H_2S and PH_3 were 100% and ~78%, respectively. The simple modification process for raw MS through adding Cu^{2+} can effectively improve the H_2S and PH_3 conversion relative to fresh MS, which can be attributed to the synergistic effect of different metal ions. However, the added Cu^{2+} in the modified MS slurry would be consumed by conversion of Cu^{2+} to CuS/Cu_2S, thereby leading to the deactivation of modified MS slurry.

Author Contributions: J.B. and X.W. contributed equally. Conceptualization, X.S. (Xin Sun) and P.N.; methodology, X.S. (Xin Sun); software, C.W. and X.S. (Xin Song); validation, X.W.; formal analysis, J.B.; investigation, X.W.; resources, X.S. (Xin Sun), K.L., and P.N.; data curation, J.B.; writing—original draft preparation, J.B.; writing—review and editing, J.B.; visualization, J.B.; supervision, X.S. (Xin Sun); funding acquisition, X.S. (Xin Sun), K.L., F.W., and P.N. All authors have read and agreed to the published version of the manuscript.

Funding: This research was funded by the National Natural Science Foundation of China (grant 51708266, 51968034, 21667015, and 41807373) and the National Key R&D Program of China (grant 2018YFC0213400, 2018YFC1900305).

Conflicts of Interest: The authors declare no conflict of interest.

References

1.   Ma, Y.; Wang, X.; Ning, P.; Cheng, C.; Wang, F.; Wang, L.; Lin, Y.; Yu, Y. Simultaneous Removal of PH_3, H_2S, and Dust by Corona Discharge. *Energ. Fuel.* **2016**, *30*, 9580–9588. [CrossRef]

2.   Zhang, C.; Cheng, R.H.; Li, S.G.; Liu, C.; Chang, J.; Liu, H. Controls on Hydrogen Sulfide Formation and Techniques for Its Treatment in the Binchang Xiaozhuang Coal Mine, China. *Energy Fuels* **2019**, *33*, 266–275. [CrossRef]

3.   Chang, S.-M.; Hsu, Y.-Y.; Chan, T.-S. Chemical Capture of Phosphine by a Sol-Gel-Derived Cu/TiO_2 Adsorbent—Interaction Mechanisms. *J. Phys. Chem. C* **2011**, *115*, 2005–2013. [CrossRef]

4.   Qu, G.F.; Zhao, Q.; Jian, R.L.; Wang, J.Y.; Liu, D.X.; Ning, P. Mechanism of PH_3 absorption by $CuILs/H_2O$ two-liquid phase system. *Sep. Purif. Technol.* **2017**, *187*, 255–263. [CrossRef]

5.   Govindan, M.; Chung, S.J.; Jang, J.W.; Moon, I.S. Removal of hydrogen sulfide through an electrochemically assisted scrubbing process using an active Co (III) catalyst at low temperatures. *Chem. Eng. J.* **2012**, *209*, 601–606. [CrossRef]

6.   Fois, E.; Lallai, A.; Mura, G. Sulfur dioxide absorption in a bubbling reactor with suspensions of Bayer red mud. *Ind. Eng. Chem. Res.* **2007**, *46*, 6770–6776. [CrossRef]

7.   Montes-Moran, M.A.; Concheso, A.; Canals-Batlle, C.; Aguirre, N.V.; Ania, C.O.; Martin, M.J.; Masaguer, V. Linz-Donawitz Steel Slag for the Removal of Hydrogen Sulfide at Room Temperature. *Environ. Sci. Technol.* **2012**, *46*, 8992–8997. [CrossRef] [PubMed]

8. Kim, K.; Asaoka, S.; Yamamoto, T.; Hayakawa, S.; Takeda, K.; Katayama, M.; Onoue, T. Mechanisms of Hydrogen Sulfide Removal with Steel Making Slag. *Environ. Sci. Technol.* **2012**, *46*, 10169–10174. [CrossRef] [PubMed]

9. Lee, E.K.; Jung, K.D.; Joo, O.S.; Shul, Y.G. Liquid-phase oxidation of hydrogen sulfide to sulfur over CuO/MgO catalyst. *React. Kinet. Catal. Lett.* **2005**, *87*, 115–120. [CrossRef]

10. Monakhov, K.Y.; Gourlaouen, C. On the Insertion of ML_2 (M = Ni, Pd, Pt; L = PH_3) into the E-Bi Bond (E = C, Si, Ge, Sn, Pb) of a Bicyclo [1.1.1] Pentane Motif: A Case for a Carbenoid-Stabilized Bi(0) Species? *Organometallics* **2012**, *31*, 4415–4428. [CrossRef]

11. Goncharova, L.V.; Clowes, S.K.; Fogg, R.R.; Ermakov, A.V.; Hinch, B.J. Phosphine adsorption and the production of phosphide phases on Cu(001). *Surf. Sci.* **2002**, *515*, 553–566. [CrossRef]

12. Chen, H.; Feng, Y.; Suo, N.; Long, Y.; Li, X.; Shi, Y.; Yu, Y. Preparation of particle electrodes from manganese slag and its degradation performance for salicylic acid in the three-dimensional electrode reactor (TDE). *Chemosphere* **2019**, *216*, 281–288. [CrossRef] [PubMed]

13. Sun, L.N.; Song, X.; Li, K.; Wang, C.; Sun, X.; Ning, P.; Huang, H.B. Preparation of modified manganese slag slurry for removal of hydrogen sulphide and phosphine. *Can. J. Chem. Eng.* **2020**, *98*, 1534–1542. [CrossRef]

14. Fleet, M.E.; Harmer, S.L.; Liu, X.; Nesbitt, H.W. Polarized X-ray absorption spectroscopy and XPS of TiS_3: S K-and Ti L-edge XANES and S and Ti 2p XPS. *Surf. Sci.* **2005**, *584*, 133–145. [CrossRef]

15. Mikhlin, Y.; Likhatski, M.; Tomashevich, Y.; Romanchenko, A.; Erenburg, S.; Trubina, S. XAS and XPS examination of the Au-S nanostructures produced via the reduction of aqueous gold(III) by sulfide ions. *J. Electron Spectrosc. Relat. Phenom.* **2010**, *177*, 24–29. [CrossRef]

16. Beattie, D.A.; Arcifa, A.; Delcheva, I.; Le Cerf, B.A.; MacWilliams, S.V.; Rossi, A.; Krasowska, M. Adsorption of ionic liquids onto silver studied by XPS. *Colloid. Surface A* **2018**, *544*, 78–85. [CrossRef]

17. Wu, H.B.; Or, V.W.; Gonzalez-Calzada, S.; Grassian, V.H. CuS nanoparticles in humid environments: adsorbed water enhances the transformation of CuS to CuSO4. *Nanoscale* **2020**, *12*, 19350–19358. [CrossRef] [PubMed]

18. Liu, H.Y.; Xie, J.W.; Liu, P.; Dai, B. Effect of Cu^+/Cu^{2+} Ratio on the Catalytic Behavior of Anhydrous Nieuwland Catalyst during Dimerization of Acetylene. *Catalysts* **2016**, *6*, 120. [CrossRef]

 *catalysts*

Communication

The Impact of Surficial Biochar Treatment on Acute H₂S Emissions during Swine Manure Agitation before Pump-Out: Proof-of-the-Concept

Baitong Chen [1], Jacek A. Koziel [1,*], Andrzej Białowiec [1,2], Myeongseong Lee [1,3], Hantian Ma [4], Peiyang Li [1], Zhanibek Meiirkhanuly [1] and Robert C. Brown [5]

[1] Department of Agricultural and Biosystems Engineering, Iowa State University, Ames, IA 50010, USA; baitongc@iastate.edu (B.C.); andrzej.bialowiec@upwr.edu.pl (A.B.); leefame@iastate.edu (M.L.); peiyangl@iastate.edu (P.L.); zhanibek@iastate.edu (Z.M.)

[2] Faculty of Life Sciences and Technology, Wroclaw University of Environmental and Life Sciences, 37a Chelmonskiego Str., 51-630 Wroclaw, Poland

[3] Department of Animal Biosystems Sciences, Chungnam National University, Daejeon 34134, Korea

[4] Department of Civil, Construction and Environmental Engineering, Iowa State University, Ames, IA 50010, USA; hantian@iastate.edu

[5] Bioeconomy Institute and Department of Mechanical Engineering, Iowa State University, Ames, IA 50011, USA; rcbrown3@iastate.edu

* Correspondence: koziel@iastate.edu; Tel.: +1-515-294-4206

Received: 11 July 2020; Accepted: 12 August 2020; Published: 16 August 2020

Abstract: Acute releases of hydrogen sulfide (H₂S) are of serious concern in agriculture, especially when farmers agitate manure to empty storage pits before land application. Agitation can cause the release of dangerously high H₂S concentrations, resulting in human and animal fatalities. To date, there is no proven technology to mitigate these short-term releases of toxic gas from manure. In our previous research, we have shown that biochar, a highly porous carbonaceous material, can float on manure and mitigate gaseous emissions over extended periods (days–weeks). In this research, we aim to test the hypothesis that biochar can mitigate H₂S emissions over short periods (minutes–hours) during and shortly after manure agitation. The objective was to conduct proof-of-the-concept experiments simulating the treatment of agitated manure. Two biochars, highly alkaline and porous (HAP, pH 9.2) made from corn stover and red oak (RO, pH 7.5), were tested. Three scenarios (setups): Control (no biochar), 6 mm, and 12 mm thick layers of biochar were surficially-applied to the manure. Each setup experienced 3 min of manure agitation. Real-time concentrations of H₂S were measured immediately before, during, and after agitation until the concentration returned to the initial state. The results were compared with those of the Control using the following three metrics: (1) the maximum (peak) flux, (2) total emission from the start of agitation until the concentration stabilized, and (3) the total emission during the 3 min of agitation. The Gompertz's model for determination of the cumulative H₂S emission kinetics was developed. Here, 12 mm HAP biochar treatment reduced the peak (1) by 42.5% ($p = 0.125$), reduced overall total emission (2) by 17.9% ($p = 0.290$), and significantly reduced the total emission during 3 min agitation (3) by 70.4%. Further, 6 mm HAP treatment reduced the peak (1) by 60.6%, and significantly reduced overall (2) and 3 min agitation's (3) total emission by 64.4% and 66.6%, respectively. Moreover, 12 mm RO biochar treatment reduced the peak (1) by 23.6%, and significantly reduced overall (2) and 3 min total (3) emission by 39.3% and 62.4%, respectively. Finally, 6 mm RO treatment significantly reduced the peak (1) by 63%, overall total emission (2) by 84.7%, and total emission during 3 min agitation (3) by 67.4%. Biochar treatments have the potential to reduce the risk of inhalation exposure to H₂S. Both 6 and 12 mm biochar treatments reduced the peak H₂S concentrations below the General Industrial Peak Limit (OSHA PEL, 50 ppm). The 6 mm biochar treatments reduced the H₂S concentrations below the General Industry Ceiling Limit (OSHA PEL, 20 ppm). Research scaling up to larger manure volumes and longer agitation is warranted.

Keywords: hydrogen sulfide; biocoal; livestock manure; agricultural safety; fertilizer; waste management; air pollution; odor; kinetics; Gompertz model

1. Introduction

Hydrogen sulfide (H_2S) is a serious safety concern in agriculture and other industries. Inhalation of H_2S can be harmful to both humans and livestock, and sometimes deadly. The Occupational Safety and Health Administration (OSHA) recommends the permissible exposure limits (PELs) concentration for H_2S at 20 ppm and an acceptable maximum peak above the acceptable ceiling concentration at 50 ppm, with a maximum duration of 10 min [1].

The mid-western United States has a significant presence of pork production. Many large swine buildings use deep-pits to store manure under the slatted floor for up to 1 year. When a pit is full, farmers pump-out most of the manure to fertilize their fields in the fall. Agitating manure prior to pump-out is required to incorporate sediments and efficiently empty the pits. This routine seasonal operation generates a high risk of inhalation exposure to gases released from manure. Agitating the manure can break the entrapped gas bubbles, which causes an instantaneous increase in H_2S concentration (Figure 1) [2]. Fatal accidents have been recorded involving a high concentration of H_2S owing to the agitation of manure in the past several years [3–6].

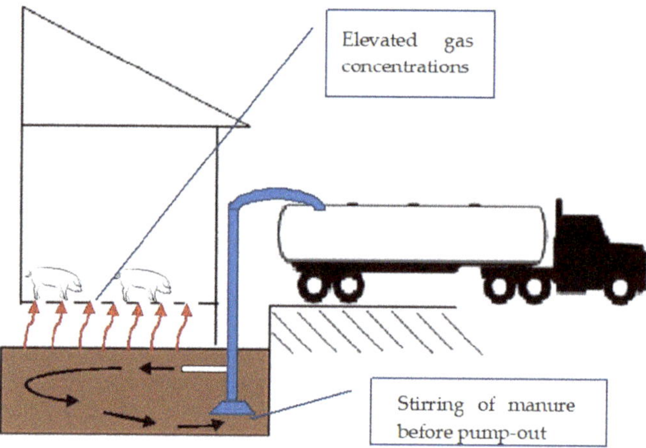

Figure 1. Schematic of the agitation process before seasonal manure pump-out from deep-pit storage under swine barn with a slatted floor. Fatal accidents are known to occur to people and livestock owing to the dangerous acute release of entrapped gases (e.g., H_2S) from stored manure during agitation.

To date, there is no proven technology to mitigate these short-term releases of toxic gas from manure. Commercial pit manure additives of the microbial mode of operation are used by some swine farmers to control gaseous emissions. Still, science-based guides, as well as more data, are needed to evaluate manure additive effect on the mitigation of gases emitted from storage [7]. Recent research on manure additives such as soybean peroxidase, zeolite, and biochar show the effectiveness of mitigating H_2S, NH_3, volatile organic compounds (VOCs), and greenhouse gas (GHG) emissions from swine manure over extended periods of time [8–13]. Additionally, we evaluated the performance of numerous commercial manure additives, but there was no overall statistically significant mitigation for gaseous emissions [14,15].

In our previous research, we have shown that 6 mm and 12 mm thick layer treatment of biochar, a highly porous carbonaceous material, can float on manure and mitigate gaseous emissions over extended periods (days–weeks). The mitigation effects on H_2S were typically the greatest on the first day of application and decreased over the duration of the trial [16]. This observation led us to explore the possibility of using surficial biochar treatment for *short-term* mitigation of H_2S emissions from swine manure. In this research, we aim to test the hypothesis that biochar can mitigate H_2S emissions over short periods (minutes–hours) during and shortly after manure agitation. The biochars tested had similar properties to those used for testing the spatial and temporal effects on pH near the liquid–gas interface owing to biochar addition to water [17] and manure surface [18].

Biochar has received considerable interest in the recent decade. It was proposed to be used as a soil amendment, an alternative source of fuel, and an adsorbent [19–21]. Biochar can be made from abundant biomass and waste through pyrolysis or torrefaction with no oxygen or a low-oxygen level [20–25]. Biochar's physicochemical properties vary as a result of differences in feedstock and its pretreatment, temperature, and time of the process [20–25]. The desired properties (e.g., pH, porosity, chemical moiety) could be explored to achieve environmental sustainability goals.

The first research question was, what biochar dose should be applied? The second research question was, how could a farm-scale system (Figure 1) be scaled down for a proof-of-the-concept experiment? The third research question was, how will the agitation of manure with added biochar influence the H_2S emission rates? Finally, will the mitigation effect be sufficient to meet the OSHA PELs recommendations, and will the results warrant scale-up research? We hypothesized that a greater biochar dose (thickness of the surficial layer applied to manure) would increase the H_2S mitigation effect; proof-of-the-concept experiments could use a shorter agitation time and a smaller amount of manure; and the mitigation effect would be significant and practical enough to warrant further scale-up research.

2. Results

2.1. Gaseous Emissions Post Biochar Application and Pre-Agitation (Stage 1)

Immediately after applying RO biochar, both scenarios showed a significant reduction in emissions. The 12 mm biochar treatment reduced the concentration of H_2S by 68.3%, and the 6 mm biochar treatment reduced 65.1% of H_2S (Table 1).

Table 1. H_2S emissions (expressed as flux) after applying red oak (RO) biochar (6 or 12 mm surficial dose) to manure surface and before manure agitation.

	RO Biochar		
Condition	Control	12 mm Biochar	6 mm Biochar
Pre-agitation H_2S $(mg{\cdot}m^{-2}{\cdot}s^{-1})$	0.00181 ± 0.000503	0.000782 ± 0.000388	0.000632 ± 0.000154

Once the HAP biochar was applied, the 12 mm biochar treatment immediately reduced the concentration of H_2S by about 99%, and the 6 mm biochar treatment reduced emissions by nearly 100% for H_2S (Table 2).

Table 2. H_2S emissions (expressed as flux) after applying highly alkaline and porous (HAP) biochar (6 or 12 mm surficial dose) to manure and before manure agitation.

	HAP Biochar		
Condition	Control	12 mm Biochar	6 mm Biochar
Pre-agitation H_2S $(mg{\cdot}m^{-2}{\cdot}s^{-1})$	0.0146 ± 0.0206	0.00014 ± 0.00011	0 *

* below detection limits.

2.2. Effect of the Dose on the Apparent Biochar Behavior Post-Agitation

After the agitation process, most of the biochar was still floating on the top of the manure. Some of the biochar was wetted and mixed with manure (as circled in Figure 2). The treatments with 12 mm biochar dose were visibly wetter and mixed more readily with manure than those treated with 6 mm biochar. Patches of open (not covered) manure were more prevalent to higher biochar dose. We observed similar dose-dependent behavior with surficially applied soybean peroxidase treatment to swine manure [11].

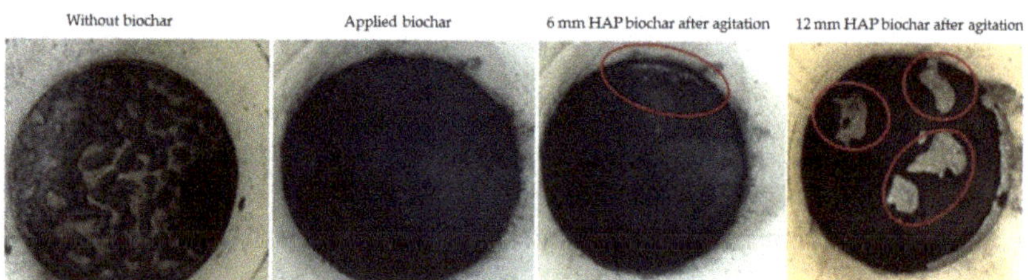

Figure 2. Swine manure surface: Control (left), highly alkaline and porous (HAP) biochar evenly spread on top of the swine manure (center left), 6 mm thick HAP biochar layer after agitation (center right), and 12 mm thick HAP biochar layer after agitation (right). Patches of open (uncovered) manure (red circles) were more apparent when higher biochar dose was used.

2.3. Gaseous Emissions during Agitation (Stage 2)

Both the 6 mm and 12 mm RO biochar treatment significantly ($p < 0.0001$) reduced the total emission of H_2S by 67.4% and 62.4%, respectively (Table 3, Figure A1). The 6 mm and 12 mm RO biochar treatment resulted in a 63.0% ($p = 0.0511$) and 23.6% ($p = 0.145$) reduction in the maximum peak flux of H_2S, respectively (Table 3).

Table 3. RO biochar treatment: the maximum peak flux and total H_2S emission *during* 3 min agitation (bold font signifies statistical significance).

		RO Biochar during the 3 min of Agitation	
	Control	12 mm Biochar	6 mm Biochar
Maximum peak flux while agitating, (mg·m^{-2}·s^{-1})	0.0504 ± 0.00078	0.0385 ± 0.0138	0.0186 ± 0.00977
% Reduction of maximum peak flux while agitating	–	23.6 ($p = 0.145$)	63.0 ($p = 0.0511$)
Total emission during 3 min agitation, (mg·m^{-2})	7.18 ± 0.644	2.7 ± 0.698	2.34 ± 0.472
% Reduction of total emissions during 3 min agitation	–	**62.4** ($p < 0.0001$)	**67.4** ($p < 0.0001$)

Both the 6 mm and 12 mm HAP biochar treatment significantly ($p < 0.0001$) reduced the total emission of H_2S by 66.6% and 70.4%, respectively (Table 3, Figure A2). The 6 mm and 12 mm RO biochar treatment resulted in 60.6% ($p = 0.05804$) and 42.5% ($p = 0.1249$) reduction in the maximum peak flux of H_2S, respectively (Table 4).

Table 4. HAP biochar treatment: the maximum peak flux and total H_2S emission *during* 3 min agitation (bold font signifies statistical significance).

	HAP Biochar during the 3 min of Agitation		
	Control	**12 mm Biochar**	**6 mm Biochar**
Maximum peak flux while agitating, $(mg \cdot m^{-2} \cdot s^{-1})$	0.0455 ± 0.0192	0.0261 ± 0.00665	0.0179 ± 0.00321
% Reduction of maximum peak flux while agitating	–	42.5 ($p = 0.1249$)	60.6 ($p = 0.05804$)
Total emission during 3 min agitation, $(mg \cdot m^{-2})$	6.36 ± 1.23	1.88 ± 0.625	2.12 ± 0.433
% Reduction of total emissions during 3 min agitation	–	**70.4** ($p < 0.0001$)	**66.6** ($p < 0.0001$)

2.4. Gaseous Emissions Post-Agitation (Stage 3)

Once the agitation stopped, the concentrations of H_2S started to decrease for both HAP and RO biochar treatments (Figures A1 and A2). The H_2S concentrations were recorded until they reached the levels before agitation and were stable. Both the 6 mm and 12 mm RO biochar treatment significantly ($p < 0.0001$) reduced cumulative H_2S emissions by 84.7% and 39.3%, respectively (Table 5).

Table 5. RO biochar treatment: the average flux and cumulative H_2S emission after agitation (bold font signifies statistical significance).

	RO Biochar after the 3 min of Agitation		
	Control	**12 mm Biochar**	**6 mm Biochar**
Duration (min)	36	36	36
Average emissions [1] $(mg \cdot m^{-2} \cdot min^{-1})$	1.37 ± 0.175	0.831 ± 0.0483	0.209 ± 0.00174
Cumulative emissions [2] $(mg \cdot m^{-2})$	49.2 ± 2.63	29.9 ± 1.74	7.52 ± 0.627
% Reduction of cumulative emissions	–	**39.3** ($p < 0.0001$)	**84.7** ($p < 0.0001$)

[1] the average emissions were calculated using the cumulative emissions divided by the duration. [2] the cumulative emissions were calculated based on the same period (post-agitation) (Figure A1).

For HAP biochar treatments, the 6 mm biochar treatment significantly ($p < 0.0001$) reduced cumulative emissions of H_2S by 64.4%. The 12 mm biochar treatment reduced the cumulative H_2S emissions by 17.9%, yet the reduction was not significant ($p = 0.2897$) (Table 6).

Table 6. HAP biochar treatment: the average flux and cumulative H_2S emission after agitation (bold font signifies statistical significance).

	HAP Biochar after the 3 min of Agitation		
	Control	**12 mm Biochar**	**6 mm Biochar**
Duration (min)	14	14	14
Average emissions [1] $(mg \cdot m^{-2} \cdot min^{-1})$	1.00 ± 0.134	0.821 ± 0.0936	0.356 ± 0.0379
Cumulative emissions [2] $(mg \cdot m^{-2})$	14.0 ± 1.88	11.5 ± 1.31	4.99 ± 0.531
% Reduction of cumulative emissions	–	17.9 ($p = 0.2897$)	**64.4** ($p < 0.0001$)

[1] the average emissions were calculated using the cumulative emissions divided by the duration. [2] the cumulative emissions were calculated based on the same period (post-agitation) (Figure A2).

The H_2S in the headspace of RO treated manure needed longer to return to the initial state compared with the HAP treatment. The H_2S release was higher in the experiment testing the RO treatment (Figure A1) compared with the experiment testing the HAP treatment (Figure A2). The control concentrations exceeded the limitations of the H_2S sensor. The apparent difference in the control concentrations is the result of the differences in manure used in RO and HAP experiments, that is, collected at the same farm, yet at two different times for the RO and HAP trials.

2.5. Kinetics of the Post-Agitation Emissions of H_2S

The kinetics modeling allowed further evaluation of the effect of biochar type and the dose. The E_0 parameter shows the potential of H_2S emission during an 'infinite' time. The cumulative emission during the post-agitation showed that there was no ($p > 0.05$) significant influence of the HAP biochar treatment on the potential maximum cumulative flux (Figure 3). However, the lack of the significance of the differences may be caused by high variability, while still, the apparent potential for lower emission is visible. The RO application of biochar significantly ($p = 0.0086$) reduced the potential of the maximum cumulative emission in the case of the 6 mm dose; however, there were no differences between the biochar dose (Figure 3, Table A1—Appendix B). For both the RO and HAP biochar, the lowest values of E_0 were determined for 6 mm biochar thickness, implying that a low biochar dose could be just as effective.

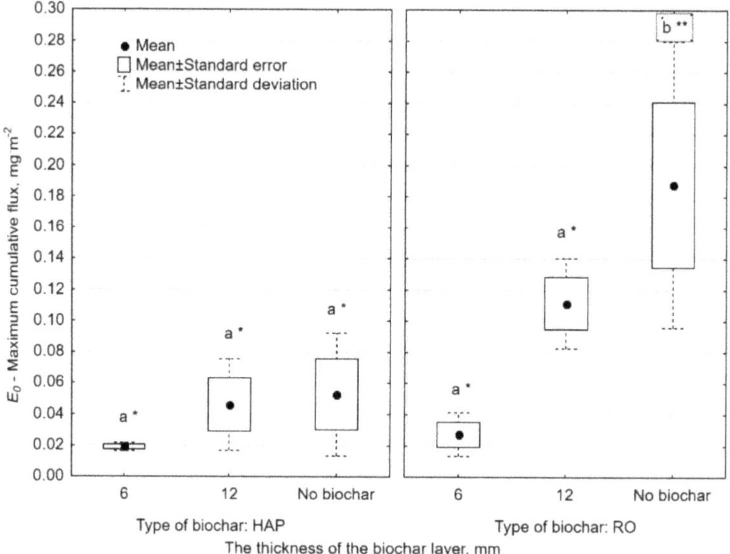

Figure 3. The differences between the maximum cumulative H_2S flux per biochar type and dose (thickness of the biochar layer). Letters indicate the significant difference within the same group of biochar types; asterisks indicate significant differences between biochar dose (Table A1). RO, red oak.

The k constant presents the rate of H_2S emission. The treatment by application of both types of biochar did not significantly influence ($p > 0.05$) the k constant (Figure 4, Table A2—Appendix B). The lack of significant differences could be caused by high values of the standard deviations. However, the influence of the biochar dose was observed only in the case of HAP, where the 12 mm biochar reduced the k value.

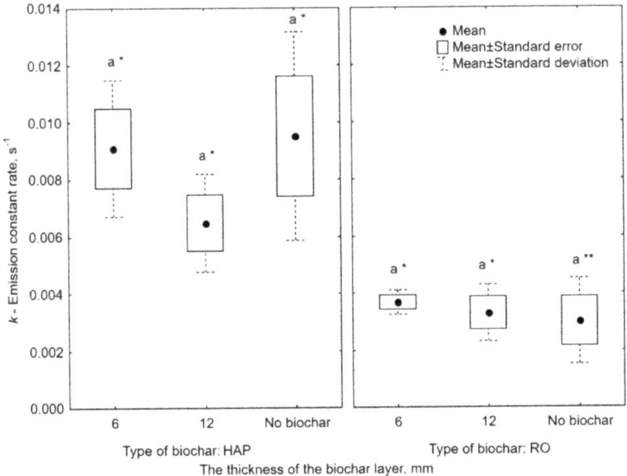

Figure 4. The differences between H_2S emission constant rates per the type and dose (thickness of the layer) of biochar. Letters indicate the significant difference within the same group of biochar types; asterisks indicate significant differences between thicknesses of the biochar layers (Table A2).

The a_1 parameter, the inflection time of the cumulative H_2S emission curve, represents the moment when the emission rate starts to 'slow' down. Similar to the k parameter, the treatment by the application of both types of biochar did not significantly influence ($p > 0.05$) the inflection time of the H_2S emission (Figure 5, Table A3). The lack of significant differences could be caused by high variability. However, in this case, the lowest a_1 values were observed for 6 mm for both biochar types.

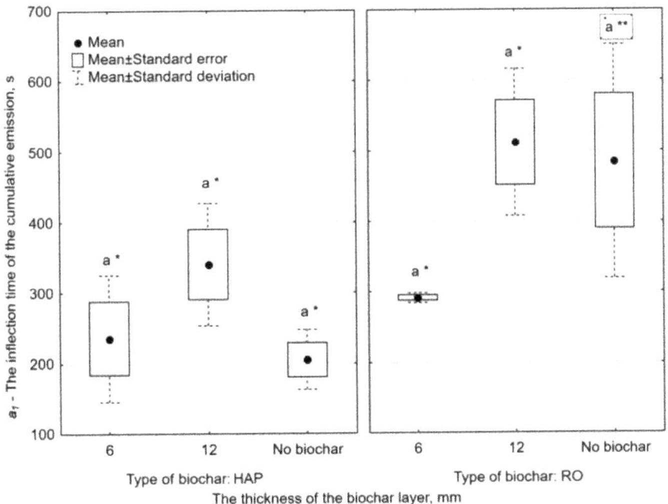

Figure 5. The differences between the inflection time of the cumulative H_2S emission per biochar type and the dose (thickness of the layer). Letters indicate the significant difference within the same group of biochar types; asterisks indicate significant differences between biochar dose (Table A3).

3. Discussion

Biochar treatments have the potential to reduce the risk of inhalation exposure to H_2S. Both 6 and 12 mm biochar treatments reduced the peak H_2S concentrations below the General Industrial Peak Limit (OSHA PEL, 50 ppm). The 6 mm biochar treatments, both HAP and RO, reduced the H_2S concentrations below the General Industry Ceiling Limit (OSHA PEL, 20 ppm) (Figures A3 and A4).

This proof-of-the-concept study shows that biochar has the potential to be an effective treatment of short-term releases of H_2S during and post-agitation of swine manure. From the kinetics of the post-agitation H_2S emissions analysis, only RO biochar has shown the significant ($p = 0.0086$) reductions on the maximum cumulative emission (E_0). Further, the smaller dosage (6 mm) worked just as well as the 12 mm dosage. The pH value of HAP was 9.2, while the RO pH was 7.5 [17]. It has been expected that HAP (more alkalic) would have a greater influence on H_2S emissions mitigation, owing to H_2S transformation into S^{2-} ions. Previously, we have found that HAP had a stronger influence on the water pH increase than RO [17]. The apparent absence of the differences between RO and HAP in the present experiment, and even (numerically) better performance of RO biochar, could be caused by the different buffering capacity of the manure used for the experiment [18]. However, the comparison between these two types of biochars was not the aim of the study.

Biochar treatments did not have much impact on the constant emission rate (k) owing to the high standard deviations, except for the 12 mm HAP biochar treatment. The high variations could be caused by high heterogenicity of the stored manure properties (i.e., stratified, biologically-active, not a well-mixed solution, with local solids aggregates, and zones with different chemical properties). Therefore, one possible solution is to work with artificial surrogate manure (if a particular mechanism behind the mitigation needs to be isolated). The inflection time (a_1) of the cumulative emissions was not influenced much by either type of biochar; the lowest a_1 values were observed for 6 mm for both biochar types, where the emission rate started to slow down after 4–5 min. These findings still need to be proven on a larger scale and optimized. Still, this initial work has implications that could potentially save people and livestock lives and reduce inhalation risks during routine seasonal manure agitation, pump-out, and land application. With further research, the optimal biochar type, dose, and form of application (e.g., pellets instead of powder), it could become an effective adsorptive 'barrier' to protect farmers, neighbors, and livestock from harmful gases and odors emitted from manure.

Surprisingly, the 6 mm biochar treatment was a numerically more effective dosage because the % reduction was slightly higher while using less biochar. The smaller amount of biochar used has an immediate impact and economics and on the feasibility of technology adoption. When the biochar is wetted, it forms 'chunks'. When manure is being agitated, the bigger chunks of biochar in 12 mm treatments started to turn over, sink, and mix with manure much faster than with the 6 mm dose. Once the physical barrier on the surface was broken, the maximum concentration of the treatment began to rise and was closer to the Control.

In future research, other kinds of biochar could be tested for their efficacy to mitigate gaseous emissions from manure. The effects of the dose and frequency of application of commercial biochars, functionalized biochars, pelletized biochar, as well as the synergy between gaseous emissions and agronomic benefits to soil should be tested. Additionally, farm-scale research is also required for the proof-of-the-concept. With more extensive farm-scale trials, researchers should consider how and where the biochar could be practically applied in order to create an effective short-term barrier to maximize the benefit of biochar treatment. Application of powdery, light, and dusty material might be hazardous itself and not be feasible in farm conditions. Pelletized biochar could be a more practical and safer mode of application. Opportunities exist to mitigate other types of gases and other applications (e.g., industrial wastewater, compost, landfill leachate) with biochar.

positively influences the process selectivity and dynamically controls the accumulation of sulfur that might compromise the catalyst performance, particularly on long-term desulfurization runs. All of this evidence taken together underline the importance of controlling the chemical and morphological properties of the N-C phase through a rational optimization of the SiC impregnation/thermal treatment sequences thus allowing a tuning of the active phase surface properties as a function of its downstream application.

Expectedly, the lower the toluene content in the stream the slower the $N\text{-}C^2/SiC$ deactivation rate under desulfurization conditions. This trend is confirmed by the catalyst deactivation measured with $N\text{-}C^2/SiC$ in the presence of variable toluene concentrations (from 1 vol.% to 4 vol.%) within a 115 h desulfurization run (Figure 3B). At the same time, selectivity follows a similar and positive trend irrespective from the toluene content but reaching appreciably higher values when the concentration of the latter increases.

As an additional trial, $N\text{-}C^2/SiC$ desulfurization capacity was investigated as a function of the reaction temperature and in the presence of variable percentages of toluene as contaminant. To this aim, $N\text{-}C^2/SiC$ was initially conditioned at 210 °C using a toluene-free sour gas stream and the process was constantly monitored throughout about 50 h. During this time, the catalyst showed a quantitative H_2S conversion (X_{H2S}) and a sulfur selectivity (S_S %) laying around 55–60% (Figure 4). The addition of toluene (4 vol.%) to the acid gas stream leads to the rapid increase of the process selectivity (up to 96%) while X_{H2S} gradually stabilizes to a constant plateau value. An increase of the catalyst temperature to 230 °C results into a rapid X_{H2S} increase that reaches values close to 80% for gradually dropping down to 72% after additional 100 h on stream.

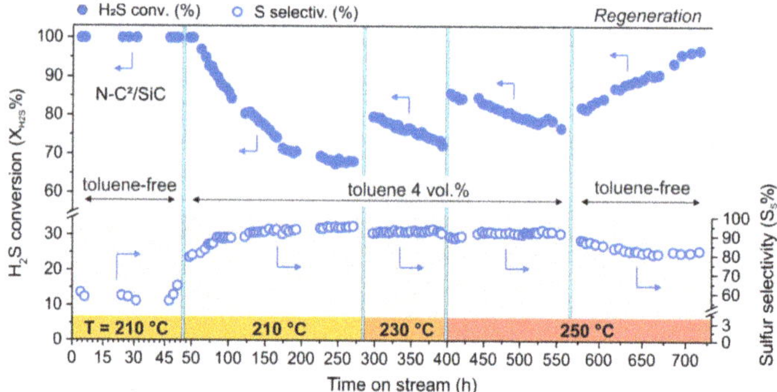

Figure 4. Effect of the temperature on the desulfurization performance of $N\text{-}C^2/SiC$ using an acid gas stream ($[H_2S] = 0.3$ vol.%) in the presence of 40,000 ppm of toluene (4 vol.%) as contaminant. The first 47 h on stream were operated under toluene-free conditions. Catalysis details: 6 g ($V_{cat} \sim 7.5$ cm^3 of $N\text{-}C^2/SiC$); O_2-to-H_2S ratio = 2.5, $[H_2O] = 10$ vol.%, He (balance); GHSV (STP) = 3200 h^{-1}. Catalyst regeneration (last 150 h on stream) occurs upon switching-off the toluene content from the acid gas stream.

As far as sulfur selectivity is concerned, the reaction temperature and X_{H2S} increase affect only marginally its mean value that slightly reduces from 96% to 95%. Similarly, an additional temperature increase from 230 to 250 °C gives rise to a further increase of H_2S conversion values that grow over 85% while S_S does no longer reduce appreciably. As Figure 4 shows, the X_{H2S} decreases again and finally stabilizes around 78% after additional 120 h on stream together with a mean S_S value of 93% that highlight the unique desulfurization properties of this metal-free catalyst while stressing again its robustness and durability under quite unconventional conditions. It should be noticed that $N\text{-}C^2/SiC$ still maintains a remarkably high X_{H2S} and S_S even after more than a 3 weeks (550 h) H_2S desulfurization run operated under continuum mode and severe operative conditions.

Notably, switching-off toluene from the acid gas stream, translates into a gradual but complete recovery of the pristine catalyst performance (Figure 4, Regeneration section).

4. Experiments

4.1. Manure, Biochars, H₂S Measurements

Fresh manure was collected twice from deep-pit storage at a local swine farm in central Iowa. The manure treated with the red oak (RO) biochar was collected in summer, whereas manure treated with the highly alkaline and porous (HAP) biochar was collected in winter. Thus, the experimental design was set up to compare the Treatment and Control of the same type of biochar. The manure properties and, therefore, baseline H_2S concentrations for control groups were different for HAP and RO trials. The proof-of-the-concept simulation of the deep pit and agitation was facilitated by 1.22 m (height) and 0.38 m (diameter) manure storage. The working volume of the manure was 103.1 L, while the headspace was ventilated with 7.5 air exchanges per hour (ACH), which is representative of the ventilation of deep-pit manure storage [11,26]. A 1/10 hp transfer pump (Little Giant, Fort Wayne, IN) was used to agitate the manure with a 1.36 m^3 h^{-1} flow rate.

Biochar physicochemical properties were described elsewhere [16–18]. Briefly, some key properties are listed below. RO biochar was pyrolyzed at 500–550 °C. It had a pH of 7.5 and a 6.75 zero-point charge, consisting of C (78.53% dry matter, d.m.), H (2.54% d.m.), N (0.62% d.m.), and volatile solids (VS, 26.38% d.m.). Fixed C and ash were 54.76% and 15.83% d.m., respectively [16–18]. The HAP biochar was made from corn stover pyrolyzed at 500 °C. The pH was 9.2 and 8.42 zero-point charge, consisting of C (61.37% d.m.), H (2.88% d.m.), N (1.21% d.m.), and VS (16.27% d.m.). Fixed C and ash were 34.98% and 46.82% d.m., respectively [16–18].

OMS-300 real-time monitoring system equipped with electrochemical gas sensors (H₂S/C-50) (Smart Control & Sensing Inc., Daejeon, Korea) was used to measure the real-time H_2S concentration [27,28]. The analyzer was calibrated with standard gas before use [27,28].

4.2. Experimental Design

The pilot-scale setup was simulating deep pit swine manure storage while manure was being agitated (Figure 6). The inlet of the pump was connected to the bottom manure sampling port; the outlet was connected to the middle manure sampling port. During agitation, the manure was pumped from the bottom to the middle zone for 3 min. Three variants per each biochar and each with triplicates experiments:

- Manure not treated with biochar—control variant.
- Manure treated with—6 mm thick layer of biochar.
- Manure treated with—12 mm thick layer of biochar.

The biochar dose was based on its volume spread over the manure surface, resulting in either 6 or 12 mm average thicknesses. The headspace H_2S concentrations were measured in the exhaust (Figure 6) continuously during the following stages:

- Stage 1: No agitation. Post biochar application and pre-agitation for all three variants.
- Stage 2: Agitation. All three variants during agitation.
- Stage 3: Post-agitation. All three variants after agitation until the headspace H_2S concentration reached its initial state.

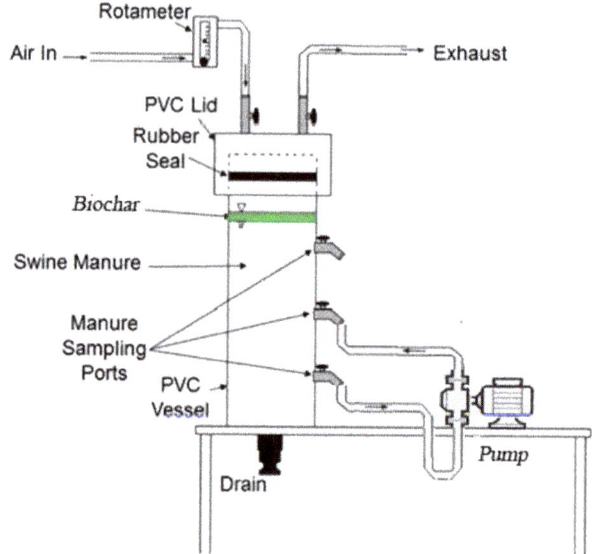

Figure 6. Pilot-scale design for simulating deep pit manure storage treated surficially with a thin layer of biochar prior to agitating.

4.3. Data Analysis, the Kinetics of Emissions

The mitigation effect was estimated by comparing measured emissions associated with the Control (not treated) and treatment (treated with biochar) manure. The % reduction was calculated as the percent ratio of (control − treatment)/control.

The one-way analysis of variance (ANOVA) and Tukey–Kramer method in JMP software (version Pro 14, SAS Institute, Inc., Cary, NC, USA) were used to analyze the data to determine the p-values of total emissions for both overall and during 3 min. The maximum levels of concentrations were used for a pooled T-test to estimate the p-values. A p-value < 0.05 was used as a statistical significance threshold. The Gompertz model was used for the determination of the post-agitation cumulative H_2S emission kinetics [29]:

$$E = E_0 \cdot e^{\left(-e^{-k \cdot (t-a_1)}\right)} \tag{1}$$

where E—H_2S emission flux, mg.m^{-2}; E_0—H_2S maximum cumulative emission flux, mg m^{-2}; k—constant rate of the H_2S emission flux, s^{-1}; t—time, s; and a_1—the inflection time of the cumulative H_2S emission, s.

The non-linear regression was used for the determination of the cumulative emission kinetics with the application of the Statistical 13 software (TIBCO Software Inc., Palo Alto, CA, USA). The R^2 determination coefficient was estimated to indicate the fitting the model to data. The kinetic analysis was completed for each variant and each repetition. The result of the regression analysis for each variant is provided in Appendix A (Figures A5–A22) and used to estimate the average values of E_0, k, and a_1 (Equation (1)). The ANOVA test was applied with post-hoc Tukey's test to indicate the statistical significance ($p < 0.05$) of the differences between average values. The calculated probabilities of Tukey's test are given in Appendix B.

5. Conclusions

The highly alkaline and porous (HAP) and red oak (RO) biochar treatments have the potential to reduce the risk of inhalation exposure to H_2S. Both the 6 mm and 12 mm RO biochar treatment significantly ($p < 0.0001$) reduced the total emission of H_2S by 67.4% and 62.4%, respectively. The 6 mm and 12 mm RO biochar treatment resulted in a 63.0% ($p = 0.0511$) and 23.6% ($p = 0.145$) reduction in the maximum peak flux of H_2S, respectively. Both the 6 mm and 12 mm HAP biochar treatment significantly ($p < 0.0001$) reduced the total emission of H_2S by 66.6% and 70.4%, respectively. The 6 mm and 12 mm RO biochar treatment resulted in 60.6% ($p = 0.05804$) and 42.5% ($p = 0.1249$) reduction in the maximum peak flux of H_2S, respectively. Both 6 and 12 mm biochar treatments reduced the peak H_2S concentrations below the General Industrial Peak Limit (OSHA PEL, 50 ppm). The 6 mm biochar treatments reduced the H_2S concentrations below the General Industry Ceiling Limit (OSHA PEL, 20 ppm). Research scaling up to larger manure volumes and longer agitation is warranted.

Author Contributions: Conceptualization, J.A.K. and B.C.; methodology, B.C. and J.A.K.; software, B.C. and A.B.; validation, J.A.K. and A.B.; formal analysis, B.C.; investigation, B.C., M.L., H.M., P.L. and Z.M.; resources, J.A.K. and R.C.B.; data curation, B.C., J.A.K. and A.B.; writing—original draft preparation, B.C.; writing—review and editing, J.A.K. and A.B.; visualization, B.C., H.M., and A.B.; supervision, J.A.K. and A.B.; project administration, J.A.K. and R.C.B.; funding acquisition, J.A.K. and R.C.B. All authors have read and agreed to the published version of the manuscript.

Funding: This research was partially funded by the U.S. Department of Energy—National Institute for Food and Agriculture, grant # 2018-10008-28616: 'Valorization of biochar: Applications in anaerobic digestion and livestock odor control (2018–2020, PI R.B.). In addition, this research was partially supported by the Iowa Agriculture and Home Economics Experiment Station, Ames, Iowa. Project no. IOW05556 (Future Challenges in Animal Production Systems: Seeking Solutions through Focused Facilitation) sponsored by Hatch Act and State of Iowa funds. The authors would like to thank the Ministry of Education and Science of the Republic of Kazakhstan for supporting Z.M. with an M.S. study scholarship via the Bolashak Program. Authors would like to thank the Fulbright Poland Foundation for funding the project titled "Research on pollutants emission from Carbonized Refuse Derived Fuel into the environment," completed by A.B. at the Iowa State University. The authors would like to express gratitude to Samuel O'Brien for his helpful corrections to grammar.

Conflicts of Interest: The authors declare no conflict of interest. The funders had no role in the design of the study; in the collection, analyses, or interpretation of data; in the writing of the manuscript; or in the decision to publish the results.

Appendix A

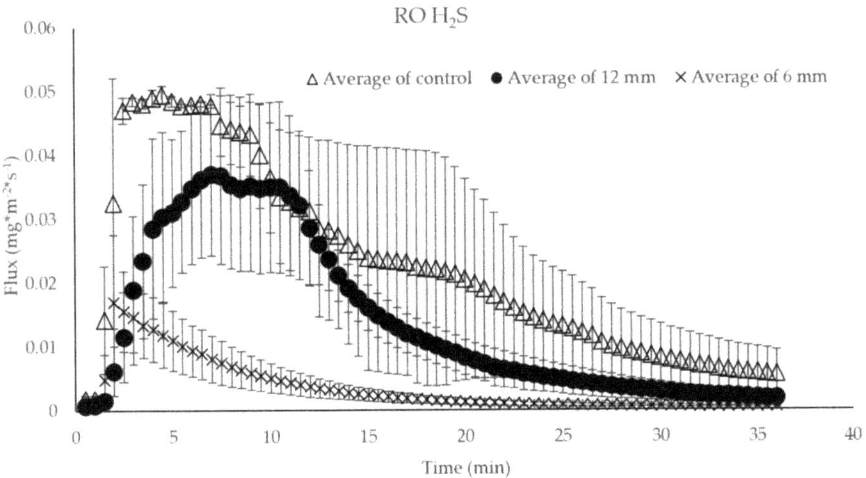

Figure A1. The short-term H_2S emissions when manure is treated surficially with RO biochar layer at two thicknesses (6 mm; 12 mm) immediately *prior to* 3 min agitation. Each data point is the average of triplicate, and the error bar signifies a standard deviation.

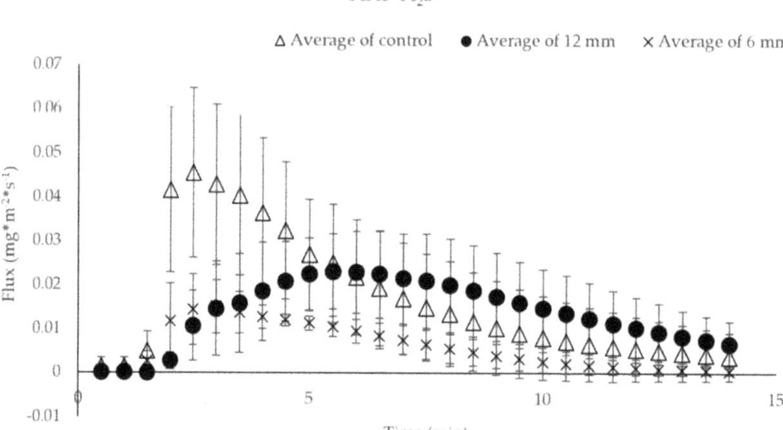

Figure A2. The short-term H_2S emissions when manure is treated surficially with HAP biochar layer at two thicknesses (6 mm; 12 mm) immediately *prior to* 3 min agitation. Each data point is the average of triplicate, and the error bar signifies a standard deviation.

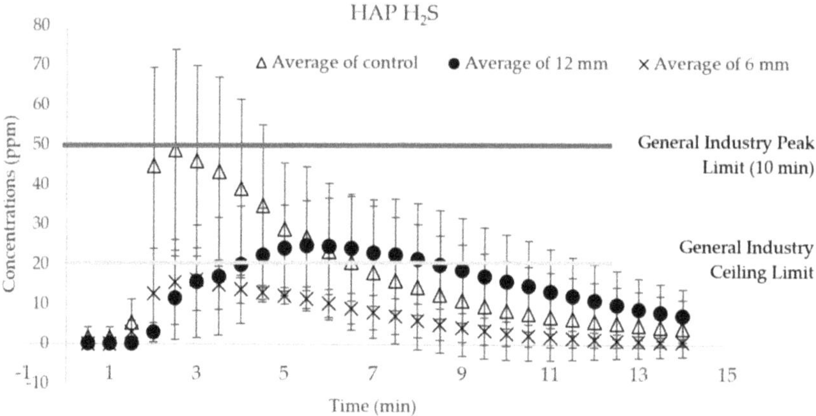

Figure A3. The short-term H_2S concentrations when manure is treated surficially with HAP biochar layer at two thicknesses (6 mm; 12 mm) immediately *prior to* 3 min agitation. Each data point is the average of triplicate, and the error bar signifies a standard deviation. Red line = the 'General Industry Peak Limit' (OSHA PEL = 50 ppm); yellow line = the 'General Industry Ceiling Limit' (OSHA PEL = 20 ppm) [1].

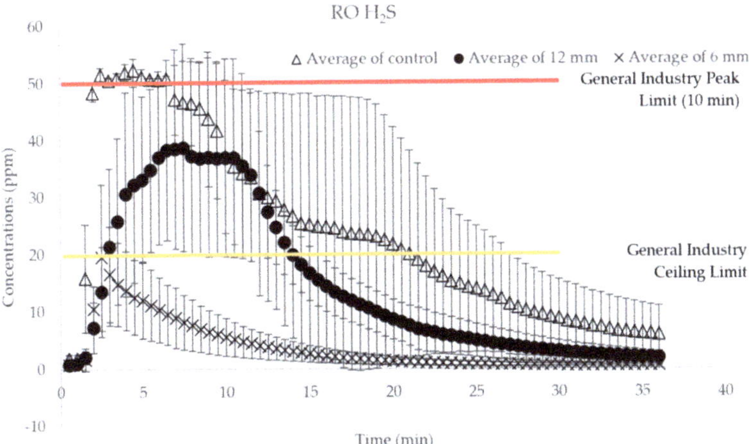

Figure A4. The short-term H$_2$S concentrations when manure is treated surficially with RO biochar layer at two thicknesses (6 mm; 12 mm) immediately *prior to* 3 min agitation. Each data point is the average of triplicate, and the error bar signifies a standard deviation. Red line = the 'General Industry Peak Limit' (OSHA PEL = 50 ppm); yellow line = the 'General Industry Ceiling Limit' (OSHA PEL = 20 ppm) [1].

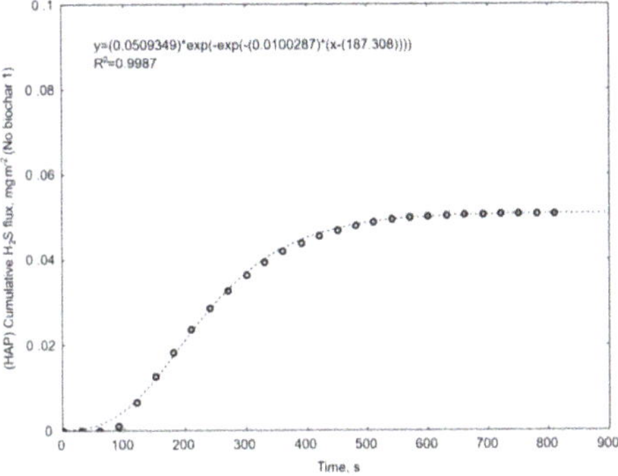

Figure A5. The cumulative H$_2$S flux. Variant with no HAP biochar, repetition 1. Gompertz equation parameters and R^2 determination coefficient.

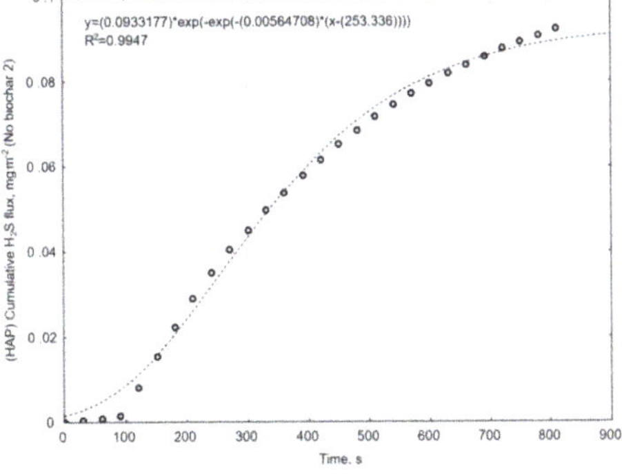

Figure A6. The cumulative H$_2$S flux. Variant with no HAP biochar, repetition 2. Gompertz equation parameters and R^2 determination coefficient.

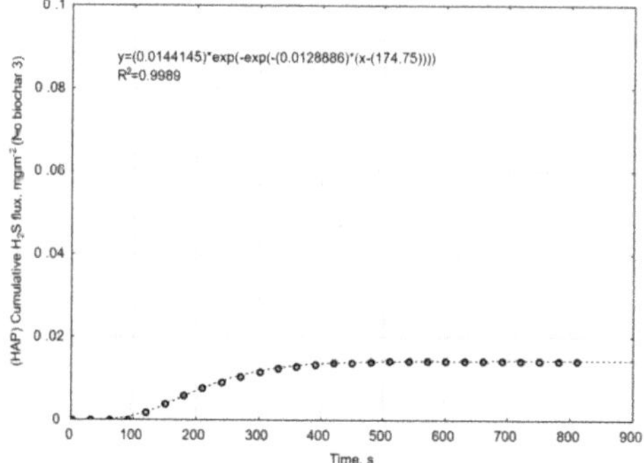

Figure A7. The cumulative H$_2$S flux. Variant with no HAP biochar, repetition 3. Gompertz equation parameters and R^2 determination coefficient.

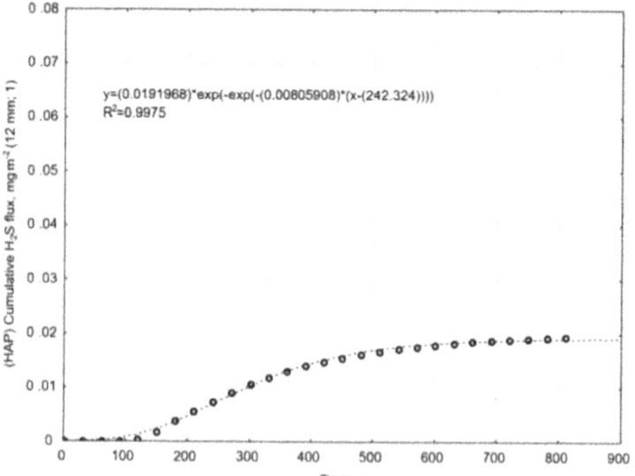

Figure A8. The cumulative H$_2$S flux. Variant with 12 mm HAP biochar layer, repetition 1. Gompertz equation parameters and R^2 determination coefficient.

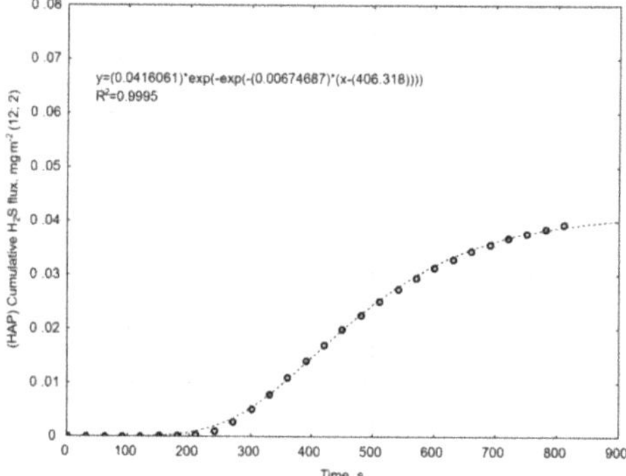

Figure A9. The cumulative H$_2$S flux. Variant with 12 mm HAP biochar layer, repetition 2. Gompertz equation parameters and R^2 determination coefficient.

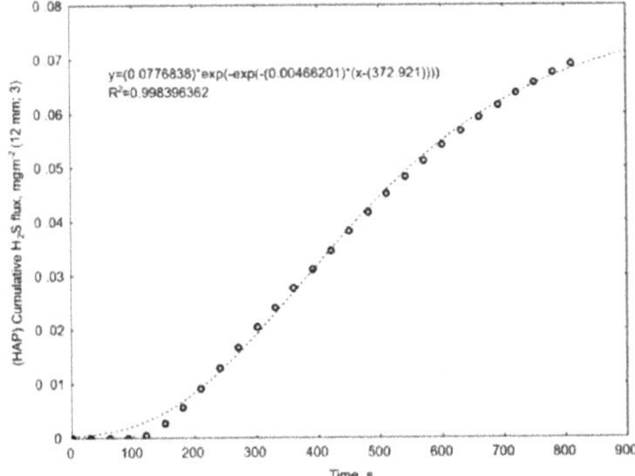

Figure A10. The cumulative H_2S flux. Variant with 12 mm HAP biochar layer, repetition 3. Gompertz equation parameters and R^2 determination coefficient.

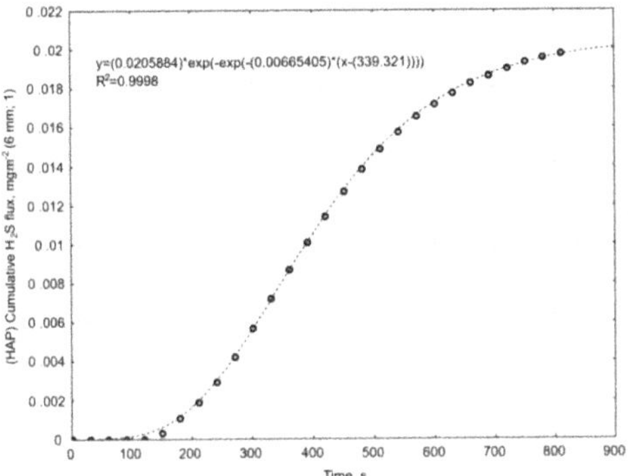

Figure A11. The cumulative H_2S flux. Variant with 6 mm HAP biochar layer, repetition 1. Gompertz equation parameters and R^2 determination coefficient.

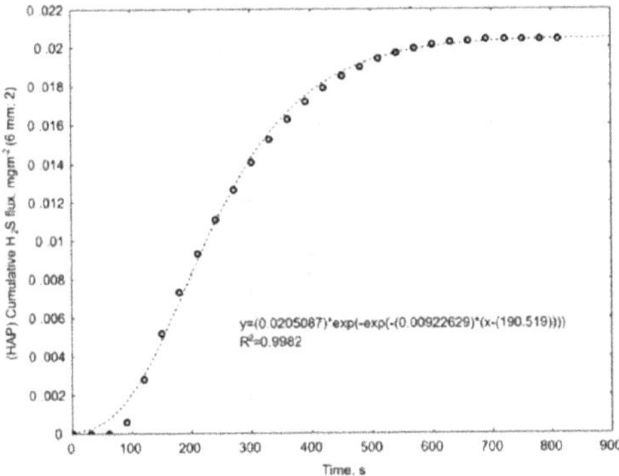

Figure A12. The cumulative H_2S flux. Variant with 6 mm HAP biochar layer, repetition 2. Gompertz equation parameters and R^2 determination coefficient.

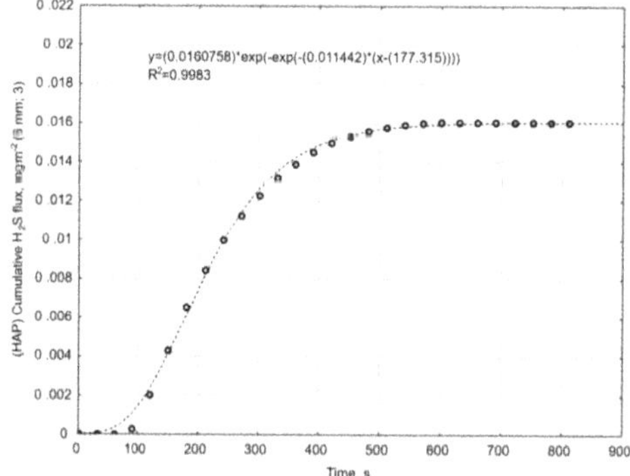

Figure A13. The cumulative H$_2$S flux. Variant with 6 mm HAP biochar layer, repetition 3. Gompertz equation parameters and R^2 determination coefficient.

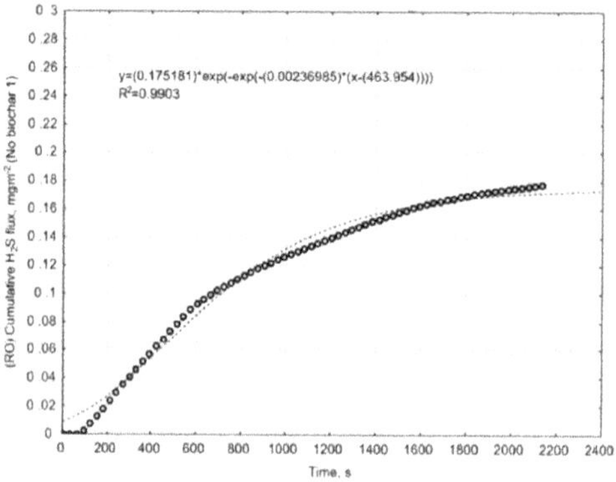

Figure A14. The cumulative H$_2$S flux. Variant with no RO biochar, repetition 1. Gompertz equation parameters and R^2 determination coefficient.

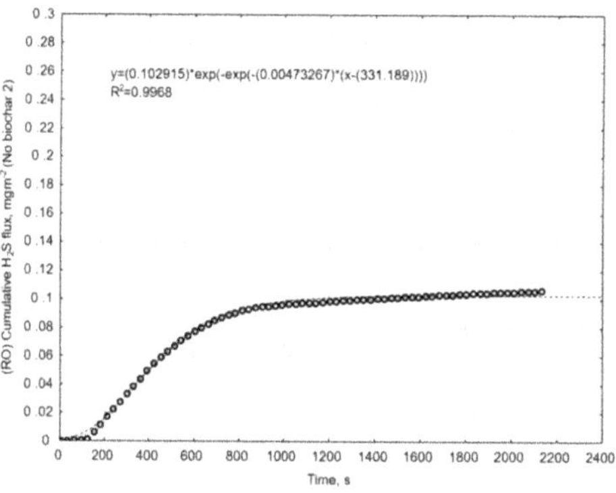

Figure A15. The cumulative H$_2$S flux. Variant with no RO biochar, repetition 2. Gompertz equation parameters and R^2 determination coefficient.

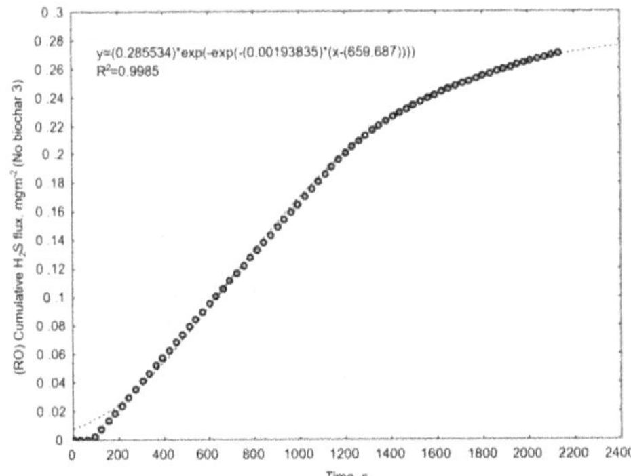

Figure A16. The cumulative H_2S flux. Variant with no RO biochar, repetition 3. Gompertz equation parameters and R^2 determination coefficient.

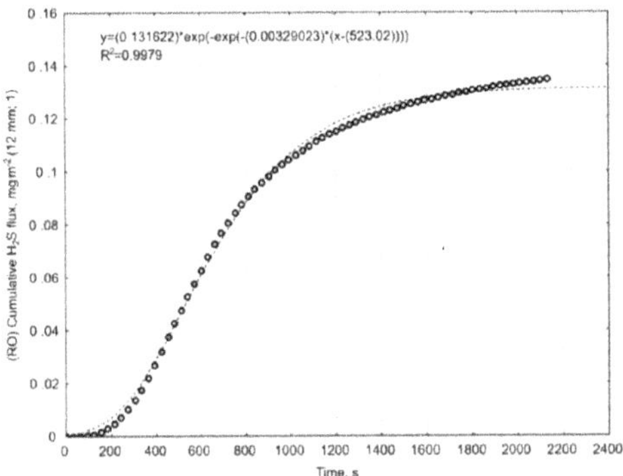

Figure A17. The cumulative H_2S flux. Variant with 12 mm RO biochar layer, repetition 1. Gompertz equation parameters and R^2 determination coefficient.

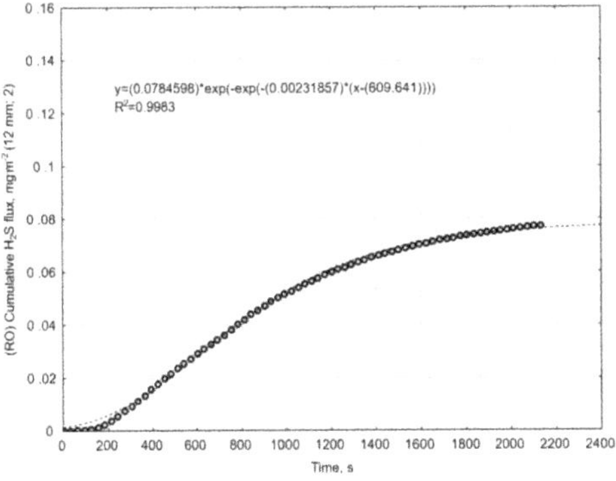

Figure A18. The cumulative H_2S flux. Variant with 12 mm RO biochar layer, repetition 2. Gompertz equation parameters and R^2 determination coefficient.

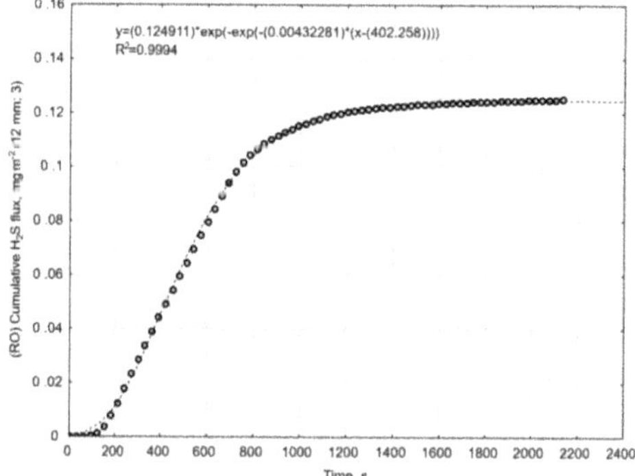

Figure A19. The cumulative H$_2$S flux. Variant with 12 mm RO biochar layer, repetition 3. Gompertz equation parameters and R^2 determination coefficient.

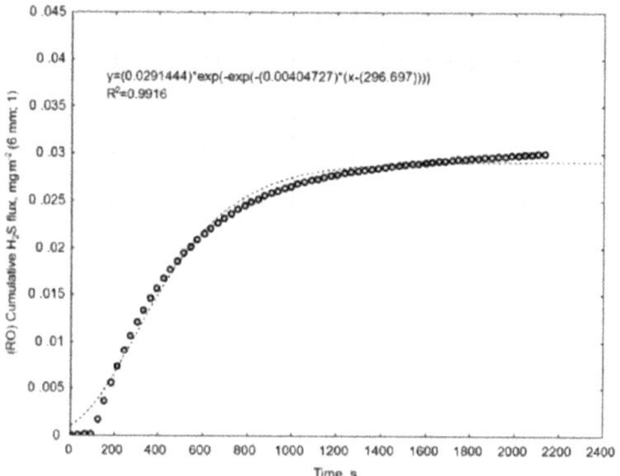

Figure A20. The cumulative H$_2$S flux. Variant with 6 mm RO biochar layer, repetition 1. Gompertz equation parameters and R^2 determination coefficient.

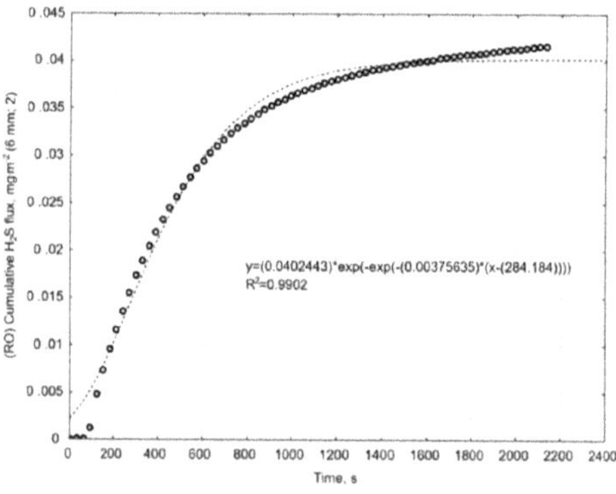

Figure A21. The cumulative H$_2$S flux. Variant with 6 mm RO biochar layer, repetition 2. Gompertz equation parameters and R^2 determination coefficient.

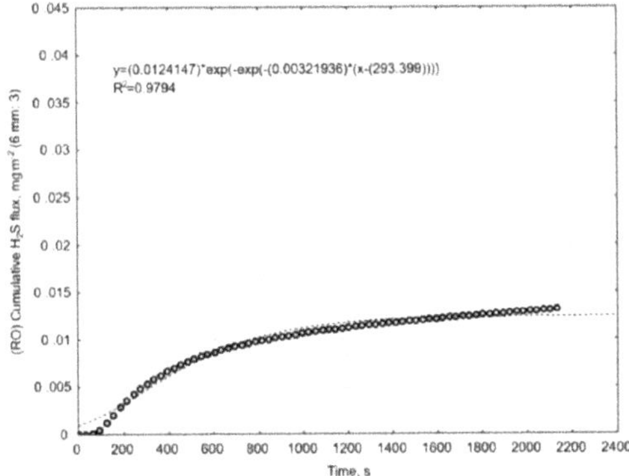

Figure A22. The cumulative H_2S flux. Variant with 6 mm RO biochar layer, repetition 3. Gompertz equation parameters and R^2 determination coefficient.

Appendix B

Table A1. Tukey's HSD test; variable: maximum cumulative H_2S flux ($mg \cdot m^{-2}$). Differences marked with red font are significant ($p < 0.05$).

Biochar		HAP	HAP	HAP	RO	RO	RO
	The Thickness of the Biochar Layer	6	12	No Biochar	6	12	No Biochar
HAP	6		0.971979	0.931431	0.999902	0.186020	0.005935
HAP	12	0.971979		0.999963	0.994297	0.500670	0.020403
HAP	No biochar	0.931431	0.999963		0.977866	0.604926	0.027845
RO	6	0.999902	0.994297	0.977866		0.258371	0.008562
RO	12	0.186020	0.500670	0.604926	0.258371		0.351435
RO	No biochar	0.005935	0.020403	0.027845	0.008562	0.351435	

Table A2. Tukey's HSD test; variable: H_2S emission constant rate (s^{-1}). Differences marked with red font are significant ($p < 0.05$).

Biochar		HAP	HAP	HAP	RO	RO	RO
	The Thickness of the Biochar Layer	6	12	No Biochar	6	12	No Biochar
HAP	6		0.637682	0.999840	0.061820	0.042644	0.031894
HAP	12	0.637682		0.496960	0.570706	0.448007	0.362320
HAP	No biochar	0.999840	0.496960		0.040832	0.028090	0.020990
RO	6	0.061820	0.570706	0.040832		0.999907	0.998455
RO	12	0.042644	0.448007	0.028090	0.999907		0.999973
RO	No biochar	0.031894	0.362320	0.020990	0.998455	0.999973	

Table A3. Tukey's HSD test; variable: The inflection time of the cumulative H_2S emission (s). Differences marked with red font are significant ($p < 0.05$).

Biochar		HAP	HAP	HAP	RO	RO	RO
	The Thickness of the Biochar Layer	6	12	No Biochar	6	12	No Biochar
HAP	6		0.762503	0.998558	0.977229	0.038806	0.068427
HAP	12	0.762503		0.543358	0.986894	0.314408	0.479691
HAP	No biochar	0.998558	0.543358		0.873116	0.020139	0.035707
RO	6	0.977229	0.986894	0.873116		0.124583	0.209978
RO	12	0.038806	0.314408	0.020139	0.124583		0.999261
RO	No biochar	0.068427	0.479691	0.035707	0.209978	0.999261	

References

1. OSHA. *29 CFR 1910.1000, Table Z-2: Toxic and Hazardous Substances*; Occupational Safety and Health Administration: Washington, DC, USA, 2017.
2. Ni, J.Q.; Heber, A.J.; Sutton, A.L.; Kelly, D.T. Mechanisms of gas releases from swine wastes. *Trans. ASABE* **2009**, *52*, 2013–2025. [CrossRef]
3. Alowairdi, L. Teenager Killed in farm accident in Clark County. 2016. Available online: https://www.weau.com/content/news/Teenager-killed-in-farm-accident-in-Clark-County-393467341.html (accessed on 9 July 2020).
4. Ballerino-Regan, D.; Longmire, A.W. Hydrogen sulfide exposure as a cause of sudden occupational death. *Arch. Pathol. Lab. Med.* **2010**, *134*, 1105. [PubMed]
5. Salem, O.H. OSHA: Ohio Farm Worker Killed by Manure Gas. 2016. Available online: https://www.farmanddairy.com/news/osha-ohio-farm-worker-killed-by-manure-gas/328366.html (accessed on 9 July 2020).
6. Mueller, C. Coroner: Farmer Died of Gas Poisoning. 2016. Available online: http://www.stevenspointjournal.com/story/news/2016/09/14/coroner-farmer-died-gaspoisoning/90365328/?hootPostID=116e126a90c5dabc8fb7604e6d681d17 (accessed on 9 July 2020).
7. Maurer, D.L.; Koziel, J.A.; Harmon, J.D.; Hoff, S.J.; Rieck-Hinz, A.M.; Andersen, D.S. Summary of performance data for technologies to control gaseous, odor, and particulate emissions from livestock operations: Air management practices assessment tool (AMPAT). *Data Brief* **2016**, *7*, 1413–1429. [CrossRef] [PubMed]
8. Maurer, D.L.; Koziel, J.A.; Kalus, K.; Anderson, D.S.; Opalinski, S. Pilot-Scale Testing of Non-Activated Biochar for Swine Manure Treatment and Mitigation of Ammonia, Hydrogen Sulfide, Odorous Volatile Organic Compounds (VOCs) and Greenhouse Gas Emissions. *Sustainability* **2017**, *9*, 929. [CrossRef]
9. Parker, D.B.; Hayes, M.; Brown-Brandl, T.; Woodbury, B.L.; Spiehs, M.J.; Koziel, J.A. Surface application of soybean peroxidase and calcium peroxide for reducing odorous VOC emissions from swine manure slurry. *Appl. Eng. Agric.* **2016**, *32*, 389–398. [CrossRef]
10. Maurer, D.L.; Koziel, J.A.; Bruning, K.; Parker, D.B. Farm-scale testing of soybean peroxidase and calcium peroxide for surficial swine manure treatment and mitigation of odorous VOCs, ammonia and hydrogen sulfide emissions. *Atmos. Environ.* **2017**, *166*, 467–478. [CrossRef]
11. Maurer, D.L.; Koziel, J.A.; Bruning, K.; Parker, D.B. Pilot-scale testing of renewable biocatalyst for swine manure treatment and mitigation of odorous VOCs, ammonia and hydrogen sulfide emissions. *Atmos. Environ.* **2017**, *150*, 313–321. [CrossRef]
12. Cai, L.; Koziel, J.A.; Liang, Y.; Nguyen, A.T.; Xin, H. Evaluation of zeolite for Control of odorants emissions from simulated poultry manure storage. *J. Environ. Qual.* **2007**, *36*, 184–193. [CrossRef]
13. Kalus, K.; Opaliński, S.; Maurer, D.; Rice, S.; Koziel, J.A.; Korczyński, M.; Dobrzański, Z.; Kołacz, R.; Gutarowska, B. Odour reducing microbial-mineral additive for poultry manure treatment. *Front. Environ. Sci. Eng.* **2017**, *11*, 7. [CrossRef]
14. Chen, B. Evaluating the Effectiveness of Manure Additives on Mitigation of Gaseous Emissions from Deep Pit Swine Manure. Master's Thesis, Iowa State University, Ames, IA, USA, 2019.
15. Chen, B.; Koziel, J.A.; Banik, C.; Ma, H.; Lee, M.; Wi, J.; Meiirkhanuly, Z.; Andersen, D.S.; Białowiec, A.; Parker, D.B. Emissions from Swine Manure Treated with Current Products for Mitigation of Odors and Reduction of NH_3, H_2S, VOC, and GHG Emissions. *Data* **2020**, *5*, 54. [CrossRef]
16. Meiirkhanuly, Z.; Koziel, J.A.; Białowiec, A.; Banik, C.; Chen, B.; Lee, M.; Wi, J.; Brown, R.C.; Bakshi, S. Mitigation of gaseous emissions from swine manure with biochar. In Proceedings of the 2020 ASABE Annual International Meeting, Omaha, NE, USA, 13–15 July 2020. [CrossRef]
17. Meiirkhanuly, Z.; Koziel, J.A.; Białowiec, A.; Banik, C.; Brown, R.C. The-Proof-of-Concept of Biochar Floating Cover Influence on Water pH. *Water* **2019**, *11*, 1802. [CrossRef]
18. Meiirkhanuly, Z.; Koziel, J.A.; Białowiec, A.; Banik, C.; Brown, R.C. The proof-of-the concept of biochar floating cover influence on swine manure pH: Implications for mitigation of gaseous emissions from area sources. *Front. Chem.* **2020**. [CrossRef]
19. Białowiec, A.; Micuda, M.; Koziel, J.A. Waste to Carbon: Densification of Torrefied Refuse-Derived Fuel. *Energies* **2018**, *11*, 3233. [CrossRef]

20. Stępień, P.; Świechowski, K.; Hnat, M.; Kugler, S.; Stegenta-Dąbrowska, S.; Koziel, J.A.; Manczarski, P.; Białowiec, A. Waste to Carbon: Biocoal from Elephant Dung as New Cooking Fuel. *Energies* **2019**, *12*, 4344. [CrossRef]

21. Pulka, J.; Manczarski, P.; Koziel, J.A.; Białowiec, A. Torrefaction of Sewage Sludge: Kinetics and Fuel Properties of Biochars. *Energies* **2019**, *12*, 565. [CrossRef]

22. Świechowski, K.; Stegenta-Dąbrowska, S.; Liszewski, M.; Bąbelewski, P.; Koziel, J.A.; Białowiec, A. Oxytree Pruned Biomass Torrefaction: Process Kinetics. *Materials* **2019**, *12*, 3334. [CrossRef] [PubMed]

23. Stępień, P.; Serowik, M.; Koziel, J.A.; Białowiec, A. Waste to carbon: Estimating the demand for production of carbonized refuse-derived fuel. *Sustainability* **2019**, *11*, 5685. [CrossRef]

24. Syguła, E.; Koziel, J.A.; Białowiec, A. Proof-of-Concept of Spent Mushrooms Compost Torrefaction—Studying the Process Kinetics and the Influence of Temperature and Duration on the Calorific Value of the Produced Biocoal. *Energies* **2019**, *12*, 3060. [CrossRef]

25. Kalus, K.; Koziel, J.A.; Opaliński, S. A Review of Biochar Properties and Their Utilization in Crop Agriculture and Livestock Production. *Appl. Sci.* **2019**, *9*, 3494. [CrossRef]

26. MidWest Plan Service. *Structures and Environment Handbook*, 11th ed.; Midwest Plan Service: Ames, IA, USA, 1983; ISBN 0-89373-057-2.

27. Wi, J.; Lee, S.; Kim, E.; Lee, M.; Koziel, J.A.; Ahn, H. Evaluation of Semi-Continuous Pit Manure Recharge System Performance on Mitigation of Ammonia and Hydrogen Sulfide Emissions from a Swine Finishing Barn. *Atmosphere* **2019**, *10*, 170. [CrossRef]

28. Lee, M.; Wi, J.; Koziel, J.A.; Ahn, H.; Li, P.; Chen, B.; Meiirkhanuly, Z.; Banik, C.; Jenks, W. Effects of UV-A Light Treatment on Ammonia, Hydrogen Sulfide, Greenhouse Gases, and Ozone in Simulated Poultry Barn Conditions. *Atmosphere* **2020**, *11*, 283. [CrossRef]

29. Hanusz, Z.; Siarkowski, Z.; Ostrowski, K. Application of the Gompertz model in agricultural engineering. *Inz. Rol.* **2008**, *7*, 71–77.

MDPI

St. Alban-Anlage 66

4052 Basel

Switzerland

Tel. +41 61 683 77 34

Fax +41 61 302 89 18

www.mdpi.com

Catalysts Editorial Office

E-mail: catalysts@mdpi.com

www.mdpi.com/journal/catalysts